Choices, Not Chances

Books by Aubrey Milunsky, M.D.

Choices, Not Chances (previously *Know Your Genes*)
How to Have the Healthiest Baby You Can

and medical texts (as author, coauthor, or editor), including
The Prenatal Diagnosis of Hereditary Disorders
The Prevention of Genetic Disease and Mental Retardation
Coping with Crisis and Handicap
Genetics and the Law (3 volumes)

Genetic Disorders and the Fetus:
Diagnosis, Prevention, and Treatment

Choices, Not Chances

An Essential Guide to Your Heredity and Health

AUBREY MILUNSKY, M.D.

Little, Brown and Company
BOSTON · TORONTO · LONDON

This is a fully revised and updated edition of *Know Your Genes,*
which was first published in 1977.

Library of Congress Cataloging-in-Publication Data

Milunsky, Aubrey.
  Choices, not chances.

  Rev. and updated ed. of: Know your genes. 1977.
  Includes index.
  1. Medical genetics — Popular works. I. Milunsky,
Aubrey. Know your genes. II. Title.
RB155.M539 1989   616'.042   88-26616
ISBN 0-316-57423-6

10 9 8 7 6 5 4 3 2 1

RRD-VA

Designed by Marianne Perlak

Published simultaneously in Canada
by Little, Brown & Company (Canada) Limited

PRINTED IN THE UNITED STATES OF AMERICA

To Jeff,
our pride and joy

If, of all words of tongue and pen,
The saddest are, "It might have been,"

More sad are these we daily see:
"It is, but hadn't ought to be."

BRET HARTE
in "Mrs. Judge Jenkins"

Contents

you're a carrier • the problem • types of hereditary diseases • harmful genes and environmental agents • why know your genes • what you should know • a right to know • a responsibility to tell • options • parents transmit genetic diseases • minor and major birth defects • inherited and acquired birth defects • having a defective child

normal and abnormal chromosomes • miscarriage and stillbirth • birth defects from too many, too few, or structurally defective chromosomes • new diagnostic techniques • Down syndrome

normal and abnormal sex chromosomes • birth defects from too many or too few sex chromosomes • mosaics • treatment

intersex • male-determining genes • genetic sex • true hermaphrodites • female and male pseudohermaphrodites • diagnosis and treatment

epilepsy • anticoagulants • anticancer drugs • acne and psoriasis medications • sex hormones • antibiotics • thyroid drugs • marijuana and cocaine

PRENATAL SCREENING AND DIAGNOSIS

ETHICS, LAW, AND EUTHANASIA

INHERITANCE OF COMMON DISEASES

TREATMENT OF GENETIC DISORDERS

APPENDIXES

INDEX

Preface

Contemplate your own history and your family's for a moment. Have there been children with birth defects or mental retardation? Stillbirths? Infant deaths? Miscarriages? Does anyone have cancer, heart disease, diabetes, or mental illness? Did anyone lose his or her mental faculties in middle age? Is there a recognized familial disorder on either side of the family? Are you aware of your ethnic origin and have you been tested to determine whether you carry a harmful gene typical of your ethnic group? Do you know which genetic diseases are especially important in your ancestral group?

If you have a genetic disorder, are you aware of how best to take care of yourself — what to avoid, and what treatment is available — and what the risks are for passing this disorder to a child? Have you been apprised of the latest gene-analysis techniques, which have spawned dramatic advances in the detection of carriers and in early prenatal diagnosis? Certain serious hereditary disorders not diagnosable before can now be detected prenatally — cystic fibrosis, muscular dystrophy, Huntington disease, and others.

Every facet of your health is regulated, modulated, or otherwise influenced by your genes. Inherited susceptibilities to many diseases are signaled by your immune (HLA) blood groups. Sensitivity or reactions to medications, as well as allergies, are genetically rooted. Your specific genetic endowment may make certain occupations or activities hazardous for your health, while many chronic though nonfatal inherited disorders may involve your vision, hearing, or learning abilities.

You are a carrier of about twenty harmful genes — as is each of us! Fortunately, among the 50,000 or more genes we each have, these few mostly cause no trouble, although they have the potential for resulting in catastrophes. Indeed, 3 to 4 percent of all children are born with a major birth defect, mental retardation, or a genetic disorder. By seven years of age, some 7 to 8 percent of children have one of these problems. "If only I had known!" is the all-too-frequent lament heard after the birth of a child with a serious birth defect, a genetic disorder, or subsequent mental retardation. Dramatic new advances increasingly make it possible to avoid such tragedies. Indeed, if you are planning to have a child, it is possible to understand, without extensive knowledge of biology or genetics, what your risks are in a given situation, what tests are available, and what options exist. I believe that you have a right to know this information and a responsibility to seek out the facts (at least for your children's sake), and that you should retain the freedom to exercise your options.

Choices, Not Chances has been written especially for people who wish to safeguard their own health, who have a genetic disease, who have or have had affected children or relatives, who are still to choose a mate, who plan to have children — in short, for those who care about themselves and their loved ones. A wealth of information on genetics is contained within these pages, and I encourage you to use the index for guidance on every subject.

For your own health and that of your children, you should know your genetic risks, their potential implications, and your associated options. By knowing your genes, so to speak, you increase your choices, limit your chances (risks), and safeguard your health. The wisdom of Henry James in "The Middle Years" should not be ignored: "A second chance — *that's* the delusion. There never was to be but one."

From responses to the first edition of this book, which was titled *Know Your Genes*, I know that the information provided enabled many individuals and families to prevent unnecessary agony, anguish, and suffering. I hope that this edition will prove even more valuable, given the new, exciting advances in understanding so many more genetic disorders. Being well informed about heredity is more important now than ever before, since so many more disorders can be prevented, anticipated, or treated. It would be a blessing to be confronted less often by anguished people with genetic afflictions or affected children and to find myself thinking sadly, "It is, but hadn't ought to be!"

AUBREY MILUNSKY

Boston, January 1989

Acknowledgments

I acknowledge with gratitude and humility the remarkable response by readers of the first edition of this book. Worldwide interest resulted in translation and publication abroad in many languages — Japanese, German, Italian, Spanish, Russian, Portuguese, Chinese, Croatian, and English. Letters from readers all over the world echoed appreciation for dearly needed and formerly ill-understood information. These responses were an important stimulus that led to this new edition.

The information presented in this book is based upon the research of thousands of people. All information provided has a scientific basis and a recognized source. I felt, however, that it was inappropriate to list the many hundreds of medical books and journals used. I pay tribute to all these workers, whose contributions collectively now provide us with knowledge about genetics that we can apply to our everyday health care.

I acknowledge with gratitude the excellent work of my secretary, Mrs. Barbara J. Thom. My thanks are also due to James C. Skare, Ph.D., at the Boston University Center for Human Genetics, for very helpful discussions on molecular genetics. And I am especially appreciative of the expert editorial work and guidance provided by Michael Brandon and Christina Ward.

It is a special joy to acknowledge the research assistance of my son Jeff, who has now begun medical school.

My wife, Babette, has once again supported and encouraged me during this arduous but enjoyable challenge. I gratefully acknowledge her love, devotion, and companionship.

Chromosomes and Genes

Hereditary Disorders

Twins

1

Why You Should Know Your Genes

You are unique. The genes you inherited from your parents, together with the effects your environment had on you, make you unique. You may be tall or short, fat or thin, fair- or dark-skinned, susceptible to allergies or other disorders, or actually affected by a hereditary disease. Regardless, everything you are is a direct result of the interaction of your genes with the environment. Certainly, your genes will largely determine your body's characteristics, your susceptibility or resistance to *all* diseases, how you react to medications, whether you carry and will transmit certain hereditary diseases or have or will fall victim to one (or more) of them.

Our collective uniqueness makes for the remarkable variation of the human species. Responses to environmental dangers have in an evolutionary way resulted in survival adaptation as advantageous genes made their presence felt. (Some examples of "good" genes are discussed in chapter 9.)

Unfortunately, *each of us* also possesses disadvantageous genes. Indeed, *you are a carrier of at least twenty such harmful genes that could cause you to have or to transmit a genetic disease.* It is likely, however, that you are unaware of your potential to transmit or even to develop genetic disease; and you may not realize the extent of the problem posed by the frequency of genetic disorders, birth defects, or mental retardation. My aim is to inform you why, for your own health and that of any children you may have, it is crucial to *know your genes.* Only by understanding your genetic endowment will you best be able to care for yourself and your children. Today there are many options available to prevent unnecessary tragedies. Remember, by knowing about your genes, you increase your choices, limit your chances, and safeguard your health.

Why the Problem Affects You

Every disease is either caused, modulated, or in some other way influenced by your genes. Genetic disorders affect at least 1 in 10 individuals in the Western world. In the United States, at least 24 million to 25 million people are affected by an inherited disorder. Of these, most suffer from diseases caused by the interaction of *multiple* genes and environmental factors, while others are directly afflicted by a disease caused by a *single* gene (such as cystic fibrosis, muscular dystrophy, Huntington disease, sickle-cell disease, and hemophilia). In the case of many of these single-gene diseases, avoidance or prevention is possible. That is why I have written this book: to elucidate the many ways in which tragedy can be averted *in advance,* or treatment initiated *in time.*

In these pages I shall emphasize the distinct hereditary diseases carried by various ethnic or racial groups: Irish, Italians, Greeks, Jews, Orientals, blacks, and whites. There is, for instance, a hereditary disease that affects 1 white child in about every 2,500: cystic fibrosis. Ashkenazic and Sephardic Jews, as another example, carry different genetic diseases. There are also many, many inherited diseases that have nothing to do with racial origins.

Certainly, it should not have to take the birth of a child with an irreparable defect to alert you to the possibility of just such an event. Science now provides many tests that help to predict and prevent genetic defects, as well as methods for successfully treating hereditary disorders obvious at birth or that crop up later in life.

Morally, there is no family, and no person planning to have a child, who can ignore the new genetic discoveries and techniques for preventing genetic disease. Your health and welfare and that of your (future) children are at stake. We all have a right and, indeed, an obligation to know about our particular genes and to consider the options available that increase our chances of having healthy children. We should also all have the freedom to exercise these options as we wish and as rationally as we are able.

Over 4,000 diseases caused by single defective genes have been identified. It is therefore not surprising that they account for 25 to 30 percent of admissions to the major children's hospitals in the

United States and Canada. This large proportion is, of course, due partly to the waning of serious infectious diseases over the last few decades, to improved early recognition of these genetic disorders, and to an increased life span for some of them.

Between 3 and 4 of every 100 babies are born with recognizable major birth defects. According to a leading National Institutes of Health study based upon analysis of 52,332 live births, by seven years of age an additional 3 to 4 percent of children are found to have major birth defects (earlier diagnoses were missed or were not possible at the time). Conditions involving the heart, muscles and ligaments, kidneys, eyes, brain, and skin were among those found with later diagnoses. Overall, then, 6 to 8 percent of all children are born with or later manifest major birth defects. There are, moreover, over 6 million persons in the United States who, for either hereditary or congenital reasons, are mentally retarded. (It is crucial to keep things in perspective, however: over 90 percent of all children are born free of genetic disease, major birth defects, or mental retardation.)

The statistics are frightening, you may say. Perhaps. But think for a moment how you would feel after having a child with a genetic disease (who suffers while you watch), and then learning that it could have been avoided. Witnessing such tragedies for almost thirty years has compelled me to share the lessons I have learned with those who would rather not say, too late, "It is, but hadn't ought to be."

Various Types of Hereditary Disorders

Because troubles that "run in the family" are apt to be glossed over, people often do not really know much about what they may have inherited. Often we do not want to know; but that is an indulgence for which we may pay dearly. Each of us may inherit a disorder directly from one parent who is affected — say, by Huntington disease, a progressive brain degeneration that causes dementia, speech defects, and purposeless muscle movements. A disease such as cystic fibrosis, which causes chronic lung infection and malabsorption of food, is transmitted by two parents who are completely healthy and may themselves never have heard about the disease, though they carry it in their genes. Tay-Sachs disease, a disorder that causes brain destruction, blindness, and early death and that is found almost exclusively among Ashkenazic

Jews and French-Canadians, is also caused by genes transmitted from both parents. Similarly, blacks transmit the gene causing sickle-cell anemia to their progeny. Hemophilia, a bleeding disease caused by a missing blood-clotting factor, is transmitted by females but occurs almost always in males. It is not special to any race (though it has been associated with the highly inbred European royalty).

Other disorders result from the interaction of multiple genes and one or more environmental factors. These include open-spine defects, which are most common among people with Irish ancestry (though they may occur in all races), high blood pressure, coronary disease, cancer, diabetes, mental retardation, schizophrenia, and skin disorders such as eczema.

Willy-nilly, we are involved in our genes. In a way, we are our genes.

Variations and Patterns of Hereditary Disorders

The nature and severity of many inherited diseases are remarkably constant. An affected child may show almost exact features, for example, as a cousin with the same disease. Frequently, however, even for the same disease in different members of the *same* family, the age of onset, the severity, and the range of signs may vary remarkably. Such variability could make diagnosis difficult. The manifestations of a genetic disease could be so subtle as to be missed, which might raise suggestions that the harmful gene "skipped a generation."

The reason for the variable expression of a single disease-causing gene is not always clear. Other genes or environmental factors could modify a gene's expression. A single gene may also cause multiple effects. In the Marfan syndrome, for example, the bones, heart, and eyes could be variably affected. Some afflicted individuals could simply be tall and have long fingers and a slightly curved spine (scoliosis), without eye or heart problems. Others may have all the signs. The basic connection in this disorder is that the harmful gene has formed weak elastic tissue common to all organs.

A single gene, which is a complex structure, may undergo just slight structural change (called mutation), with the resulting disease (or effects) varying between different families. Hence, a particular condition, such as profound childhood deafness, could occur in unrelated persons and be caused by a slightly altered or even a different gene in each person. The result, deafness, would nevertheless be the same. Moreover, the exact same birth defects could result from a single harmful gene, a chromosome defect, or medication or alcohol taken in the earliest weeks of pregnancy.

Harmful Genes and Environmental Agents

You may believe you are in perfectly good health, and it is to be hoped that you are. Many of us, however, through our genetic endowment — our genes — are predisposed to react in a specific way, a possibly fatal way, to certain environmental agents. For example, you may appear to be perfectly healthy but nevertheless may have a particular enzyme — a substance involved in the normal making and breaking-down of body chemicals — missing from your red blood cells. Consequently, when you are exposed to certain drugs (even aspirin or sulfa drugs), a severe reaction may occur, or a hemolytic anemia, which reduces the life span of red blood cells, may develop. Greeks, Italians, Orientals, blacks, and other groups may be especially prone to such reactions because of a deficiency in this enzyme (glucose-6-phosphate dehydrogenase). Fatal reactions to penicillin are probably also genetically determined, but in a different way.

Another alarming phenomenon is the potentially fatal reaction to general anesthesia of apparently healthy individuals who have a particular hereditary muscle disease (malignant hyperthermia) that is not ordinarily evident. During or after their operation under general anesthesia, they may suddenly develop an extraordinarily high fever (such as a body temperature of 108 degrees Fahrenheit) and die from a complication unrelated to the disorder for which they had undergone surgery.

Of particular special interest is an enzyme that may be at high levels in some persons and that may act harmfully by potentiating the effects of carcinogens (cancer-causing chemicals). Such enzyme levels are under genetic control. A person's enzyme may, for example, potentiate the effects of carcinogens from cigarettes, thereby raising his or her particular risk of lung cancer. By the same token, those who have inherited a low level of such enzyme activity have a much lower risk of developing cancer. This explains why some who smoke heavily all their lives escape lung, and perhaps other, cancer. (They do not, however, escape the ravages of heart disease and emphysema.)

Why Know Your Genes?

Knowledge about many hereditary diseases has been available for years — so why the new urgency about these matters? Until relatively recently there were more pressing problems that took priority over genetic diseases, and this is still the case in many underdeveloped parts of the world. Malnutrition and infectious disorders are the primary problems that require attention before a society can turn to more sophisticated considerations, including the management and prevention of hereditary diseases.

Unfortunately, many genetic disorders manifest themselves in our adult years for the first time. It is to your advantage to know about genetic disease to secure your personal health, to recognize your reproductive options, and to obtain the best care for your offspring. From your personal health perspective, even though you may have a genetic disorder, awareness of precipitating or aggravating factors can be extremely valuable, even lifesaving. Furthermore, recognition of an inherited susceptibility may enable a safer choice of occupation or avoidance of certain medications (such as barbiturates in porphyria).

Knowing your genes helps you exercise your options, which include having genetic counseling to discover what risks you have yourself, what risks there are for having defective offspring, and what risks to consider when choosing your mate; having tests to determine if you are a carrier of a genetic disorder; having prenatal-diagnosis tests to detect fetal defects; requesting artificial insemination by donor; selecting in vitro fertilization ("test-tube baby" method); or choosing not to have children and having a vasectomy or tubal ligation. To obtain the best care for your child will require information about appropriate tests for your newborn baby to avert mental retardation by initiating immediate treatment. Knowledge about screening tests in pregnancy could facilitate the safest outcome for your child.

By *not* knowing your genetic background, you risk having a child with a genetic disease, who is likely to become a burden on your healthy offspring. I know of many couples over thirty-five years of age who decided against prenatal diagnosis and subsequently had a child with Down syndrome — a child who eventually will have to be cared for by his or her siblings or the state.

What You Should Know

It's true: you're unique. But like each of us, you harbor about twenty disease-causing genes. You should therefore be fully aware of your genetic background — your family history. If you took up a pencil this moment and drew a diagram of your family pedigree (appendix A explains how to do it), would you be able to account for all instances of birth defects, hereditary disorders, different types of cancers, mental retardation, and stillbirths or recurrent miscarriages in your brothers and sisters, parents, uncles and aunts, first and (preferably) second cousins? This is the first step in the genetic-counseling process — to determine the genetic nature of a disorder and its risk of recurrence.

Knowledge of your ethnic (ancestral) origin is also important, since certain serious genetic disorders typically affect particular ethnic groups. If you and your partner have the same ethnic background, tests to determine if you carry a certain harmful gene may be advisable (see chapter 9).

For your own and your family's health you need to know what factors will precipitate or aggravate illness in the face of a genetic condition and which dietary or other factors will result in disease if you are genetically susceptible.

You must know your risks in childbearing so far as birth defects are concerned. They relate to many different factors, including your family history, your ethnic group, previously affected children, your age and that of your mate, your health, and exposure to infections, drugs, or illness during pregnancy — to name the key items explored in this book.

You should be aware of the options you have to avoid genetic disease in your offspring. Moreover, remarkable advances in medical genetics will continue, and you should remain in touch with a medical school–based program so as to benefit early from new developments.

A Right to Know

You have a right to know if your risks for having defective children are higher than normal, if you are a carrier of specific hereditary disorders, and what tests or other options are available to you. True, you also owe it to yourself and your future children to know your family history, to seek consultations from specialists, and to take advantage of the recent advances in medical technology that allow for the prevention of genetic diseases.

Nobody wants to discover that he has a hereditary disease or that one of his children is affected — especially when a particular disorder was avoidable. Certainly, there are individuals who, because of their religious beliefs or other reasons, choose not to interfere with what they regard as their destiny, or God's will, and would not consider induced abortion of a defective fetus. This is their free prerogative, and such action and belief must be safeguarded. There are those, however, who select this option to avoid serious or fatal genetic disease; their rights, too, must be protected.

The responsibility to one's offspring before procreation is self-evident. Surely, children have the right to be born free of birth defects and serious or fatal hereditary disease whenever possible. Indeed, the Supreme Court of Rhode Island, among others, has spoken explicitly to this point: "Justice requires that the principle be recognized that a child has a legal right to begin life with a sound mind and body." In order to ensure this right, all prospective parents have to act in a responsible way by determining if they are indeed carriers or are personally at risk for having a genetic disease.

Society, too, has a stake in your actions and how you perceive your responsibilities. Suppose that you choose simply to have children with serious hereditary disease — children who, for example, also have mental retardation. The state will inevitably end up being responsible — sooner or later, and in one way or another — for the care of your affected offspring. I will explore the morality and ethics of some of these problems and approaches later.

A Responsibility to Tell

The death of a child, owing to a grave birth defect, soon after birth is a painful tragedy that lives on in eternal memory. Years later, when the healthy brothers and sisters of the child are planning their own families, facts about their deceased sibling become important because of possibly increased risks of similar birth defects in their own offspring. Time and again I have encountered well-meaning parents who have not told their children about such tragedies — or about a severely retarded, institutionalized maternal uncle (which can be important in sex-linked diseases), or about a stillbirth, or about birth defects or retardation in other family members, or about recurrent miscarriages. Protectiveness, shame, culpability, superstition, unwillingness to relive the pain in telling, and ignorance of the importance the facts bear have been the main motivating factors I have observed. Parents simply must understand their responsibility to tell. Equally important is the need for couples who have a defective child to inform their siblings, uncles, aunts, and cousins who are still in their childbearing years. Given the sad state of modern family relationships, it is necessary to remind families of these responsibilities.

Avoidable births of severely mentally retarded children to near relatives because of failure to inform about a chromosome defect is unacceptable and irresponsible. *Families that share and tell, care.*

Minor and Major Birth Defects

There is, of course, great variation in the severity of birth defects. Minor ones are very common, occurring in up to 40 percent of live births. Minor "defects" may often simply be normal variations, or have no medical, surgical, or cosmetic significance. You may well have some yourself. Do you, for example, have a single crease across your palm? While this so-called simian crease occurs very frequently in Down syndrome (which is no longer called mongolism), a single transverse palmar crease is also found on one or both hands in about 1 percent of healthy, normal newborns — who, strangely enough, are most commonly male firstborns. Are your second and third toes webbed or bunched? Perhaps one of your parents has a similar minor defect. Are your fifth fingers slightly incurved? Do you have flat feet? How many birthmarks? Do your ears have an unusual shape? Is your palate high-arched? Each of these minor birth defects usually has no significant meaning if it occurs in an isolated fashion. However, when three or more minor defects are noted, there is a 20-percent likelihood that a major birth defect is also present. For example, if one of your ears is shaped *very* differently from the other, the kidney on the side of the abnormally shaped ear may also be abnormal — and that could be very important to your health!

In contrast, major birth defects are serious. They disfigure the body, impair health, or threaten life. Heart defects, very small heads (microcephaly), mental retardation, blindness, deafness, dwarfism, and many others are in this category.

Inherited or Acquired Birth Defects

Both major and minor defects may occur "out of the blue," without any cause ever being determined. Certain medications taken during pregnancy (anticonvulsants, for example) or a suspected viral illness in early pregnancy (such as German measles) may be considered in trying to evaluate the cause. The condition, of course, may be hereditary. Not infrequently, problems are encountered in trying to differentiate an acquired defect, such as one caused by a drug or virus, from one that has genetic origins, even though no previous cases have occurred in the entire family. Individuals with hereditary diseases may be born with gross disfiguring abnormalities or may show no evidence of the particular disease for months or even decades (as with Tay-Sachs disease and Huntington disease). They may die within hours or days from irremediable biochemical faults in metabolism, or may live for months before irreparable mental retardation becomes apparent.

An infant born with an abnormality may merely have suffered from some trauma in early intrauterine life. Babies who suffer from lack of oxygen or are brain-damaged in some way during the hours before, during, or immediately after delivery may later be found to have cerebral palsy, mental retardation, or epilepsy. These very sad instances of damage occur in what up to that moment has usually been a normal fetus. These acquired conditions that cause brain injury will not be discussed specifically in this book; the main thrust here is a consideration of hereditary disease and defective development.

Options

In the past, the only possible approach by a physician — because so little was known — was simply to await the birth of a child with a serious genetic disease. Only at that point was it possible to counsel the parents about the exact risk of recurrence in subsequent pregnancies. Recent advances make it possible to diagnose more and more hereditary disorders in the fetus. It is therefore critically important that families with serious problems of this kind remain in contact with major medical centers that can provide sophisticated genetic counseling, in order to benefit from continuing new discoveries.

The following is a good case in point:

> Pam and Ernie were twenty-four and twenty-six years old, respectively, when they married. Both were perfectly well themselves. Pam's brother, however, had died at the age of twelve from muscular dystrophy. At the time of her brother's death, her parents were told that Pam was indeed a possible carrier of muscular dystrophy and was therefore at risk for having male offspring with this disease. No effort was made to determine if she actually was a carrier, and as the years rolled by and Pam's marriage approached, no one suggested genetic counseling again. Her very first pregnancy yielded a beautiful boy who appeared entirely normal. When the son was three years old, however, Ernie noticed that the child had difficulty climbing stairs and even getting up from a sitting position on the floor. The diagnosis of muscular dystrophy was made during the third month of Pam's second pregnancy.

> Pam's obstetrician advised the couple that Pam indeed was a carrier (this was confirmed by a blood test) and that there was a 50-percent risk that any future male offspring would be affected by muscular dystrophy. The physician explained, however, that it had just recently become possible to detect muscular dystrophy early enough in pregnancy to offer the parents an elective abortion if an affected male fetus was indeed found. Pam and Ernie opted for an amniocentesis test, which showed that the fetus was a male with muscular dystrophy, and they elected to terminate the pregnancy. Using prenatal-diagnosis studies, they subsequently had a normal son.

There is much that is new in clinical genetics today, and it takes major efforts by practicing physicians to keep up with the latest recommendations stemming from advances in genetics (as well as in all other fields of medicine). For example, only since 1986 could Pam and Ernie have been able to use the sophisticated DNA tests that can determine whether or not a male fetus actually has muscular dystrophy. It is not, however, possible for any physician to know the answers to all the various complex problems in medicine, though he or she can reasonably be expected to seek and to find expert consultation for patients. This applies equally well to medical genetics. Should you at any time feel that you would be happier with another opinion, then it would be entirely reasonable to seek one in a major medical school or hospital center. All physicians should be sufficiently sensitive and secure to offer the very anxious or extremely concerned person an opportunity for a second opinion. Sadly, this is not always the case.

Paradoxically, the patient can play an important part by bringing to the attention of a physician a recent advance in medicine culled from the latest weekly news magazine, or learned from this book or other sources. The parents, relatives, and friends of children and adults with various genetic diseases have established valuable organizations dedicated to support, education, and research. Those in the United States are listed in appendix B. These and other charitable organizations have played critical roles in making people aware of the need for genetic counseling.

Having a Defective Child

After the initial shock of learning that their child has a major birth defect or is likely to be severely mentally retarded, a complex set of reactions and adaptive mechanisms generally appear in the parents. Depression and feelings of guilt are natural, despite the fact that in the majority of cases no culpability can be ascribed to either parent. On occasion, either the mother or the father or both may refuse to accept the likelihood that their child will be severely retarded. Indeed, their defensive rejection of the probable reality may be so strong as to block reception of any genetic counseling that is provided soon after the birth of the affected child. The result is that even when the risk of having another similarly affected infant is high, another pregnancy is apt to follow rapidly.

Bitterness is common and is occasionally accompanied by envy of close relatives who have normal children. Either or both parents may indulge in self-pity or seek solace in alcohol. Initial anger directed at the physicians for not having prevented the tragedy often gives rise to a sense of increasing frustration over the inadequacy of care or the lack of a specific diagnosis, treatment, or cure.

The presence of a child with serious birth defects in the home becomes a chronic emotional and physical drain on the parents, and often leads to a severe state of exhaustion affecting all avenues of their life. Economic hardship may follow, which almost invariably increases marital conflict. The sex life of the couple becomes a major casualty and these problems further feed the fires of anger and frustration. Separation and divorce are only too well known in families in which such tragedies have occurred. The enormous drain on the energies of the parents frequently leads to a relative neglect of their *unaffected* children. Generally speaking, it is simply not possible for parents in this situation to have sufficient energy and time left over to attend to their normal children as they would otherwise have done. As a consequence of this neglect, which is often not recognized, emotional, behavioral, and psychological problems develop in the normal siblings.

These major difficulties can become chronic and provide lifetime complications of one sort or another for all members of the household, though not every family having such a child is as devastated as described. Those sufficiently wealthy to employ full-time help are often the most vocal about how families can and should cope with these tragedies. Try as they might, middle- and lower-income parents often find it impossible to do their best for both the affected child and their other children. It is a very rare family that "benefits" from such catastrophes. True, the qualities of compassion, patience, and love can be brought out by caring for the abnormal child or adult — a healthy adaptation, but, unfortunately, unusual in most families. All things considered, it is a heavy burden, and feelings of anger, guilt, and frustration are especially bitter when parents realize too late that the tragedy could have been prevented.

Science has now made us responsible through knowledge and has provided the means to take the fear out of old superstitions and aversions.

I have witnessed the deep despair of a mother and father when they are told that their tragedy was not inevitable; for, all the time, deep inside, they really did know it, but . . .

All of us have a tendency to postpone action on a health problem until it becomes urgent, and often it is then too late. This book is sending you a personal message about a slightly different, equally real, and even more compelling urgency. In the words attributed to Robert Louis Stevenson:

> Let me do it now
> Let me not defer it nor neglect it
> For I shall not pass this way again.

2

Chromosomes

"But why us? What are chromosomes? Why does *our* baby have an extra one? Does it come from either of us? Can it happen again?"

Distraught and bewildered young parents in my consulting room for the first time, still caught in the dismal fog of disbelief and unreality caused by my diagnosis, invariably ask me such questions as soon as they are able to speak at all.

Time and again, couples say that they have no idea what chromosomes are, how they differ from genes, or even what the difference is between congenital and hereditary defects. I try to answer as clearly as possible, while assuring them that they need no prior knowledge of biology to understand. We begin with the cell.

Cells

Our bodies are made of billions of microscopic units called *cells*, many of which have very special functions: brain cells for memory and intelligence, heart cells for rhythmic contraction, intestinal cells for making mucus, and so on. The cells of our bodies live for variable periods of time, depending upon their organs of origin. While we cannot grow new brain cells (in fact, we steadily lose brain cells as we get older), the cells lining our intestines are lost and replaced by new ones every twenty-four hours or so. It has been estimated that about 50 million cells in our bodies die every second and are rapidly replaced by about the same number. Sperm cells in the testes may live only a few months, while ova (eggs) in the ovaries may live longer than fifty years. The implication for birth defects is therefore that a woman's ova are exposed to environmental influences, such as X rays and drugs, all through her childhood and childbearing life.

Despite their specialized functions, all cells have similar basic component parts. The tiny center of operations within each cell is called the nucleus. It not only controls the functions of that cell, but also contains the messages, or blueprints, inherited from our parents that determine what the cell will actually do, and what characteristics will be passed along to each of our children. Each cell nucleus contains tiny threads of chemical compounds called *chromosomes*, the most important component of which is *DNA* (deoxyribonucleic acid).

Normal Chromosomes

It was known almost a century ago that when a special dye was added to a cell at a critical point in its formation, these threadlike structures of protein would take up the dye and stain, becoming easier to see. They were therefore called chromosomes, from the Greek words *chroma*, meaning "color," and *soma*, for "body." Late in the nineteenth century, the chromosomes were already considered to be the likely carriers of hereditary factors.

All the necessary information required to direct the formation and function of a human being — or of any other living thing, from bacteria, to plants, to elephants — is contained by these complex, thin threads. The chromosomes are, in turn, composed of *genes*, which are the units of heredity. Single genes are so small that they remain invisible even when looked at by the most modern instruments, including the electron microscope.

You receive half your chromosome complement from your mother and half from your father. The genes constituting the chromosomes are therefore equally contributed to you by each parent. In turn, you will pass along half your chromosomes and genes to each of your own children. A look at what normally happens to chromosomes as they pass from parent to child helps us understand what can happen to chromosomes when things go wrong.

NUMBER, SIZE, SEX, AND IDENTITY

The number of chromosomes and their structure vary greatly among living organisms, ranging from 4 to 500 in each cell. Chimpanzees and gorillas have 48 chromosomes per cell. In 1956, humans were shown to have 46 chromosomes in every cell (except sperm and ova, which have 23 each) and not 48 as was previously thought. The chromosomes in a cell can be viewed through the microscope and photographed; they appear as shown in figure 1. Each chromosome in the photograph can be cut out, paired with a corresponding chromosome, arranged in order with the largest first, and pasted on cardboard — as shown in figure 2. There are 22 pairs (a total of 44), which are numbered as shown. One chromosome of each pair is from the father, and the other is from the mother.

The remaining two chromosomes in every cell are called *sex chromosomes*, since they carry the message that determines which sex the individual will be. Note that the two sex chromosomes are

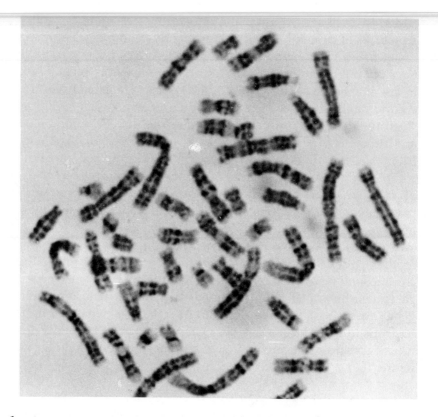

Figure 1

Human chromosomes within a cell as seen through a microscope

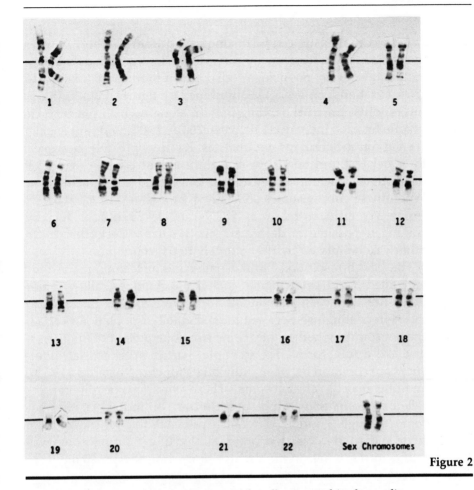

Figure 2

Normal human chromosomes from a single cell, arranged in descending order of size (except, by convention, for the number-22 pair and the sex chromosomes). Staining reveals the cross-striations, or bands. Note the two female sex chromosomes (XX).

arranged separate from the other 22 pairs in the pasteup. Each parent normally passes along one sex chromosome to his or her offspring. The two sex chromosomes in females are denoted as XX and in males as XY. The mother can contribute an X chromosome only, while the father can contribute an X or a Y. When the mother contributes a single X, and the father also contributes an X, then the offspring will be a girl (XX). The presence of the Y chromosome always dictates that the offspring will be male (even in abnormal states when there are two or even more X chromosomes together with the Y chromosome).

Each chromosome can be distinguished from another, in ways even more accurate than by size alone. Using various staining techniques, it is possible to distinguish horizontal bands, or cross-striations, on every chromosome (see figure 2). As with our fingerprints, the total banding pattern along each chromosome is unique for each individual. Between 200 and 400 bands are usually seen in routine chromosome analysis. It is now possible, however, to "catch" chromosomes at a slightly earlier phase during cell division, at which time they appear longer under the microscope. When these chromosomes are stained, at least 800 to 1,400 bands (some say even as many as 5,000) can be visualized, thereby making it possible to detect extremely small defects that could otherwise be missed by the routinely used techniques.

These high-resolution chromosome-banding techniques (as they are called), and even regular staining techniques, allow minor differences between individuals to be recognized. Experts can therefore distinguish between them and tell, especially for certain chromosomes, whether they come from one person or another — just like a fingerprint. For example, particular structural differences are common between individuals for chromosomes 1, 9, 16, and the male Y chromosome. These chromosome differences, called *polymorphisms*, can be used to trace an individual chromosome through a family. I recall a prenatal-diagnosis case from some years ago in which a rather odd-looking Y chromosome was found in the fetus. In an attempt to exclude the possibility of a potential structural abnormality, I requested a blood sample from the husband to check his Y chromosome. Study of his chromosomes showed a completely different Y chromosome. This gentleman, who was a sailor and who spent much time away from home, was clearly not the father!

Chromosomal polymorphisms have also been used to distinguish the origin of the extra chromosome that appears in conditions such as Down syndrome (discussed below). From such studies we now know that the extra number-21 chromosome in Down syndrome originates from the mother in about 75 percent of cases and from the father in about 25 percent.

SPERM AND EGGS

In the formation of both the sperm and the ovum, or egg, the usual 46 chromosomes per cell are reduced to half that number: 23 chromosomes. When they fuse at fertilization, they form one cell that again contains 46 chromosomes. How do the sperm and egg form with only 23 chromosomes each? And once fertilization occurs, how do the chromosomes divide as we grow from a single cell into a whole person? The sequence of events is best followed by observing figure 3 carefully as we go along. Let us start with one cell nucleus in the testis. (The same process happens in the ovary.) To make it easier, we will follow only one of the 23 pairs of chromosomes in that cell nucleus. The same process, called *meiosis,* occurs for each of the 23 chromosome pairs in every cell from which sperm or eggs originate.

Step 1: Shows one cell with a pair of chromosomes.

Step 2: The chromosomes split longitudinally and begin to pair off.

Step 3: The cell nucleus begins to divide.

Step 4: The cell nucleus (and the cell that it occupies) has divided into two new nuclei, each containing a pair of chromosomes.

Step 5: The two chromosomes in each new nucleus now begin to move apart as the cells and their nuclei divide.

Step 6: New cells and nuclei are formed, each with only one

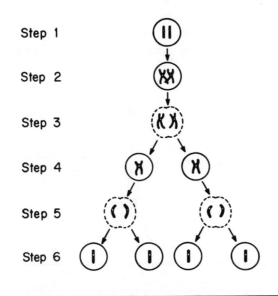

Figure 3

Chromosome division (meiosis), step by step

chromosome from the preceding cell. We can see that, from the original cell with a pair of chromosomes, there are now four cells each with a single chromosome. These are the sperm cells (or eggs, if in the ovary), and they each obviously contain 23 chromosomes, which is half the original number. When a sperm with 23 chromosomes and an egg with 23 chromosomes meet in fertilization, a single cell is constituted with 46 chromosomes. We thus receive half our chromosomes (and, therefore, half our genes) from our father and half from our mother.

MULTIPLICATION BY DIVISION

We all start like this, as a single cell with 46 chromosomes. Let us follow that single cell (figure 4) as it divides, a process called *mitosis,* and again focus for the sake of simplicity on only one pair of chromosomes.

Step 1: Shows the single cell with the pair of chromosomes enclosed by the nuclear membrane.

Step 2: The chromosomes split longitudinally, making a total of two pairs, beginning the stage called *metaphase.*

Step 3: The chromosome pairs separate; the cell nucleus and the cell itself begin to divide.

Step 4: One chromosome member of each pair is now found in a new cell. Note that there are now two cells, each generated from the original single cell. This whole process of

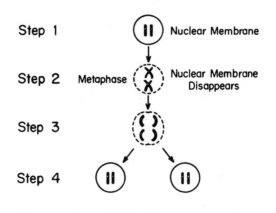

Figure 4

The process of cell division (mitosis)

cell division continues an infinite number of times to constitute, finally, all the cells in the human body.

One very important phenomenon, with implications for the risks of inheriting any genetic disease, can occur between any pair of chromosomes. A variable-sized piece of one chromosome — a piece as small as a single gene or one as long as thousands of genes — can cross over, or interchange, with a segment on the other member of the pair. This crossing-over mechanism, like the game of musical chairs, results in a chromosome bearing "unexpected" genes, which, when transmitted to an offspring, could be the cause of a specific genetic disorder. This process of *recombination* can occur both in meiosis and mitosis. Crossing-over of genes during meiosis yields about 8 million different combinations in the egg or sperm of a single individual. Recombination therefore is a process that heavily contributes to genetic variation, or how different we are from one another.

A similar crossing-over process, called *sister-chromatid exchange*, seems to occur more frequently in certain genetic diseases in which there is an increase in chromosome breakage (discussed below).

Abnormal Chromosomes

MISCARRIAGE AND STILLBIRTH

It is quite startling to realize that up to one-third of all recognized pregnancies end in miscarriage; as many as 22 percent are lost before or at the time of the first menstrual period after conception has occurred and even before pregnancy has been detected. (It has been calculated that a remarkable 78 percent of all conceptions do not result in a live birth.) Early miscarriage is therefore very common. In the United States, 4 percent of married women report having two miscarriages; 3 percent report more than three such losses. (I have extensively reviewed the causes and treatment of miscarriage in *How to Have the Healthiest Baby You Can* [Simon and Schuster, New York, 1987].)

Among the most common causes of miscarriage are disorders of the chromosomes of the developing embryo or fetus. Recent estimates imply that 1 in 10 sperm or eggs carries a chromosome abnormality. And recently, Swedish researchers, in the first study of its kind, demonstrated that about 50 percent of the ova of women with fertility problems have chromosome defects. Not surprisingly, chromosome defects occur in some 8 percent of all pregnancies (about 1 in every 13 conceptions); the body is able to detect and reject more than 99 percent of them! Some 50 to 60 percent of the embryos or fetuses lost in the first three months of pregnancy have some kind of chromosome disorder. Fortunately, a woman's body has the remarkable ability to detect chromosome defects in the embryo or fetus and reject the vast majority.

In couples who experience miscarriage repeatedly, there is an increased likelihood that one of the partners carries (but is usually not affected by) a chromosome defect. After two miscarriages, there is a 1 to 3 percent likelihood that one member of the couple carries a chromosome defect that is being passed on to the embryo. After three miscarriages, that likelihood increases to 3 to 8 percent; it is therefore expected medical practice to offer, at least after three miscarriages, to arrange for blood chromosome analysis of *both* partners. If a chromosome abnormality is in fact detected in one partner, the risks of actually bearing a chromosomally defective child can be determined. This risk is rarely more than 20 percent and is usually less than 10 percent. Fortunately, prenatal diagnostic tests are available (chapter 22); they are recommended in all future pregnancies when one partner carries a chromosome defect.

Given the body's remarkable mechanism of rejecting chromosomally defective embryos and fetuses, the frequency of such defects drops from a high of about 78 percent at about two weeks of pregnancy, to about 3 percent at sixteen weeks, and to 0.6 percent in live-born infants. Stillborns and babies who die in the first few weeks of life have a 6 to 11 percent frequency of chromosome defects. This, incidentally, makes it critically important to check the chromosomes of every stillborn baby, not only to determine the cause, but to recognize any future risks in order to be able to offer specific options, including prenatal diagnosis.

Fortunately, although chromosome disorders are among the most common causes of birth defects, the remarkable efficiency of the body in recognizing and rejecting defective embryos or fetuses results in very few babies actually born with these disorders. Let us consider those that actually do survive and remain the source of major trouble and concern.

BIRTH DEFECTS DUE TO CHROMOSOME ABNORMALITIES
Many specific birth defects or groups of abnormalities (called *syndromes*) result from certain defects of individual chromosomes.

Chromosomal defects occur as frequently as 1 in every 156 live births (0.65 percent). In the United States alone, over 20,000 infants are born each year with such defects. About 0.4 percent of live-born babies have a *serious* chromosome defect, while an additional 0.2 percent carry a chromosome defect that may later have repercussions when that individual begins childbearing. The vast majority of serious birth defects due to chromosome abnormalities are naturally aborted by the body; these include fetuses or stillborn infants with such defects as extremely small heads, deformed brains and faces, cleft lips and palates, single nostrils, and abnormal ears and eyes, as well as heart and other defects. It should be remembered, however, that virtually every recognizable chromosome defect can be *diagnosed* in the fetus sufficiently early in pregnancy for the parents to consider the option of elective abortion (see chapter 22).

TOO MANY OR TOO FEW CHROMOSOMES

Sometimes there are too many chromosomes. Most commonly such abnormalities arise during the process of cell division. This process is easier to understand if you follow figure 5. It is the same basic diagram as figure 3; steps 1, 2, 3, and 4 are identical. But the crucial difference occurs in the following steps:

Step 5: The cell nucleus and the cell that it occupies divide into two, but this time the chromosome pair fails to separate.

Step 6: Both remain in one cell, which is now, say, the egg; the other cell ends up with all the other chromosomes but lacks this particular one. Most commonly this is the number-21 chromosome. When the egg with the extra chromosome is fertilized by a normal sperm, the resulting single cell has an extra chromosome.

The mechanism of chromosome separation could go wrong earlier in the process, for example between steps 3 and 4. The sperm or eggs formed would again end up with either one too many or one too few chromosomes. Any person born with an extra number-21 chromosome (see figure 6) in all or many of the cells will have all the features of Down syndrome, or trisomy 21 (previously called mongolism). This is the commonest kind of Down syndrome, and it constitutes about 94 percent of all such offspring. The remaining 6 percent of progeny with Down syndrome are caused by abnormal rearrangements of the chromosomes or by mixtures of normal and trisomy-21 cells (see below).

The phenomenon of one chromosome sticking with another during cell division — *nondisjunction* — is not confined to the extra number 21 in Down syndrome, but may occur with number 13, number 18, or any other chromosome, almost invariably causing serious defects in the child. This tendency to have "sticky" chromosomes increases with a woman's age; hence, so do the risks of bearing a child with a trisomy, such as Down syndrome. In addition to the mother's advanced age, there are other unknown influences that make the extra chromosomes stick in these trisomic disorders. Many factors have been suggested as causes, including a "stickiness" gene, X-ray exposure *before* pregnancy, virus infections, and diabetes or thyroid disease of the mother. In support of a hereditary factor is the fact that individuals with a near relative (sibling, aunt, uncle, or first cousin) with Down syndrome or another trisomy have a slightly increased risk (less than 1 percent) themselves of having a trisomic child.

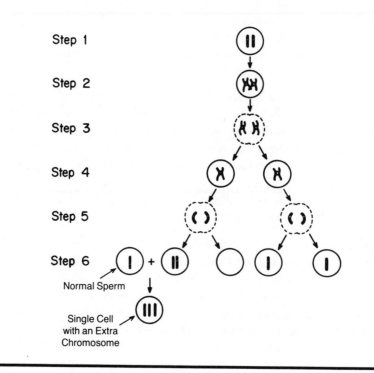

Figure 5

Abnormal chromosome division (meiosis) — for example, in the ovary — can result in one cell with an extra chromosome.

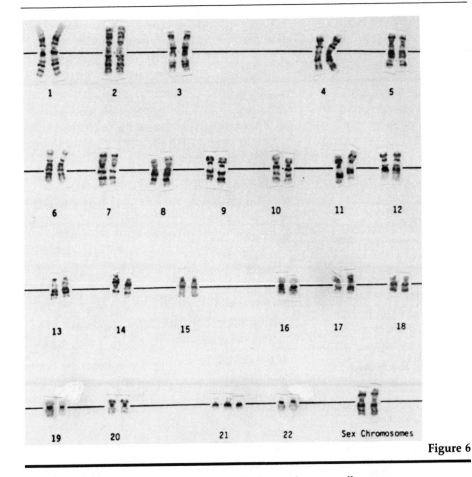

1   2   3   4   5

6   7   8   9   10   11   12

13   14   15   16   17   18

19   20   21   22   Sex Chromosomes

Figure 6

An abnormal number of chromosomes (47) from a human cell, as seen
in Down syndrome. The extra number-21 chromosome is typical of this
syndrome.

It has also been repeatedly observed that some of the major
chromosomal abnormalities, including Down syndrome, may
occur in clusters. A group of affected children may, for example,
be born in the same fall or winter. This clue has raised questions
about the role of viruses in causing chromosome "stickiness"
during cell division.

The presence of the extra chromosome number 21 was observed
for the first time in 1959 in France. Such a discovery in a fetus or
child means that the parents ultimately will have to face mental
retardation and the other features of Down syndrome (see table 1),
as well as various medical, emotional, and social problems. Those

Table 1. The Most Common Clinical Features in Three Trisomies (Disorders Caused by an Extra Chromosome) That Are Associated with Advanced Maternal Age

Trisomy 21 (Down Syndrome) 1:770 live births	Trisomy 13 (Patau Syndrome) 1:5,000 live births	Trisomy 18 (Edward Syndrome) 1:6,600 live births
Mental and physical retardation	Mental and physical retardation	Mental and physical retardation
Short stature	Undescended testes	Failure to thrive
Flat face	Widely spaced eyes	Undescended testes
Large tongue	Low-set and malformed ears	Difficulty feeding with poor suck
Oblique, slanting eyelids	Jitteriness and spells of no breath (apnea)	Heart defects
Cross-eyes	Small head (microcephaly)	Low-set, malformed ears
Flat back of head	Small, abnormal eyes	Elongated skull
Short limbs	Cleft lip or palate	Short sternum
Short, broad hands; short fingers, especially the fifth	Six fingers and toes	Single umbilical artery
High-arched, narrow palate	Heart defects	Contracted overlapping fingers
Short fifth finger (middle phalanx)	Presumptive deafness	Poor or excessive muscle tone
Poor muscle tone	Small chin	Short neck
Flat bridge of nose	Excess skin at nape of short neck	Heart defect
Abnormally shaped and low-set ears	Strawberrylike birthmarks	Groin or umbilical hernia
Incurved fifth finger	Long, curved nails	Small chin
Heart defects (in about 50 percent)	Flexible thumbs	
Single crease across palm	Contraction deformity of fingers	
Excess skin at nape of neck	Single crease across palm	
Intestinal (duodenal) obstruction	Prominent heel	
Premature aging	Poor or excessive muscle tone	

NOTE: An affected individual might not have every listed feature.

with Down syndrome (see figures 7–10) invariably are mentally retarded, short in stature, have typical flat faces with slanted eyes, are floppy (lack muscle tone) in infancy, and have some of the other common features or associated defects listed in table 1. In none of the chromosome disorders does an affected individual have every feature. Moreover, some of the individual features (for example, a single crease across the palm) may be present in perfectly normal persons.

The IQ level in Down syndrome is mostly around or below 50. IQs of up to 80 have, however, been reported. Lifetime care is required for all, and this is not made easier by premature aging — an unexplained feature of this chromosome disorder (see figure 10). Frequent loss of intellectual ability after thirty years of age in Down syndrome is probably related to degenerative changes in the brain similar to what has been found in the dementing disorder called Alzheimer disease (discussed in chapter 32). In fact, there seems to be higher frequency of Alzheimer disease in families with Down syndrome.

Life expectancy in Down syndrome, now that most patients are cared for at home, or at least not in large institutions, has increased dramatically. Just a few years ago, more than half had died by five years of age. Today, even though some individuals may reach seventy years of age, it is estimated that 79 percent of those without heart defects will survive to at least thirty years of age. Even with repaired heart defects, survival until at least fifty is now expected. Projections of a Scandinavian study estimate that *at least* 40 percent will survive to sixty years of age. Longer life expectancy has meant that brothers and sisters are almost invariably left with the burden of caring for those with Down syndrome, and where there are no siblings, or they simply refuse and one or both parents have passed on, the state almost always ends up with this responsibility.

Early deaths in Down syndrome used to be due mainly to respiratory infections and serious congenital heart defects. Now these are vigorously treated and other complications, such as leukemia (twenty times more common than in normal people), other cancers, diabetes, and common illnesses, claim their lives.

The presence of the extra number-13 chromosome usually produces a disorder called Patau syndrome, or trisomy 13, with profound mental retardation, a small head, malformations of the ears and eyes, and multiple other defects as listed in table 1.

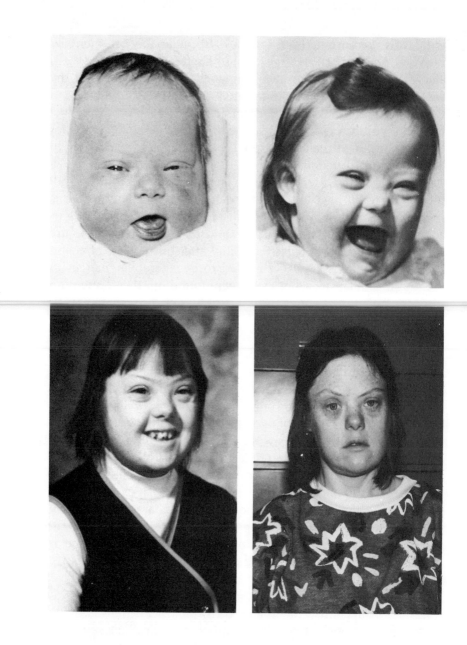

Figures 7, 8, 9, 10

Down syndrome in the same person as a newborn (*top left*), a toddler (*top right*), an adolescent (*bottom left*), and an adult of thirty years. Note the effects of premature aging.

Associated defects are frequently severe, and usually 86 percent have died by the end of their first year of life. Only an occasional child has survived beyond five years.

The presence of the extra number-18 chromosome — a condition called Edward syndrome, or trisomy 18 — also results in profound mental retardation and severe defects as listed in table 1. More than 30 percent of these profoundly retarded infants die during the first two months of life. Few survive childhood and only an occasional child has reached adolescence.

Very rarely, an infant is born with one too few of the regular (that is, nonsex) chromosomes. In most cases this abnormality is sufficiently serious as to be incompatible with life; if born alive, the babies have serious birth defects and invariably die soon after birth.

MIXTURES OF NORMAL AND ABNORMAL CELLS

During the process of cell division there can be mixture of normal and abnormal cells, with two different cell types emerging: one with normal chromosomes and one with an extra chromosome. The resulting individual may finally be made up of a thorough mixture of cells, or only certain organs or tissues may have abnormal cells. For instance, only the brain, the genital organs, the blood, and the skin may have cells with too many chromosomes, with all the other organs having normal cells. Affected individuals are described as having *chromosomal mosaicism*.

If, for example, 40 percent of the cells have the normal 46 chromosomes, while 60 percent have the extra number-21 chromosome, then the features of Down syndrome will be present but are likely to be milder, depending upon which organs have normal cells. As few as 1 to 4 percent of the cells of a person could have too many or too few chromosomes. Such individuals will usually be healthy, or possibly have infertility or miscarriage problems. The difficulty in these situations is trying to determine whether in fact such low degrees of mosaicism are actually causing the infertility or recurrent miscarriage. Generally, mosaicism of the regular chromosomes is much less common than with the sex chromosomes.

CORRECT NUMBER OF CHROMOSOMES, BUT DEFECTIVE STRUCTURE

Even when all the chromosomes are present, something may go wrong with one or two of them. These structural defects, generally the result of breakage, occur commonly: approximately 1 in every 500 live births. A whole variety of changes in chromosomal structure may result from chromosomal breakage. For example, two small pieces may break off the end of two different chromosomes and exchange positions. This process is called *translocation,* and about half the time it occurs spontaneously at the time of conception. Its other occurrences are inherited and passed down through the generations. Hereditary Down syndrome is the consequence of translocation between a number-21 chromosome and another, most often a number 14. When the chromosome pieces change places without any piece being lost in the exchange, the process is called a *balanced translocation.* The vast majority of balanced translocations are *not* associated with birth defects. However, for those balanced translocations that are not inherited from one parent, there is a *slightly* greater risk than random for having a child with mental retardation, birth defects, or both. The reason for this is unknown, but the additional risk may reflect damage to a gene at the site of breakage, malalignment of the exchanged piece, or disturbance in controlling functions from neighboring genes. Moreover, there is also an increased risk of chromosome "stickiness" (nondisjunction, mentioned above) and hence a slightly greater risk than random of having a child with some type of trisomy (for example, Down syndrome).

The chromosome with the translocated piece is vulnerable, so during fertilization or at the very time one cell (the fertilized ovum) splits into two, duplication of the whole or part of an involved chromosome may occur. This results in almost a whole extra chromosome being present (effectively a trisomy), yielding a triplet or an extra segment (called *partial trisomy*). This is termed an *unbalanced translocation* and it is this situation that is associated with serious birth defects and with a significant likelihood of recurrence.

There are a fair number of us who, without knowing it, are balanced carriers of various chromosomal abnormalities. Each year in the United States alone over 7,000 children are born with balanced or unbalanced chromosomes. Those of us who have a balanced-chromosome exchange are at risk of having defective offspring: mostly a 10- to 20-percent risk for prospective mothers and about a 2- to 4-percent risk for prospective fathers, depending upon which chromosome is involved. The lower risk for fathers with unbalanced chromosomes suggests that their abnormal sperm do not make it to fertilization. A rare situation in which a translocation occurs between a pair of number-21 chromosomes results in a child with Down syndrome in every pregnancy: 100 percent risk!

Recognition of these translocation situations is obviously of critical importance, as the following case shows:

> Jim and Barbara had tried for five years to have a child. Barbara had become pregnant three times during that period, but each time suffered a miscarriage during the second or third month of pregnancy.
>
> Because of the recurrent miscarriage, their obstetrician wisely suggested that they both have their chromosomes studied. The results showed that Barbara herself was a balanced-translocation carrier for Down syndrome and it was inferred that at least some of the miscarriages were due to defective embryos. Barbara and Jim were advised that they could still have unaffected children through prenatal genetic studies. This they did in a subsequent pregnancy; the prenatal studies showed a fetus with a balanced translocation who was therefore, like Barbara, perfectly healthy.
>
> Discovery of the translocation abnormality in Barbara led to a search for the same abnormality in other members of her family. Tests determined that her mother as well as two aunts and an uncle all had the same balanced-chromosome abnormality. Even more important, the continuing search revealed that four of Barbara's cousins were also balanced carriers of this chromosomal abnormality. Indeed, one of the cousins had already had a child with Down syndrome due to an unbalanced translocation. Another cousin, one of the balanced-translocation carriers, was pregnant at the time. The call from Barbara came in the nick of time. She had an immediate amniocentesis and prenatal study that showed she was carrying a fetus with Down syndrome, and she and her husband elected to terminate that pregnancy.

After Jim and Barbara had suffered three miscarriages, it was important for them to have chromosomal studies performed. After it was established that Barbara was the carrier of the translocation, the couple wisely availed themselves of the valuable prenatal studies. While their risks of having an affected child were about 11 to 15 percent, the knowledge they gleaned from the prenatal studies provided tremendous emotional relief.

Barbara also acted responsibly — and I wish this were the rule — by calling or writing every member of her mother's family. Through her intelligent behavior not only were other carriers detected, but one cousin and her husband were spared the sadness of having a child with serious birth defects.

MORE STRUCTURAL PROBLEMS

It is possible for a chromosome to break in two places somewhere along its course and for the two ends of the severed segment to reattach after turning upside down (see figure 11). Since the function of the genes, to some extent at least, depends upon their location along the chromosome, this process, called *inversion*, may (or may not) have untoward consequences. Usually, no abnormalities result in a child born with an inversion, especially if the chromosome in question was directly inherited from one otherwise healthy parent. If, however, that parent had a previous child with birth defects ascribed to the inversion, the risk of this recurring in a future pregnancy is about 7.5 percent for women (with the inversion) and 4 percent for men.

Couples who experience repeated miscarriage and among whom one partner is found to have an inversion *probably* do not have an increased risk of eventually bearing a child with birth defects due to the inversion. Nevertheless, it would be judicious to discuss and consider undergoing prenatal diagnostic studies

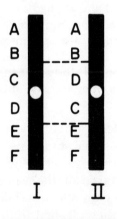

Figure 11

Inversion in one of two paired chromosomes. Breaks have occurred in two places, and the separated piece CD has reattached after turning upside down.

(discussed in chapter 23) each time pregnancy is successfully established. Depending upon the chromosome involved, a new inversion may result in such defects as severe mental retardation, a small head (microcephaly), congenital heart disease, and other major birth defects. In such cases a duplication or a loss (or both) of a chromosome segment involved in the inversion is the source of the problem. Other complicated structural abnormalities of chromosomes occur, but generally are rare.

DELETED CHROMOSOMES

On occasion a piece of a chromosome may simply break off and disappear — a process called *deletion*. Again, depending upon the size of the piece that is missing and which chromosome is involved, various birth defects may occur. The nature of these birth defects may vary from minor to extremely serious. I vividly remember one striking case in which a deletion had occurred along the number-5 chromosome.

Mary, then twenty-seven years old, had sought a consultation with me for a rather unusual reason. She had called the plumber in for repairs two weeks before her consultation. Her baby was then four weeks of age. While repairing her sink in the kitchen, the plumber asked her if she had acquired a new kitten. She was initially indignant, since what the plumber had heard was her baby crying. For the next two weeks, however, her attention began to focus more and more on the baby's cry, which she realized did indeed very much resemble the cry of a cat. Since the baby had also been extremely difficult to feed, she decided that a visit to the doctor was necessary.

She began by complaining about the feeding problem, but it soon became apparent that her main concern was the abnormal cry. She had had a perfectly normal pregnancy, and had no family history of hereditary diseases.

When I examined the child, I found that he was underweight. The cry indeed was remarkably similar to that of a cat. In addition there was a certain roundedness of the face, wide-spaced eyes, and some incurving of the fifth fingers of each hand. A heart murmur was also audible.

The clinical diagnosis I reluctantly and sadly made was the cri-du-chat (cry-of-the-cat) syndrome, which had at that time barely been recognized. Analysis of this baby's chromosomes confirmed the clinical diagnosis and also the poor prognosis of grave mental and physical retardation, which became apparent in the next few months and years.

Only recently, with new advances in technology, has it been possible to detect reliably a deletion of a tiny segment along the course of a chromosome. These minute deletions are not always visible, even under the high-powered microscope. Examples of disorders in which deletions are frequently (but not always) visible include Duchenne muscular dystrophy, the Prader-Willi syndrome, retinoblastoma, and Wilms' tumor.

BROKEN CHROMOSOMES

Many environmental agents may cause breakage of chromosomes, as frequently happens in circulating white blood cells. X rays (as when used for outlining the bowel in a GI series), viral infections, or various medications are common examples. These changes are usually transitory and cause no hereditary or future ill effects.

There are, however, a few disorders resulting from autosomal recessive inheritance (discussed in chapter 7) in which chromosome breakage is typical. Affected individuals share a tendency to infection, because their immune systems fail to function properly, as well as a tendency to develop cancer. The three most important conditions — Fanconi syndrome, ataxia telangiectasia, and Bloom syndrome — are discussed in later chapters.

Virtually all the discussion in this chapter has been confined to consideration of the 44 regular chromosomes. Unfortunately, problems with the remaining pair, the sex chromosomes, occur even more commonly, and these form the substance of the next chapter.

3

Sex Chromosomes

The function of the sex chromosomes is more than simply to "dictate" the sex we become. Sterility, absence or abnormality of menstrual periods, and speech, language, and learning problems, to name but a few difficulties, may stem from disorders of the sex chromosomes. Indeed, for reasons not understood, abnormality of the sex chromosomes often has a deterrent effect on intellectual development and function.

As noted in chapter 2, each of our cells normally contains 46 chromosomes, one pair of which pertain to sex. The female sex chromosomes are called X, and a woman normally has two of them (XX). The male's sex chromosome is called Y; he normally has one Y and one X (XY) in each cell. However, either male or female may be born with too many X or too many Y chromosomes; also, an X or a Y may be missing. The mechanism in the embryo that produces this error is the same one that can occur with all the other chromosomes — the mechanism described in the preceding chapter.

Individuals possessing the Y chromosome will always appear to be males even if they have two, three, or even four X's as well. A male who appears to have no Y chromosome is an extremely rare exception. Very recent elegant work has shown, however, that this individual (called an XX male) has a segment of the Y chromosome inserted into another chromosome, not easily visible, and the power of the sex-determining gene on that tiny piece of the Y chromosome is still operative (see "Rarer Additions," below).

Sex chromosomes can suffer the same breakage or deletion as the other chromosomes. A person can be born with some cells containing normal sex chromosomes and others containing abnormal ones. Such a condition produces what has already been designated as a mosaic individual — a fairly common occurrence for sex-chromosome disorders. Indeed, sex-chromosome abnormalities generally occur at least once in every 500 births!

Too Many Sex Chromosomes

MALES WITH AN EXTRA X

In 1942 research was done on nine male patients who had abnormal breast development, very small testicles, and no sperm production. Certain sex hormones found in their urine were at the high level usually found in castrated males. Their body build was feminine and resembled that of the traditional Mideastern eunuch.

This condition, called *Klinefelter syndrome,* is not at all rare, occurring in 1 in 1,000 males. It usually becomes apparent and is mostly diagnosed only after puberty. Males so affected are apt to be taller than normal, frequently have learning difficulties in school (especially language-based dyslexia), and are mildly to moderately retarded in up to 15 percent of cases. Personality and behavior problems are common. (Table 2 lists additional common features.)

Males with Klinefelter syndrome are often remarkably indolent about their own care and future, and are prone to alcoholism and other antisocial behavior. Generally, they tend to be reticent, passive, and low-keyed. Mental illness, both neurosis and psychosis, are common, and depression as well as periods of mania may occur. Sexual behavior is normal, even though some individuals have been convicted for sexual offenses and petty crimes and have ended up in mental institutions or prisons. While their sexual behavior is normal, their libido (sexual drive) tends to be low. Many have managed to lead a normal married life.

Breast cancer is very much more common among men with Klinefelter syndrome. Life expectancy for men with this syndrome appears to be shortened by about five years, while their overall mortality rate is increased 50 percent. Strokes, brain hemorrhage, chest infections, disease of the aortic heart valve, and cancer of the breast account for almost all of their increased mortality rate.

Sterility in this disorder is irreversible: the small testicles produce no sperm. Breast development, which often becomes obvious at puberty, can be quite embarrassing but can be treated with male hormones such as testosterone or surgically corrected by mastectomy. Such hormones deepen the voice, cause facial hair to grow, and perhaps slightly increase the size of the penis. Some men with the syndrome claim that testosterone enhances their libido. Frequently, even though they claim the hormone helps them, they tend to give it up.

In 1956, some fourteen years after the recognition of the clinical signs of this disorder by Dr. H. Klinefelter, several groups of researchers working in different countries discovered that these males had an extra X chromosome: their sex-chromosome pattern was XXY instead of simply XY, as in the normal male.

I recall Joey, who was my first patient with the XXY, or Klinefelter, syndrome.

Joey, age sixteen years at the time, was brought for consultation by his parents, who were concerned because he had been devel-

Table 2. The Most Common Clinical Features in Two Disorders That Are Caused by an Extra Sex Chromosome and Are Associated with Advanced Maternal Age

Klinefelter Syndrome (XXY Males) 1:1,000 males	Triple-X Females (XXX Females) 1:1,000 females
Small testes, normal penis	Normal appearance
Sterility/infertility	Menstrual irregularities
Mental retardation in up to 15 percent	Mental retardation (in a few cases)
Female body build	
Excessive breast development in 50 percent	Mental illness
Tall stature (especially long legs)	Relative infertility
Personality/character trait disturbances	Early menopause
Emotional/behavioral problems	Learning disorders
Psychiatric/sexual problems	
Various nervous-system abnormalities	
Osteoporosis (thin bones)	
Bone anomalies	
Delayed language acquisition	
Obesity in adulthood	
Learning disorders	

NOTE: An affected individual might not have every listed feature.

oping rather prominent breasts. They told me that Joey had always been a quiet, withdrawn child who seemed to have difficulty getting along with other children, and had emotional and behavioral disturbances both in school and at home. He had been kept back at certain stages of his school career and was now a full three years behind his peers.

He had always grown faster than normal and at the time of the consultation was much taller than boys his own age. The only abnormal physical sign that the parents mentioned was curving in of his fifth fingers. My examination also revealed his prominent breasts and small testicles. He seemed extremely immature, with borderline intelligence.

Chromosomal studies showed that he had three sex chromosomes, one of them an extra X. This was an important diagnosis to make, since some therapy was available. He was given male hormones as a long-term treatment. His breasts returned quite soon to normal male size. He became more aggressive and was able to lead a normal sex life, though he remained sterile.

MALES WITH AN EXTRA Y

Individuals born with an extra Y chromosome are, of course, always male. The XYY syndrome has been associated in the past with a tendency to criminality. This whole matter has generated considerable controversy and is discussed in detail in chapter 5.

FEMALES WITH AN EXTRA X

Once misleadingly called "super females," women with three X chromosomes (instead of two) usually appear to be entirely normal physically. This common disorder, affecting 1 in 1,000 females, is usually associated with no obvious serious physical birth defects. (Table 2 lists common features of the disorder.) Minor signs, such as an incurved fifth finger, might be present, and later in childhood and in adulthood these women may tend to be taller than most, occasionally outstandingly so. In contrast, their heads in general tend to be somewhat smaller than average. Even though XXX women are found two to ten times more often among the mentally retarded and mentally ill (psychotic), the actual frequency of these disorders among such women is not known exactly. Epilepsy, too, seems to be more frequent in XXX women, in whom brain-wave (electroencephalographic) abnormalities have often been seen.

Careful studies following young girls diagnosed from birth

clearly show that XXX girls have lower IQs than their siblings. *Very few,* however, appear to have mental retardation; they may have a normal or low-normal IQ. Speech and language development is frequently delayed and learning disorders are common.

Menstrual difficulties are not unusual and include late onset of periods, scant to absent periods, infertility to sterility, and early onset of the menopause. Most XXX women lead a normal sexual life, and despite their chromosome complement, most of their children have normal chromosomes. In view of their theoretical risks, however, XXX women are advised to have prenatal genetic studies.

A significantly increased mortality rate among XXX women has been reported. This conclusion was based on a study of 94 triple-X women, done by a Medical Research Council unit in Edinburgh, Scotland, where so much of our knowledge on sex-chromosome abnormalities originated. The researchers followed these women for an average of sixteen years, during which time 24 died — double the expected death rate. Although this observation was highly significant, no specific single cause of death was recognizable.

RARER ADDITIONS

Truly bizarre sex-chromosome numbers occasionally do occur. For example, there have been persons born with 48 or 49 chromosomes in each cell instead of the usual 46. All these individuals have been severely or profoundly mentally retarded. Women with four X chromosomes instead of two in each cell have been noted to appear physically sound and to menstruate normally. They have, however, been severely retarded, with IQs of less than 50.

An unexpected "rare addition" has emerged as the explanation for a disorder of males born only with two female sex (XX) chromosomes. About 1 in 25,000 newborn boys is an XX male having essentially the same features as described above for Klinefelter syndrome. Two distinguishing features are that these males are shorter and are invariably of normal intelligence compared with males with Klinefelter syndrome. Recent discoveries using DNA technology (see chapter 8) have elegantly demonstrated that male-determining genes (originally from a Y chromosome) were incorporated in one of the X chromosomes of these XX males!

Too Few Sex Chromosomes

A deficiency of sex chromosomes can also wreak havoc. A female born with a missing X chromosome will invariably develop a well-recognized group of physical abnormalities collectively called Turner, or XO (monosomy-X), syndrome (see table 3). At birth the infant appears to be a female, and frequently is noted to have striking puffiness of the back of the hands and feet. This puffiness disappears slowly in the first year of life. In adulthood these women are usually short (under 5 feet), and characteristically have webbing of the neck and a typical facial appearance. While they do develop underarm and pubic hair, their menstrual periods, almost invariably, do not appear. The ovaries are internally underdeveloped and are replaced by a streak of connective tissue. Except for rare exceptions, these women are unable to have children. (However, new advances and successes do give hope — see chapter 19.)

IQs are considered to be within the normal range in Turner syndrome. Some females who appear to perform tasks at a

Table 3. The Most Common Features of Turner Syndrome

Turner Syndrome (XO Syndrome) 1:3,500 females	Turner Syndrome (cont.)
Short stature	Kidney or urinary-tract anomaly (or both)
Absent menstrual periods	
Sterility	Underdeveloped, curved nails
Infantile sex organs	Short fingers and toes
Lack of breast development	Birthmarks
Transient swelling (hands and feet)	High-arched palate
	Short neck
Webbed neck	Cross-eyes
Increased carrying angle at elbow	Defective hearing
Low hairline in neck	Small chin
Shield-shaped chest	Bony defect of chest wall
Cardiovascular anomaly	Learning disorders
Hypertension (high blood pressure)	Thyroid disorders
Osteoporosis (thin bones)	

NOTE: An affected individual might not have every listed feature.

lower-than-expected level have a characteristic of Turner syndrome in that they exhibit difficulties with spatial perception and mathematics. Their personalities are sometimes characterized as immature and childish. Psychological problems, if they arise at all, may become severe in adolescence, when lack of growth and breast development in these women become obvious to them in comparison with their peers.

One key problem is short stature. Hence, early diagnosis of this disorder makes it possible to initiate treatment with human growth hormone (now made by recombinant DNA technology — see chapter 8), which enables these children to grow taller than they would have without treatment. (Human growth hormone derived from the pituitary gland should no longer be used, because in at least three cases recipients have died from a rare brain virus.) Administering certain female hormones when the greatest height has been achieved facilitates reasonable development of the external genitals and breasts, and leads to the appearance of pubic and underarm hair as well as menstruation. Cyclical hormone treatment is usually recommended until forty to fifty years of age. This therapy is critically important during the teens for the development of emotional stability and good mental health in women with Turner syndrome.

An underactive thyroid, high blood pressure, and osteoporosis may be notable in some women with this syndrome. Cyclical estrogen hormones are taken by these patients for breast development and to allow them to have menstrual periods, although they are usually doomed to sterility. However, in 1986 two such women, with the aid of hormone therapy and in vitro fertilization (see chapter 19), were able to bear their own children!

As noted earlier, in about 50 percent of all early miscarriages the embryos or fetuses have chromosomal abnormalities. For reasons that are presently still unclear, the commonest abnormality is Turner syndrome, which occurs in about 20 percent of all miscarriages. It is reassuring that nature is so efficient in ridding the body of abnormal fetuses that only about 2 percent of Turner-syndrome conceptions actually complete pregnancy with a resulting live birth! It is also curious that there appears to be a five- to tenfold increase in twinning among the brothers and sisters of these affected patients. Strangely, too, most of these twins are identical.

Females born with only part of one of their X chromosomes

missing show features similar to Turner syndrome, but usually these traits are less prominent.

Mosaic Chromosomal Patterns

Mosaicism, earlier described as a mixture of cells each with normal or abnormal chromosomes, may be extremely difficult to diagnose. Affected individuals vary, depending upon which organs are involved. There are many parents who appear on testing to be completely normal but have a child with a sex-chromosome disorder. Subsequent studies may reveal that one of the parents is a "mosaic" for that particular disorder, an individual who because of the presence of normal cells in vital organs appeared not to be affected.

It can be extremely difficult to confirm a suspected diagnosis of mosaicism. The most common tissues that are studied are blood cells, skin cells, and bone marrow cells. Generally, if no mosaic pattern is found in either blood or skin, the person is very unlikely to have a mosaic abnormality. It is nevertheless impossible to rule out a mosaic pattern in any particular person, since it is conceivable that all cells in blood, bone marrow, and skin may have a normal constitution, while the brain, ovary, or testes cells (for example) may have an abnormal set of chromosomes.

There are some other disorders that involve the sex of an individual but are most often not reflected by any abnormality, inherited or otherwise, of the sex chromosomes. In most of those affected, something went awry during very early fetal development, when the sex hormones were influencing the formation of the sex organs. Some of these conditions are characterized by unexpected findings, such as apparent males with two X chromosomes. This produces an ambiguity that is beset with problems.

It should be made clear at this point that research has *not* found that homosexuality, transvestism, and other psychological sexual disturbances are caused by sex-chromosome abnormalities. So-called *intersex* states, in which the genitals may be ambiguous, pose special problems, which the next chapter examines in some detail.

4

The Remarkable Conditions of Intersex

"It's a boy!" "It's a girl!" The ringing cry of the obstetrician or midwife as the baby emerges through the birth canal is never forgotten by a mother or father. Happily, this declaration of gender is almost invariably correct. There are, however, certain disorders, which may or may not be inherited, that serve to confuse interpretation of a child's real sex. Doubt most commonly arises when it is unclear from examination of the genitals whether the child is male or female. This situation, described as *ambiguous genitalia*, may arise as a result of various disorders.

All parents are extremely upset when ambiguous genitalia are discovered. They may have already announced the birth of their "son" or passed the word following results of prenatal diagnosis as early as eleven weeks of pregnancy. Anxiety, embarrassment, and worry about the cause and the future all compound the difficulties of coming to grips with this problem.

Most people are not aware of how complicated assigning a person's sex can actually be. First there is *genetic sex*, which is determined by a male-determining gene on the Y chromosome. This gene could be translocated to another chromosome, but as long as it is present, the individual will become a male. Hence, even when a male is born with two "female" X chromosomes (an XX male), this male-determining gene has been detected on one of the X chromosomes. *Chromosomal sex* is reflected by the presence of two X chromosomes in a female and an X and a Y present in males. Mixtures of male and female cells (mosaicism) may occur in an individual and affect the appearance of the genitals. *Gonadal sex* refers to microscopic examination of what is thought to be the testis or ovary. In certain conditions, microscope study reveals

45

that neither ovary nor testis is present, that both an ovary and a testis is present in the same individual, or that the ovary and testis are combined as one organ. The sex of the genitals may vary in the circumstances.

Genital sex is the usual way we assign the sex of a baby. However, when ambiguous genitalia are present, this method is unreliable in determining the genetic sex, but is critically important in determining the appropriate *rearing sex*. Gender identity — that is, how an individual personally establishes his or her sexual identity — is heavily influenced by how the child is reared. Parents largely influence gender identity by the name they give the child, the style of dress, the haircut, the toys used, and their own psychological attitude. Hence, it should be clear that a person could have female (XX) chromosomes, a male-determining gene on one of them, and male or female genitals.

The matter can be further understood by considering how normal sexual development occurs from the time of conception, when the genetic sex of the embryo is determined. For the first few weeks in the life of the embryo, the developing sex organs are neither male nor female. If the male-determining gene on the Y chromosome is present, the indifferent developing sexual organs transform into testicles in the seventh week of pregnancy (except in extremely rare conditions). If this male-determining gene is not present, the tissue develops into ovaries instead, identifiable by the tenth week of pregnancy. Male hormones made by the developing testes are responsible for making the male sex organs. If these hormones are absent or are functionally abnormal, the child's genitals will appear as female. Male hormones also influence the development of the male's internal genitalia, which include the various tubes and canals that transfer the sperm from the testes through the penis. Absent or functionally abnormal male hormones will result in abnormalities of these channels.

The presence of another hormone inhibits development of a uterus in a male. The absence of this hormone allows the full development of a uterus and Fallopian tubes in normal females. Clearly, abnormalities in this inhibiting substance will seriously influence the presence or absence of a uterus — even in a genetic male. In summary, the apparent sex of an individual as seen by genital examination reflects the consequences and interactions of genes, chromosomes, and fetal hormones. Hormones taken by the mother (such as certain oral contraceptives) may override the fetal hormones and result in genital masculinization of a female fetus.

With a clear view of the elements that contribute to identifying an individual as male or female, we can consider the three main groups of disorders in which ambiguous genitalia occur. Appropriate treatment is contingent upon understanding the cause. The term *intersex* describes disorders in which internal or external genital organs of both sexes are present. A person may have a penis as well as a vagina and have one ovary and one testis. Such an individual may or may not be sterile, and the sexual organs may remain infantile or may mature fully. The penis, for example, may remain the size of a clitoris.

True Hermaphrodites

The hermaphrodite has always been a symbol of tragedy as well as a subject of humor. A true hermaphrodite possesses both testicles and ovaries, either as two separate organs or as a single, combined organ. The external sex organs in these patients vary tremendously and do not help in the diagnosis. They may appear perfectly female, perfectly male, or have external evidence of both male and female sex organs. Most of these individuals (over two hundred have been reported) have normal female chromosomes. Some, however, have normal male chromosomes, while others are mosaics of mixed female and male chromosome sets. Rarely, a person may have two distinct sets of cells, both having female sex chromosomes (XX) or one group with male (XY) and another group with female chromosomes. This remarkable situation of a person having two entirely distinct sets of (all) chromosomes — called a *chimera* — is the result of either two sperm fertilizing an ovum with a double nucleus or two fertilized ova fusing together and developing into a single person rather than twins. Such persons have two different sets of blood groups! They are otherwise invariably normal (sometimes with short stature).

Rita and Tom had been happy to announce the birth of their first son. At the very first examination after birth, the pediatrician noted that instead of the hole, or meatus, being situated at the end of the penis, it was located beneath and at the base of the organ (a condition called hypospadias). One testicle was located in the groin. No other abnormalities were noted, and the child underwent surgical repair of his penis. He grew and developed normally and reached puberty without any problems.

At puberty, however, he began to develop very prominent breasts. At the same time, the child complained that he was passing blood in the urine for a day or two each month. The parents brought him in for an immediate consultation.

On examination the boy looked particularly feminine in build, with large breasts; the penis was rather small; one testicle was in the groin, and the testicle on the other side did not feel normal and was associated with a groin hernia on the same side.

I suspected an intersex condition and proceeded with various studies. The boy's chromosomes turned out to be female (two X chromosomes). Exploratory surgery later revealed that he had a poorly developed testicle on one side, and on the other side, in association with the hernia, a combined, poorly developed ovary and testis. In addition, he had a small vagina connecting at the base of the bladder, which explained the bleeding in the urine: it was a menstrual period! A tiny rudimentary uterus was connected to the small vagina. The diagnosis of a true hermaphrodite was made on the basis of his having both an ovary and a testis.

Because of the risks of malignancy, both the functioning testis and the ovary-plus-testis on the other side were removed, together with the rudimentary vagina and uterus, and the hernia was repaired. The use of male hormones made the breasts regress to normal male size and allowed the boy to continue growing up as a male, albeit sterile.

Pseudohermaphrodites

In this condition there is a discrepancy between the sex of the genital organs and the chromosomal sex or gonadal (ovary or testis) sex. The external genitals are not necessarily ambiguous.

FEMALE PSEUDOHERMAPHRODITES

Females with this condition have external genitals that resemble a male's, although their chromosomes are female (XX). A number of disorders, all sharing a mechanism in common, can result in a female pseudohermaphrodite. In essence, this mechanism interferes with the normal production by the adrenal gland of certain cortisone-like hormones (resulting in deficiency of that hormone), while other hormones (such as masculinizing hormones) are produced in excess, which results in the virilization of the female genitals. The clitoris enlarges and looks like a penis and the labia fold over to resemble a scrotum.

By far the most common cause of female pseudohermaphroditism is an inherited disorder called *congenital adrenal hyperplasia* (*CAH*). Surprisingly, the incidence of CAH in the United States and Europe is as common as 1 in 5,000 to 1 in 15,000. About 1 in 50 people are thought to carry the gene for this condition, which is transmitted as an autosomal recessive trait (explained in chapter 7). In this disorder the deficiency of an enzyme called 21-hydroxylase is the reason why critically necessary cortisone-like hormones are not formed. More important than the masculinization of female genitals is the life-threatening vomiting and collapse of the child in the first few weeks of life. A rapid diagnosis (easily made, if considered) will save the child's life. Lifetime treatment to replace the missing hormone will be necessary for the child, who will otherwise develop normally.

The gene responsible for the deficient enzyme happens to lie very close to genes for certain important "blood groups" called the HLA complex (see chapter 9). This close linkage can be used to determine if a person carries the gene for CAH and in fact, used together with hormone measurements of amniotic fluid, makes prenatal diagnosis possible in early pregnancy.

While CAH is the most important and commonest cause of ambiguity of the female genitals, another, less common disorder, called *gonadal dysgenesis*, may have similar effects. This condition, characterized by normal female chromosomes, is also inherited as an autosomal recessive disorder and is usually called "true gonadal dysgenesis." A similar condition, also inherited as an autosomal recessive disorder, appears in individuals with female genitals but whose chromosomes are normal male (XY).

Besides some other rare inherited disorders that cause female pseudohermaphroditism, certain hormones with male effects (some oral contraceptives, for example) may override the effects of female hormones produced by the fetus. The result could be masculinization of a female child — something that happens in about 0.3 percent of women taking such hormones during very early pregnancy. In rare cases, mothers with a tumor that is producing hormones with a male effect could also deliver a masculinized female child.

MALE PSEUDOHERMAPHRODITES

An affected individual of this type has testicles and normal male chromosomes (XY), but incompletely masculinized, female-appearing genitals. This disorder, called the *testicular feminization syndrome*, is transmitted as an *X-linked* condition: the gene responsible resides on one of the X chromosomes of the mother. The problem is not one of hormone production but rather a failure of the genital tissues to react normally to male-hormone (testosterone) stimulation during development of the embryo. The result is that the external genitals develop as a female while the internal genitals are male. Consequently, the usual way such a female-appearing individual comes to medical attention is at puberty, when the complaint is absent menstrual periods; on examination, a groin hernia is frequently found and surgical repair reveals a testicle within the hernia sac!

Male pseudohermaphrodites often come to medical attention soon after birth or in infancy, however. This usually happens if the tube that transports urine from the bladder through the penis exits somewhere along the undersurface of the penis between its tip and base, and there are no testicles in the scrotum. The testes in these individuals are usually found during surgery, either in the groin or within the abdomen. The tissue is either that of a normal testis or is simply fibrous in nature. The abnormalities found reflect a failure in the production of the male hormone testosterone or inability to convert testosterone to another hormone responsible for normal formation of the external male genitals. In essence, a male pseudohermaphrodite comes about because of a failure to manufacture or promptly utilize the male hormone testosterone. A whole host of disorders are recognized in which such faulty chemistry occurs.

Male pseudohermaphrodites may have all normal male chromosomes (XY) or may be born as mosaics, having some cells with normal male chromosomes mixed with others containing only one X but no Y chromosome (X/XY mosaics). Their testicles are not normally formed and fail to produce adequate testosterone. Even though the chromosomes are male (XY), the external genitals appear to be female and the clitoris may be enlarged. Internally, a tiny uterus and attached Fallopian tubes may be present. Tissue in the testicles is mainly fibrous. A critically important potential complication is that 15 to 20 percent of such individuals will develop a malignant tumor in a "testicle." Surgical removal of both "testicles" in those with this condition is recommended immediately after diagnosis.

Male pseudohermaphrodites may also result from at least five different enzyme deficiencies that interfere with the normal formation of testosterone. Some of these conditions are inherited as autosomal recessive disorders and require very careful diagnostic investigation to distinguish one from another. They may be characterized by ambiguous genitals at birth, breast development in a male at puberty, absent development of a beard, or other disturbances in the development of sex characteristics. Another, separate disorder in which there is an autosomal recessively inherited inability to convert testosterone to another hormone means that the following will not occur: beard growth, receding hairline, acne, and enlargement of the prostate gland.

For all these intersex states, various combinations and permutations occur. Consultation with a medical geneticist or endocrinologist is very important to sort out the correct diagnosis. Clearly, this is important not only for appropriate treatment but also because of hereditary implications related to future childbearing.

Diagnosis and Treatment

Diagnosis of one of the intersex states arises mostly in the newborn period or at puberty. Occasionally, the question of sexual ambiguity may arise in the late teens or even early twenties. Certainly, the appearance of ambiguous genitalia in the newborn constitutes a medical emergency, since congenital adrenal hyperplasia could be life-threatening. Given the complicated set of possible diagnoses, parents are generally advised to seek help from specialists (an endocrinologist and a medical geneticist) in a medical school–based hospital. Tests that can be anticipated are likely to include hormone measurements of blood and urine, blood chromosome and biochemical tests, pelvic ultrasound studies, and possibly X-ray studies to determine the presence and normality of internal genital organs, and surgical exploration for the same purpose.

Once a definitive diagnosis has been made, a treatment plan can be formulated in consultation with the patient's parents. Depending upon the diagnosis, recommendations might include correction of the external genitals by plastic surgery, removal of abnormal internal genital organs (such as an abdominal testis), initiation of hormone treatment for life, or, beginning near puberty, X rays of the urinary tract and kidneys (because of the frequent association of urinary-tract abnormalities).

Evaluation may take about one week, but only after a definite diagnosis has been made should the next critical issue — gender assignment — be tackled. The medical and nursing staff must display the most sensitive insight possible in dealing with the anxieties of parents, not to mention teenaged patients. Extremely careful consideration is required when deciding whether the child will be reared as a male or reared as a female. Important considerations include whether puberty will conform to the assigned sex as well as to the individual's potential for normal sexual function and future ability to bear children. One major question that requires immediate resolution is whether an affected infant, despite having normal male chromosomes as well as testicles, will have an adequate penis to function as a male. It might be necessary to remove a tiny penis and abnormally located testes surgically, while reconstructing the external genitals by creating a vagina that could at least function for sexual intercourse. One critical recommendation is that all reconstructive surgery should be done before 2½ years of age to avoid the psychological problems associated with confused gender identity.

5

The Myth of the Criminal Chromosome?

Has heredity anything to do with crime, or are all the causes environmental? Given the appalling and increasing amount of crime and violence, this is more than a simply theoretical question. Men, for example, have been consistently found to be criminal, antisocial, or alcoholic more often than women, regardless of their race, social history, or cultural background. Female criminals have an excess of *both* sons and daughters who are criminals. This in fact may be due to their own fathers' criminality. Some studies on identical and nonidentical twins have shown that in the case of identical twins more often *both* are involved in criminal acts. The implication is that some hereditary factor is involved. Unfortunately, however, none of these studies was performed in a way that would take into account familial and environmental factors; hence, no satisfactory evidence can be adduced from them to suggest a hereditary criminal factor.

The same point is made by a more common example. Consider for a moment families you may know in which the children and the parents alike are markedly overweight. Any initial impression that their obesity may be due to inherited factors could be readily dismissed after realizing the remarkable quantity of food consumed by the entire family!

Distinguishing hereditary from environmental factors that lead to crime is difficult. A whole host of social factors is obviously involved: broken homes, child abuse, psychological and physical deprivation, poverty, overcrowded life in ghettos, and so on. These and other blows of fate during childhood may be sufficient cause for the development of criminal or deviant behavior. Yet it is quite possible that a predisposition to mental illness or psychopathic personality may play a strong if not major part.

While it is well known that child abusers are frequently individuals who were abused in their own childhood, there has been no evidence to suggest that the practice is hereditary. Violent crimes may be perpetrated by different members of the same family. Take the case of one Ohio family. When the youngest son was four years of age, his father murdered the boy's mother. As a boy of eighteen, this son also committed murder. Did this boy inherit his father's excessive aggression? Was this simply a coincidental occurrence in the same family? Or was it simply a reflection of early environmental trauma?

Swedish researchers have approached the problem by studying adopted individuals, their biologic parents, and their adoptive parents. Adoptees whose biologic *and* adoptive fathers were criminals had a remarkable 36-percent rate of criminality. Prolonged institutional care and adoption in an urban setting were noted to be the most important environmental factors contributing to criminality in female adoptees. Petty female criminals more often had biologic parents with petty criminality than their male counterparts. While antisocial personality is less common among women, those so predisposed have more affected relatives than do antisocial men. Maladjusted girls, who may also exhibit misconduct in school, hyperactivity, and attention deficits, more often had biologic fathers with criminality than did similarly affected males.

While *some* genetic effect has emerged for *petty* criminality, we are clearly not "prisoners of our genes." Environmental factors interacting with our genes may explain some predisposition to crime. This may even be true in the case of a commonly occurring disorder — the XYY male — which we need to examine in detail.

The Male with an Extra Y Chromosome

In the early 1960s, research scientists in Edinburgh questioned whether abnormal sex chromosomes might predispose individuals toward deviant behavior. To determine whether the presence of an extra Y chromosome had anything to do with unusually aggressive behavior, the researchers studied the chromosomes of male patients who were mentally subnormal and being held for treatment in special high-security institutions after committing various crimes of violence. Some 6.1 percent of the 196 males studied showed abnormal chromosomal patterns. Of these males, 3.6 percent had an extra Y chromosome — the sex-chromosomal pattern called XYY.

Scientifically reliable information about expected development and prognosis for XYY males is still very limited. There are two main reasons. The first is that there are no studies yet that carefully document the progress and development of XYY males from birth to adulthood or beyond. The second reason reflects the effects of a self-fulfilling prophecy (discussed below) that could influence a child's behavior. Most researchers report that in early childhood these boys seem to have lower intelligence than their peers (they later catch up), as well as educational and behavior problems that are especially related to mild learning disorders, delayed speech, and some impaired language skills. Physically, most are tall for their age, and they commonly develop severe acne in their teens.

A few children with an extra Y chromosome have been reported to exhibit an extremely defiant nature, destructiveness, outbursts of temper when frustrated, and an inclination to climb to dangerous places — all evident by four years of age. You do not have to be an expert in psychology, however, to recognize that there must be very many children whose behavior fits this description but whose chromosomes are normal. We will need decades' more study before specialists can accurately inform a couple with a newborn (or a fetus) with XYY chromosomes about expectations and prognosis.

Characteristics of XYY Males

Men with the XYY sex-chromosome pattern are usually tall, tend to have severe acne, often have lower IQs than their siblings, and at least 15 percent of the time have learning disabilities (see table 4). IQs in perhaps 40 percent of all XYY males range between 70 and 89. Verbal IQ is usually more affected than performance IQ. Over half of these men have perfectly normal IQs, and one registered a genius IQ score of 145. They tend to be impulsive, to react poorly to frustration and adverse circumstances, and may exhibit wild tempers. They are *not*, however, more violent or aggressive than other males. Their sexual behavior does not appear to be abnormal. Antisocial behavior, together with their lower IQ and emotional lability, lands them in trouble with the police at least ten times more often than males with normal chromosomes.

Studies of XYY males in the 1960s were initially done only on tall prisoners who had committed violent crimes. We now know that an XYY male may be entirely normal in most respects or may have mild to severe behavioral disturbances associated with an immature personality. Other characteristics may include a weak concept of self, a serious risk of social maladjustment, a borderline-normal intellect, or nonspecific physical features (such as incurved fifth fingers).

It is *very* important to realize that all the above traits have been described in XYY males who have come to medical attention. Many men with the XYY pattern, if not most, have not been recognized, and they may have managed their lives no differently from other males. To understand the true nature and frequency of problems XYY males could have, researchers need to study them from birth.

XYY Fathers

Since XYY males are fertile, they make sperm that each contain either only one X, only one Y, an X and a Y, or two Y chromosomes. If one of these sperm fertilizes an ovum with one X chromosome, the possibilities are that the child will be a normal girl (XX), a normal boy (XY), an XYY male, or a male with

Table 4. The Most Common Features of XYY Males

XYY Males 1:1,000 males	XYY Males (cont.)
Tall stature	Large teeth and long ears
Normal appearance	Severe acne in adolescence
Personality disorder in some	Occasional undescended testes
Antisocial/criminal behavior in some	or small penis
Tendency to low IQ	Learning disorders

NOTE: An affected individual might not have every listed feature.

Klinefelter syndrome (XYY). There are few reports that document chromosome abnormalities in the children of XYY fathers. The likely explanation is that sperm with an abnormal number of chromosomes are less apt to achieve successful fertilization. Nevertheless, the partner of every XYY male should have prenatal diagnostic chromosome studies in every pregnancy. While the exact risk for bearing a child with a chromosome defect is unknown in these circumstances, it is higher than random chance. Moreover, while advanced maternal age has not been associated with having an XYY male, the phenomenon of "sticky" chromosomes (nondisjunction — discussed in chapter 3) further increases the risks for such couples. In fact, the very first XYY male described (in 1961) was the father of a Down-syndrome (trisomy-21) child.

The Extent of the Condition

About 1 in every 1,000 males born has the XYY chromosome pattern. Studies in the United Kingdom showed that the incidence of XYY males among the men committed to four maximum-security institutions in England, Wales, and Scotland in 1972 and 1973 was 2.1 percent. During that two-year period, 70 percent of these XYY males were aged fifteen to twenty years. This meant that 1 in every 16 incarcerated males of that age group had an extra chromosome. A few studies have noted the striking youth (nine or ten years of age) of convicted XYY boys. Careful calculations have suggested that there is a chance of about 1 in 1,000 that a normal male will be committed to a high-security prison during his lifetime. In contrast, the likelihood of an XYY male being committed is 1 percent — at least a tenfold greater risk! The increased number of XYY males in mental or penal institutions is now well established.

Other Chromosomal Abnormalities and Deviant Behavior

It is interesting that surveys done on mentally subnormal, dangerous criminals have shown a higher frequency of males with other sex-chromosome syndromes. A fair number have Klinefelter syndrome (XXY) or variants thereof (XXYY).

Prenatal Diagnosis of the XYY Fetus

In a few score of reported cases, an XYY fetus has been unexpectedly diagnosed during prenatal genetic studies that were initiated because of advanced maternal age or other reasons. What would you do if suddenly confronted by the news that the fetus you or your spouse was carrying had been diagnosed as having XYY chromosomes? Would you dwell on the optimistic side, knowing that there are XYY males who exhibit no deviant behavior and lead perfectly normal lives? Or would you worry that the risk of eventual committal to a maximum-security institution could approximate 1 in 100? In the United States, an informal survey by one of the leading researchers on XYY males, Dr. Arthur Robinson of Denver, showed that about 50 percent of parents elected to terminate such pregnancies. In contrast, parents in Scandinavia have been making that choice about 80 percent of the time.

The Self-fulfilling Prophecy

Imagine that you and your spouse are the proud new parents of a normal-appearing male baby. You had agreed prior to the birth of your son to participate in a newborn-screening program in which a tiny skin prick enables blood analysis of the infant's chromosomes. Unexpectedly, the analysis reveals that the baby has an XYY chromosomal pattern. Would you rather the doctor *not* tell you this information? After all, it is clear that the final answers about the ultimate outcome for intellect and behavior in XYY males can not yet be forecast with any degree of certainty. Given this real uncertainty, would you and your spouse elect not to have the information?

Your knowledge about your son's abnormal chromosome pattern may easily influence the way you rear him. For example, knowing his chromosomal abnormality may make you more anxious, and either more or less demanding of him; you might insist on more discipline, or you might be more permissive, and so on.

Physicians who study children with an extra Y chromosome thus may be unwittingly responsible, via the parents, for influencing the child's behavioral development. Indeed, the criticism goes, this awareness of the extra Y chromosome may be harmful to the development of the child, as the parents might interpret bad behavioral characteristics as indicative of the child's future criminal disposition and overreact accordingly. For example, a child may be fascinated by the dissection of an insect or a frog. The parents, knowing about their child's extra Y chromosome, may interpret their son's activity as demonstrating a clear propensity toward murder. A more rational and reasonable explanation would be that he has a well-developed curiosity about biology. Their efforts at disciplining him in such a situation, perhaps repeated in other contexts, may destructively influence his childhood years.

If you were about to adopt a child and were informed that he had an XYY constitution, would you complete the adoption process? This possibility is quite real, since in a study at Johns Hopkins University, 3 out of 23 XYY males were adopted. Despite the possible effects of the self-fulfilling prophecy, would you insist on knowing this information prior to adoption? In this situation, it would in fact be illegal for this information to be withheld from you — if it were known.

A Plea of Insanity

There have been at least six criminal trials in which the defendant has raised his abnormal XYY chromosomal pattern as a basis for a plea of insanity. This defense refers to the legal concept that defines the extent to which accused or convicted individuals may be relieved of criminal responsibility because of mental disease at the time the crime was committed. The defense attorney must first demonstrate that the accused was indeed suffering from a "mental disease." Moreover, a relationship must be established between this mental disease and the alleged criminal act.

The insanity defense based on the XYY chromosomal pattern was first raised in April 1968 in France. The defendant, Daniel Hugon, was accused of murdering a sixty-five-year-old prostitute in a Paris hotel. Chromosome studies, made after he attempted suicide, indicated that Hugon had an XYY pattern. Nevertheless, he was found legally sane and was convicted of murder. The prosecution requested a five- to ten-year sentence rather than the fifteen years normally imposed for similar crimes. Hugon received a seven-year sentence.

In the same year, twenty-one-year-old Lawrence Hannel came to trial in Australia charged with the stabbing death of his seventy-seven-year-old landlady. Again, the plea of insanity based on the defendant's XYY chromosomes was offered. The jury, after only eleven minutes of deliberation, delivered a not-guilty-by-reason-of-insanity verdict, and Hannel was committed to a maximum-security hospital until "cured."

Ernest D. Beck, a twenty-year-old farm-worker with the XYY chromosome pattern, was tried in Bielefeld, West Germany, in November 1968. The court in this case accepted the prosecution's argument that Beck was fully aware that he was committing murders, even though he might not have been able to control his impulses to kill. Beck received the maximum sentence of life imprisonment for the murder of three women.

In April 1969, Sean Farley of New York, a six-foot-eight-inch, twenty-six-year-old male, pleaded not guilty on the grounds of insanity to the alleged brutal murder and rape of a forty-year-old woman in an alley near her home. On cross-examination, the prosecutor established that it was possible to live a normal life although possessing XYY chromosomes. The jury found Farley guilty of murder in the first degree.

Other convictions have been obtained despite the individual's condition of having XYY chromosomes. In one celebrated case, Richard Speck, the convicted murderer of eight Chicago nurses, was *incorrectly* characterized as an XYY male. He actually had a normal chromosome constitution.

Tall Men

Because of the difficulties in getting at the truth about the extra Y syndrome, a unique, major collaborative study was initiated. Danish and American researchers set out to study, in an unbiased fashion, males with an extra Y chromosome living in the community. They chose Denmark because of the excellent social records kept there.

The researchers decided to study males born in Copenhagen during a particular period — January 1, 1944, to December 31, 1947. Out of 31,436 men in this group, they were able to account for some 28,884 whose heights were also known. They elected to study only those who were more than six feet tall, and they obtained chromosome analyses on 4,139 such men.

The investigators found 12 XYY males and 16 males with an extra X chromosome (XXY). Some 41.7 percent of the XYY males and 18.8 percent of the XXY males were found to have been convicted of one or more criminal offenses, in contrast to 9.3 percent of males with normal chromosomes. The difference in rates of conviction between the XYY and XXY males was not statistically significant. Moreover, crimes of violence were *not* more frequent among XYY versus XXY males, or when compared to males with normal chromosomes (XY).

This study therefore confirmed that XYY males had a much higher rate of criminal convictions than normal males, but statistically not much greater than XXY males. Both the XYY and XXY males had achieved lower educational levels and had much lower IQ scores than normal males. No evidence was found to suggest that XYY males were particularly aggressive. Indeed, the essence of the study was the recognition that the antisocial behavior exhibited by XYY males was in all probability a reflection of a low intelligence level rather than any genetic criminal propensity.

A major reservation to bear in mind is that this study focused *only* on tall men. Since we do not know for sure how many XYY males are less than six feet tall, a full understanding of the frequency of all physical, intellectual, and mental problems in such men will have to await the results of long-range studies.

Conclusions

There is now little doubt that both XYY and XXY males end up in mental or penal institutions at a higher frequency than normal males. The idea that the extra Y chromosome has a special propensity to induce criminal behavior is a myth that seems to have been exploded. The most likely reason for some XYY males ending up in trouble is their low intelligence associated with antisocial behavior. Whether features such as being impulsive, being a "loner," and being prone to an uncontrollable temper, inattention in school, and so on, are symptoms isolated to the XYY syndrome is something that will, we hope, be settled by further study.

Meanwhile, what about those parents who discover that they have an XYY male fetus and have to decide for or against abortion? Some are likely to be swayed by the evidence of low intelligence, learning disabilities, and antisocial behavior; others may take their chances that their son, like some others, will be a normal, nonviolent XYY child.

6

The Fragile-X Syndrome: The Most Common Inherited Cause of Mental Retardation

The fact that mental retardation is more common in males has been known for about a century. The major reason for this excess became clear in the mid-1970s, when studies from Australia focused attention on an unexpectedly common disorder with striking features: the *fragile-X syndrome* (previously referred to as the Martin–Bell syndrome). This disorder, caused by a single defective gene on the X chromosome, has highly variable signs that usually include mental retardation, large testicles, and distinctive facial features. Special studies have revealed the location of the defective gene on the X chromosome — a vulnerable spot that tends to break; hence, the term "fragile-X syndrome." Because of the remarkable variability of the physical, behavioral, and developmental features of fragile-X syndrome and the delayed appearance of some major features, definitive recognition of this disorder eluded researchers for many years. Confusion was also generated by the fact that although males were primarily affected, within the same families mildly affected females were also observed. Indeed, we now know that about 1 in 1,060 males is born with the fragile-X syndrome, and that the disorder accounts for about 25 percent of all cases of mental retardation in males and about 10 percent of mild to moderate mental retardation in females.

The Main Signs

The cardinal features of the disorder are listed in table 5. Since the vast majority of affected persons have never been diagnosed, you may wish to study the features in table 5 carefully and examine the

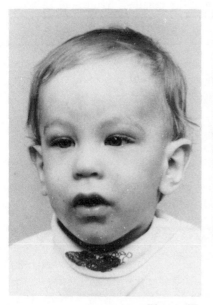

Figure 12

Fragile-X syndrome in a 2½-year-old boy. Note the high forehead, prominent ears, long midface, and comma-shaped folds of tissue at the inner corners of both eyes.

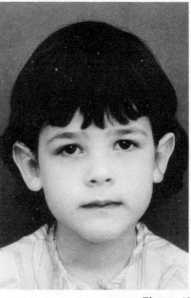

Figure 13

Fragile-X syndrome in a five-year-old girl who was studied because of serious behavior problems. The prominent ears and strong chin are characteristic.

Figure 14

A severely retarded, autistic man with the fragile-X syndrome. Note the long midface, prominent ears and chin, high forehead, and his constant preoccupation with his hands and a piece of string. (Figures 12, 13, and 14 originally appeared in the *Journal of Pediatrics* [110:821; 1987] and are reprinted courtesy of Dr. A. E. Chudley, Children's Hospital, Winnipeg, Manitoba, Canada, and with the permission of the C. V. Mosby Co.)

Table 5. The Most Common Features of the Fragile-X Syndrome

Characteristic	Comments
Nervous System and Skull	
Mental retardation	Varies from mild to profound; about 20 percent of males are normal
Learning problems	—
Speech and language problems	Unclear, repetitive, echoes others; poor language content; usually, delayed speech
Hyperactivity	—
Shyness early on	Later often cheerful, friendly, and cooperative
Features of autism	Examples: hand flapping, hand biting, repetitive mannerisms
Seizures	Only in some
Large head	In most
Facial Appearance	
Prominent ears	Often large, soft, and protruding
Long midface	—
Prominent lower jaw	—
Prominent forehead	—
Overcrowded teeth with overbite	—
High-arched palate	Usually a mouth-breather
Genitals	
Large testicles	After puberty; in most adults; occasionally in children
Muscles, Ligaments, and Bones	
Floppy	As infants
Loose-jointed	—
Stiff walk	Only in some
Flat feet	—
Long, thin arms	—
Curved spine	Occasionally
Prominent shoulder blades	—
Caving-in of breastbone	Occasionally; mild to moderate

Table 5 (continued)

Characteristic	Comments
Downward slope of the shoulders	—
Protuberant abdomen	—
Hernias of groin and diaphragm	—
Skin	
Velvety soft texture	Often
Hand calluses from self-abuse	Occasionally
Abnormal palm creases and fingerprints	—
Heart	
Prolapsed mitral valve	—
Slightly dilated aorta	Occasionally

NOTE: An affected individual might not have every listed feature.

accompanying photographs closely (figures 12–14). If you conclude that this is a possible diagnosis in a *maternal* relative, consult with a clinical geneticist in a medical school and request a fragile-X chromosome study.

A wide range of intellectual function occurs in males with this disorder. While intellectual deficits are mostly in the moderate-to-severe range, they may vary from learning disability to profound retardation. Remarkably, however, about 20 percent of males with this condition may have entirely normal intelligence and unwittingly transmit the disorder to their offspring. A male with a normal IQ thus may transmit the fragile-X syndrome to half his daughters, who may then have an affected son with mental retardation.

Boys with this disorder usually have enlarged testicles, which becomes obvious around the time of puberty. Curiously, when examined under the microscope, testicular tissue appears normal; the extra size is largely due to an increase in fluid within the tissues. Sexually, these males function normally; they produce sperm and may father offspring. The penis may occasionally be large but is usually normal. There is some evidence that affected males have congenital anomalies more frequently than other males do — conditions such as undescended testes (testicles that fail to exit the abdomen and enter the scrotum) or a penis opening that is situated somewhere along the undersurface of the penis rather than at the tip (hypospadias).

The appearance of the face tends to be quite characteristic (see figures 12–14). A usually large head is bordered by long and protuberant ears associated with a prominent forehead and prominent jaw. The midface area between the eyes and the mouth tends to be long and flattened. The eyebrows (or ridges above them) tend to be prominent and the nose rather long. The distinguishing facial features become increasingly obvious between eight and twelve years of age.

Only in 1983 were abnormalities of connective tissue suggested as an integral part of the fragile-X syndrome. Hyperextensible joints, flat feet, and many other features listed in table 5 signal involvement of the tissue that binds the body together.

Delayed speech is common in retarded fragile-X boys. Later, however, the quality and nature of the speech problem becomes highly characteristic. Affected males have difficulties in articulation and leave out words, which results in grammatically poor sentences that tend to sound like the recitation of a litany. This speech pattern is so characteristic that it has alerted physicians to consider a fragile-X diagnosis. These males also have a tendency to repeat words, phrases, and sentences continually, and to echo someone else's words automatically when spoken in their presence. Frequently, they may simply talk continuously and compulsively.

A wide range of behavior has been observed among fragile-X males. Typically, when young, they are very shy and anxious, though later they may be friendly, cheerful, and cooperative. Unfortunately, distressing behavior is not uncommon; it varies from simple hyperactivity to outright psychotic behavior in some. Since 1980, autistic behavior has been noted in fragile-X males. Indeed, 2 to 5 percent of all males with infantile autism probably have the fragile-X syndrome. Some Scandinavian reports suggest that the frequency could be somewhat higher. The psychotic activity occasionally resembles paranoid schizophrenia.

In addition to the characteristics listed in table 5, other features include brown birthmarks that resemble coffee stains (called café-au-lait spots), cleft palate, and wryneck (torticollis), which results in an odd position of the head. A number of twins with fragile-X syndrome have been reported, as well as a set of triplets who all had this syndrome.

Looking Backward for the Future

An enormous number of families need to know about the fragile-X syndrome. If in your own family there is a history of unexplained (undiagnosed) mental retardation in males on your *mother's* side, genetic counseling is advisable. Since no effective treatment exists, efforts at prevention are paramount.

The case history that follows illustrates the importance of diagnosis.

Harry and Beatrice had been high-school sweethearts. After getting married, they waited for about five years, until they were financially secure, to have their first child. Jeremy, their first son, appeared fine at birth and seemed to be developing normally throughout the first year of his life. Speech delay became obvious by the end of his second year, during which time the child's pediatrician referred them to a neurologist.

After a thorough evaluation, including multiple X rays and blood tests as well as chromosome analysis, the specialist concluded that Jeremy was developmentally slow and that the cause was unknown. "More time would have to pass," said the specialist, before he could make a more accurate estimate of the boy's degree of retardation. Harry and Beatrice were worried during this time about the fact that Beatrice had a moderately retarded uncle and a mildly retarded aunt on her mother's side of the family. At the time, twenty years ago, the doctor thought that this family history was simply coincidental.

For the first five years of Jeremy's life, Harry and Beatrice took him to various specialists, including those at university hospitals. The fact that he was moderately retarded became increasingly obvious, and their constant disappointment and distress at finding no additional explanation was punctuated only by the joyful birth of their daughter. Encouraged by her perfectly normal development over two years of age, they embarked on a third pregnancy.

George was born and followed a course of development almost identical to his brother. He, too, was eventually found to have moderate mental retardation, and once again an absolutely certain cause was not determined. However, this time, with two sons affected, the doctors advised that a hereditary cause for mental retardation in boys (*sex-linked* mental retardation) was the most likely diagnosis, albeit without a specific name.

I saw Harry and Beatrice in consultation (over twenty years after the birth of their first son) because their daughter was concerned about her own risks for having a mentally retarded child. By this time, the physical features of the fragile-X syndrome and the characteristic appearance of the X chromosome were already known. The boys both had large testicles, a condition that had not been observed previously since their evaluations had ceased many years before puberty.

I subsequently saw the daughter for genetic counseling and advised her that while there was theoretically a 40-percent risk of her being a carrier of the fragile-X chromosome (rather than the usual 50 percent for X-linked disorders), her tests were all negative. Nevertheless, because the technology was not (and still is not) good enough to be certain that she was not a carrier, I counseled that she should keep in touch with a clinical geneticist each year before planning a pregnancy. I indicated that new research involving DNA probes (see chapter 8), still in progress, would likely yield a more precise diagnosis of the carrier state in the near future, and that prenatal diagnosis was already possible in specialized centers.

The Inheritance Pattern

Transmission of the fragile-X disorder was initially thought to conform to patterns seen with other sex-linked disorders, such as hemophilia or Duchenne muscular dystrophy (see chapter 7). Quite unexpectedly, a unique pattern that does not conform exactly to sex-linked inheritance has been discovered only recently. Our current knowledge allows certain risk predictions:

1. An intellectually normal female who inherits the fragile-X gene from her carrier mother has a 50-percent risk of having an affected son, whose risk of being retarded is 40 percent. Half her daughters will carry the gene, but only 16 percent will be retarded.
2. If such a daughter is retarded, her risk of having an affected *and* retarded son is 50 percent. If she has a daughter herself, the risk is 28 percent that the child will also be mentally impaired.

3. Men who are seemingly entirely normal and do not even show the fragile-X chromosome when tested may nevertheless transmit the gene to all their daughters. These females are usually intellectually normal. However, when they reproduce, 50 percent of their sons will be affected, and 40 percent will be retarded. Half their daughters will be carriers, among whom 16 percent will be retarded.
4. Normal-but-transmitting males may account for 20 percent of all cases of the fragile-X syndrome. Unfortunately, they will remain undetectable until new technology reveals their ominous burden or until one of their children or grandchildren is diagnosed as having this fateful flaw.
5. Curiously, women carriers who bear a son who is a normal-but-transmitting male have a 50-percent risk of having an affected male, who has only a 9-percent risk of being retarded. This carrier female also has a 50-percent risk of having carrier daughters, and these girls have only a 5-percent risk of being intellectually impaired.

The reasons for these seemingly complicated differences in risks are not yet completely understood. Moreover, these figures are likely to change a little with further experience. If current speculation is correct, loss of fertilized eggs that were destined to become carrier females or affected males, or the weakened or variable effects of the transmitted gene, may explain the differences in risks for this unique disorder.

The latest analyses suggest that fragile-X mutations are entirely confined to sperm, and that this mutation rate is the highest single-gene change detected in man (and other mammals), if X-linked inheritance (see chapter 7) is assumed. New data suggest that normal males who transmit the fragile-X carry a partially changed (mutated) gene that really becomes fully mutated only when the X chromosome carrying it passes through a female. It seems that the fragile-X change comes about through a stepwise progression; but much more work is needed to understand precisely how this remarkable disorder is inherited.

The Fragile-X Chromosome

Located on the long arm of one X chromosome is a very constricted site resembling a gap — the site of the defective gene — which constitutes the typical appearance of the fragile-X chromosome.

After staining of the chromosomes, this characteristic X chromosome can usually be demonstrated only after the cells have been grown in special broth containing no folic acid. Various chemicals can be added to help demonstrate the fragile-X site. Notwithstanding all current techniques available, even in affected males, rarely more than 50 percent of cells studied reveal this fragile site. Indeed, an affected male may have as few as 4 percent of cells studied showing the typical chromosome feature. Not unexpectedly, it might not be possible to demonstrate the presence of the fragile-X site even in known female carriers. Among those female carriers in which this chromosome can be demonstrated, frequently only a small percentage of studied cells reflect this typical feature.

Certainly, in those families with mentally retarded males among whom no diagnosis has been made and for whom normal chromosome reports have been issued, it would be advisable to submit another blood sample for chromosome analysis, but this time specifically for fragile-X study. Since not all laboratories are expert at demonstrating this typical chromosome, it is important first to determine the best laboratory to which this important sample can be sent.

Females and the Fragile X

About two-thirds of the females known to have inherited the fragile-X chromosome are entirely normal both intellectually and physically. Among the other one-third, who are intellectually below par, about 56 percent have borderline-normal intelligence, about 24 percent are mildly retarded, and about 20 percent are moderately retarded. Among all female fragile-X carriers, it is presently possible to demonstrate the fragile X only in those with intellectual impairment. It is estimated that about 1 in 677 women are carriers of the fragile-X chromosome.

While the matter is still under study, it seems that most females with the fragile-X chromosome show no facial or other features (other than the mental retardation) seen in affected males. Some undoubtedly have a few of the typical facial features (see figure 13). Of particular interest is an observation by Israeli researchers (in only one case thus far) that the ovaries of a fragile-X female carrier seen by ultrasound study may be larger than normal. There is also some evidence showing that the older the carrier female is, the more difficult it has been to demonstrate the presence of the fragile-X chromosome. New techniques using DNA probes will ultimately resolve this question.

Prenatal Diagnosis

The basics of prenatal diagnosis are outlined in chapters 21 through 23. At present, because of technical limitations, it is possible to achieve only about a 92-percent certainty for the prenatal diagnosis of the fragile-X syndrome using amniotic-fluid cells. This figure only applies to the detection of affected male fetuses. There is insufficient information to determine how often a carrier female could be detected. Not only would the figure be considerably less, but until new DNA technologies emerge, considerable numbers would be missed.

Using fetal-blood sampling, however, it is likely that close to 100 percent of affected males would be detectable. Unfortunately, this technique, called fetoscopy (see chapter 23), may be associated with a 4- to 6-percent risk of pregnancy loss because of the procedure. Of interest is a recent observation that an affected male fetus had larger-than-normal testicles.

Fragile Sites on Other Chromosomes

The location at which a chromosome is likely to break is described as a fragile site. These sites become visible during the process of cell division (metaphase) and may appear spontaneously or be demonstrated by the use of chemicals that uncover their location. Many of these sites, including the fragile-X site, can be induced by growing the cells to be studied in *broth* (culture medium) that lacks folic acid. More-complex compounds are also being used routinely to detect the presence of these fragile sites. While the most common and serious genetic disorder associated with a fragile site is the fragile-X syndrome, there are some other, rare conditions involving mental retardation that are due to fragile-site disorders on other, nonsex chromosomes.

In the final analysis, the fragile-X chromosome involves a gene at the visible constriction site. This syndrome represents one of a relatively few genetic disorders in which a visible chromosome defect reveals the site of a defective gene. This is the right moment, therefore, to consider you and your genes in some detail.

7

You and Your Genes

"What is a gene? How do you know we carry this gene when you haven't even tested us? How could this disorder be inherited when no other family member has had it? How sure can you be that the medicines I took in early pregnancy didn't cause my child's defects? Could my other children be affected in the future? Could this happen again? Isn't it possible there's been a mistake?"

This couple's questions echoed those of so many other upset parents upon learning that they have both transmitted a harmful gene to their now defective child. They were seeking clarification, explanations, disclaimers, relief from guilt, and sympathetic counsel and treatment. Just like so many other couples, they had read newspapers about genes and disease but never identified themselves as potential carriers of any harmful genes. It came as a shock when they realized that each of us carries at least twenty harmful genes.

"Genes," I have often explained, "are chemical structures within our cells that carry the blueprints for our body's structure and function. Interference with our genes can result in diseases or defects."

For some, reducing our heredity to chemicals may be a difficult concept. But it is a concept that can be understood. The chemicals that make up the universe have been discussed for many years, and we know that chemicals are the basis of all living things, not to mention the atmosphere, rocks, and soil.

Given the astonishing progress in human genetics and the anticipated tremendous surge in knowledge over the next few years, everyone should understand some of the more important facts about genes, especially those with direct implications for our own health and that of our children.

Let us first examine what a gene is, how it functions, and what can go wrong when it is abnormal. Then we can consider how an abnormal gene is transmitted through a family. In the next chapter, I will discuss the new gene-analysis techniques, their already established value for diagnosis, and the exciting future these advances portend.

What Are Genes?

Each of our 46 chromosomes in every cell is itself made up of genes, just like a strand of pearls. In each cell we have 50,000 to 100,000 *working* genes and many more whose function (if any) is unknown. Each gene is contiguous with its neighbor along the length of a chromosome and is constructed like a ladder or spiral staircase — the *double helix* (see figure 15). The two strands, or "sides of the ladder," are composed of long chains of molecules containing nitrogen, sugar, and phosphate, and the surrounding infrastructure is bolstered by calcium, magnesium, and certain proteins, including a type known as *histones*. About 100 different histones are known that themselves influence gene action.

The long chains of molecules, called *nucleotides,* are made up of sugar, phosphate, and nitrogen-containing molecules called *bases.* Together, these substances constitute *nucleic acids,* of which there are two. The one with the sugar ribose is called *ribonucleic acid,* or *RNA.* The other, with a different sugar, deoxyribose, is referred to as *deoxyribonucleic acid,* or *DNA.* In cells, RNA is mainly found in the *cytoplasm,* the fluid that bathes the nucleus, while DNA is largely confined within the nucleus itself. The "rungs" of the ladder are made up of the nitrogen-containing bases, which are always connected to each other in the same way. For example, DNA base A (called *adenine*) always pairs up with base T (called *thymine*), and G (*guanine*) always pairs up with C (*cytosine*), as shown in figure 15. In RNA there is a slight difference in that base A pairs up with another base called *uracil* (U).

There are an incredible 3 billion of these base pairs. The largest chromosome (designated number 1) has about 249 million base pairs of DNA, while the smallest (number 21 — *not* number 22, as would be expected) has about 48 million. The X chromosome, of great importance because of its role in many severe sex-linked

Figure 15

At left: a coiled strand of DNA made up of many "rungs," representing many genes; *at right:* the "copying," or replication, process as DNA duplicates itself. The long, twisting DNA strand, or "ladder," looks as though it unzips; then it joins chemical substances from the cell to form two new strands.

diseases, has about 154 million base pairs of DNA. Individual genes seem to range in size from 1,000 to more than 200,000 base pairs.

Genes are made up of DNA. They constitute the inherited blueprints that make us what we *are* physically and mentally. (What we *become* reflects our interaction with the environment.)

When a cell divides — for example, for growth, or to make a sperm or egg — the chemical bond holding the two "sides of the ladder" unzips, and each single side, or strand, attracts molecules from within the cell to duplicate itself exactly (see figure 15). In this way, every new cell gets a copy of the original DNA strand, or gene. The DNA strands, bolstered by proteins, are tightly coiled and make up the length of a chromosome. To give you some idea of how tightly our genes are coiled — if they were unraveled, they would measure a few yards. The total length of all chromosomes in a cell together measure less than half a millimeter — less than 0.02 of an inch!

How Do Genes Function?

Exactly how a gene — or several genes, separately or together — determines hair color, facial structure, or a specific disease is not exactly understood. We know that the main function of DNA (genes) is to make *proteins*. Proteins are built by linking simpler chemical substances, called *amino acids*, together. A gene first makes amino acids by using three bases — a triplet, or *codon* — as a "mold," or "pattern." A particular *order* of bases specifies a single amino acid. For example, a triplet (codon) made up of two uracil bases (UU) and one cytosine (C) base — in this exact order — codes (UUC) for making the amino acid phenylalanine, so important in the treatable disease of phenylketonuria. A codon made up of guanine (G) and two adenine bases (AA), in this order (GAA), specifies formation of glutamic acid, whereas the amino acid valine is specified by the triplet guanine, uracil, and adenine (GUA). The various amino acids are then forged together to make a protein, which is simply a chain of amino acids.

To give some idea of their complexity: One chain of the protein hemoglobin, which carries around our oxygen, has 146 amino acids. When one specific amino acid in that chain, glutamic acid, is replaced by the one called valine, the hemoglobin protein malfunctions, and sickle-cell disease occurs! Instead of the gene codon being made up of guanine and two adenine bases (GAA), the single base change to uracil gives a triplet of guanine, uracil, and adenine (GUA), which makes valine instead. The change in this gene is called a *point mutation* (discussed further below).

The particular order of the codons, or triplets, constitutes the genetic code, which will be different for all individuals except identical twins. Human genes can code for only twenty different amino acids, which in turn link up in various numbers and combinations to form chains that have the potential to make 4 million proteins. In fact only about 20,000 such proteins are known thus far. There are sixty-four codons, three of which are called "nonsense" or "stop" codons because their function is literally to terminate a coding sequence from a gene.

Through a complex mechanism, an RNA molecule — called *messenger-RNA* — transcribes the code and transmits the information, enabling the message to be "translated" so that protein synthesis can occur. Remarkably, more than 80 percent of DNA does *not* code for protein. Some of the noncoding DNA is interspersed *within* genes. These intervening sequences are called *introns* and are distinguished from "working" DNA sequences, called *exons*. What function the large amount of "nonworking" DNA introns have, if any, is a mystery. When proteins are to be made, mechanisms in the cell exist to cut out the nonworking introns and splice together the exons from which the message to make a protein is translated. Actual transcribing is also initiated and controlled by specific genes.

Once the messenger-RNA has the message, it migrates out of the nucleus into the surrounding cell fluid (cytoplasm), where proteins are synthesized. The detailed knowledge of how DNA makes RNA that makes protein has been used to link base pairs together in the laboratory and actually synthesize a gene that works in yeast cells. Implications for gene engineering in humans are obvious and are discussed in the next chapter.

JUMPING GENES AND OTHER PHENOMENA

As if things are not complicated enough, we know that genes may move spontaneously from one chromosome to another. These so-called *transposons* can actually become integrated into another gene, producing a mutation. Such phenomena — "jumping genes" — are well recognized in maize, bacteria, yeast, and flies. Whether jumping genes cause mutations in man is still uncertain. Copies of existing genes that no longer function have been discovered. How and why these so called *pseudogenes* were "silenced" is unknown.

Thus far, it should be clear that the genetic message to make protein is transcribed from DNA to RNA and thence translated into protein. From studying viruses, we now realize that genetic information may flow in a reverse direction — from RNA to DNA! In theory, then, a mutation in the messenger-RNA could actually be converted and incorporated into DNA.

GENE CONTROLLERS

The bacteria *E. coli,* which so often cause travelers to experience diarrhea, have been a boon to geneticists (except when they travel!). Studies with *E. coli* — each has a single chromosome — have taught us that one gene may regulate or control the effects of another. An operator gene, lying adjacent, controls expression of a structural gene (which initiates protein synthesis). In turn, the flanking operator gene is itself controlled by a regulator gene on the same or on a different chromosome. The regulator gene makes a substance, called a *repressor*, that controls the operator gene (at least in microorganisms). Hereditary disorders in which more than one enzyme is deficient may occur when this gene control mechanism goes awry.

ARCHITECTURAL GENES

Why is it that your eyes are set in their locations? Why not much farther apart? Why are your ears where they are? What mechanism stops our fingers from growing twelve inches or more? Why are your spleen and your liver tucked under the left and the right side of your lower rib cage, respectively?

Complex, interacting, and interdependent organizing mechanisms exist about which we still know little. However, from genetic studies of lowly fruit flies, worms, frogs, mice, and recently man, researchers have discovered clusters of genes that act as master planners.

These architectural genes seem to be in charge of a body plan and appear to control spatial organization during embryonic development. These or other controlling genes may order cells to move in one direction or another, to stop growing, or even to die. The architectural genes, in flies and other species, have a DNA sequence in common, termed the *homeobox*. Replacement of one body part by a different one is called *homeosis*. Hence, mutation of a gene in the homeobox may result, in the fly, for example, in legs growing on the head, where antennae should be. Such homeotic mutations seem to disrupt the controlling function of the architectural genes.

It is likely that similar functioning genes are subject to mutation in humans and that at least some birth defects may result. Remarkably, the DNA sequence common to several homeotic genes in the fruit fly is also found in various species, including man. Is there, perhaps, a universal genetic master plan for body organization common to all species?

GENE FAMILIES

Groups or clusters of genes with similar structure and function exist on the same or on different chromosomes. It is thought that they have a common origin in evolution. Such gene families code for closely related proteins that may make their appearance at different times as the embryo develops. The best example are genes (on chromosomes number 16 and 11) that make globin proteins progressively in the embryo and fetus, until a final adult form is synthesized.

GENES OUTSIDE THE NUCLEUS

To complicate matters further, genes exist outside the nucleus in tiny structures called *mitochondria*, which are dispersed in the cell fluid around the nucleus. Mitochondria, of which there are 4 in a sperm and about 1,000 to 1,500 in a liver cell, contain their own DNA, or genes, which codes for enzymes and proteins that yield energy for the body. In fact, because the proteins and fats we eat are "shoveled" into the mitochondrial "furnace," the mitochondria are often called the power plants of the body.

Needless to say, genes within the nucleus exist that are necessary for making mitochondrial enzymes. The genes within each mitochondrion are the same and code for amino acids in a similar fashion to DNA within the nucleus. There is a slightly different genetic code and other technical variations, which I will not explore here. Suffice it to say, however, that the mitochondria in sperm do *not* enter the egg at the time of fertilization. Obviously, then, the entire set of genes in mitochondria in all cells of an individual comes from the mother! A few very rare diseases due to mitochondrial malfunction are known. In such cases a mother would transmit the disease to *all* her children!

How Single Genes Cause Hereditary Disorders

Among the 50,000 to 100,000 working (structural) genes known are some 4,000 that cause the number of thus-far-recognizable different genetic defects. While individually uncommon or rare, single-gene defects cause considerable ill health. A normal gene in a human egg or sperm may, of course, undergo change — mutation — and that person will either be born with a genetic disorder or simply carry that abnormal gene without obvious effect. Mutations are common causes of single-gene defects that suddenly appear in a family for the first time. Once a person is born with a mutated gene, that gene can be transmitted to offspring. How do such mutations occur?

While X rays and chemicals can cause mutations, we mostly do not know why they occur. Certainly, any interference with the normal gene structure (DNA), the transcribing of messenger-RNAs, or the enzymes responsible for cutting out the unnecessary introns or splicing the exons together, may alter the formation (translation) of proteins, their transport, or their function, thus resulting in a genetic disease. Gene structure, or DNA, can be changed by a single base (such as cytosine) being altered in the triplet code, and this in turn can result in the synthesis of a defective protein and a specific genetic disease. This single-base substitution is referred to as a point mutation and accounts for many known genetic diseases. Deletion or even insertion of one or more bases into the DNA strand can alter the whole base sequence, resulting in non- or malfunctioning proteins and possibly a genetic disease.

A key example in which a change in the triplet mold occurs involves the gene that causes hemophilia. The structure of this gene is now known and it consists of 2,351 amino acids! The complete gene, including "working" sequences (exons) and intervening segments (introns), consists of 186,000 base pairs. There are twenty-six exons — the "real stuff" of the gene — and mutation anywhere along any exon could result in formation of an abnormal sequence or absence of amino acids and consequently a defective protein. Hemophilia results when a protein (called Factor VIII) that is responsible for blood clotting is abnormal in structure and function. With such a long gene, it is easy to understand that mutations, including deletions, could, and do, occur at various sites along the length of the gene, each resulting in an abnormal Factor VIII protein, and hence the bleeding disease, hemophilia. Similarly, the gene for Duchenne muscular dystrophy is long and hence subject to more possible mutations.

Another good example of a long gene in which point mutations or actual deletions of gene segments — called *sequences* — occur is thalassemia. There are many different forms of this hemolytic anemia, which varies from mild to fatal. Certain types are more characteristic among Italians from Sardinia than among Orientals or Indians, and so forth.

When there is a defective gene, the code may be read as nonsense and no amino acid may be produced, or the code may be read as missense and a different amino acid made, resulting in a defective protein. Missense mutations probably occur in many biochemical genetic diseases, such as phenylketonuria. The next question, then, is how are these defective genes transmitted through a family?

How Genes Are Inherited

Among all these tens of thousands of genes, we usually do not know which harmful genes we, as parents, carry, or which of our own parents was the carrier for a specific disease. Recent scientific advances have made it increasingly possible to find out.

We know, as discussed in earlier chapters, that normally we get exactly half our genes from our mother and half from our father. If you have fair skin and suffer badly in the sun, it may be a characteristic of only one side of your family. A very high intelligence may be a recognizable family trait: there are some famous cases, such as the Darwin family, which produced outstanding scientists for five generations, and the Bernoullis, who generated altogether nine eminent mathematicians or physicists.

On the other hand, genius often springs from completely average families of no particular intellectual distinction, as in the case of Newton, Keats, and Einstein.

Inheriting Harmful Genes from One Parent Only

As we frequently take after one parent more than the other — even though we inherit height, hair, body-build, and other genes from both — we can infer that in certain cases the gene we inherited from one parent was "stronger" than its equivalent from the other parent. Dark hair, for instance, is said to be *dominant;* dominant genes governing such traits are harmless. But there are also destructive dominant genes that cause disease or abnormality. Statistically, the parent possessing a dominant gene will pass it along to half of his or her children, usually regardless of sex, as in Huntington disease (see figures 16–18). The case history of a family I saw ten years ago gives a good idea of the possible problems dominant inheritance can create.

> Ann was twenty-one years old, engaged to be married soon. Her father had had some kind of disease affecting the brain and nervous system for some years, the name of which had not been told to her. The significance of this condition, unfortunately, had

also not been communicated to her. Since the disease was, in fact, Huntington disease, serious questions arose when she became informed of all the facts just a few weeks prior to marriage.

Her fiancé had not known that Ann had a 50-percent risk of actually having Huntington disease, nor that this disease could slowly appear within a few years after marriage. If his intended wife indeed had the disease (even though it was not yet apparent) and they eventually had children, then there would be a 50-percent risk that each child would develop the same disease. In this case, realization of all the facts led the girl's fiancé to cancel the marriage and break off the relationship. Five years later, tragically, Ann developed Huntington disease.

There are just over two thousand other recognized dominant disorders. Most dwarfs have a condition called achondroplasia, which is characterized by a prominent head, normal intelligence, and a waddling gait. Achondroplasia is a dominant disorder; therefore, should such a dwarf marry, he or she has a 50-percent risk of having offspring with the same condition. It turns out that about seven-eighths of achondroplastic dwarfs have normal parents, which means that this condition mainly occurs "out of the blue," the result of a spontaneous gene mutation. Once affected, that dwarf will pass on the gene for achondroplasia to half of his or her children, on average. In contrast, mutations either do not occur or are very rare in Huntington disease.

A single gene may have several effects. For example, in Marfan syndrome (a dominantly inherited disorder in which the elastic fibers of the body's connective tissues are defective and weak), bone, blood vessels, and eye defects occur, all as a consequence of one harmful gene. In contrast, several other, different harmful genes may have the same effect. A good example is profound deafness from birth. Dominant, recessive (see below), and sex-linked genes can all result in congenital deafness. In fact, there are as many as eighteen types of recessive deafness alone, and about 1 in 10 persons carries one of these deafness genes.

MUTATION

How does mutation happen? Earlier, I described the gene as a ladder of DNA, which by splitting rungs acts like a mold, thus allowing, by copying, the formation of identical genes. When some change or fault occurs in this copying process, a gene that is just slightly different is formed. A good analogy is the common

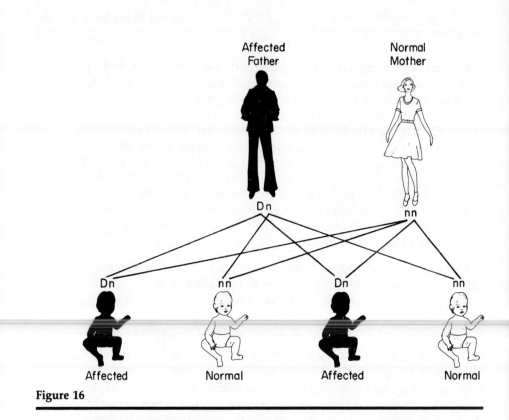

Figure 16

Dominant inheritance. One affected parent (the father, in this case) has a defective gene (D) that dominates its normal counterpart (n). Each child has a 50-50 chance of inheriting the defective gene from the affected parent, and thus of having the disease.

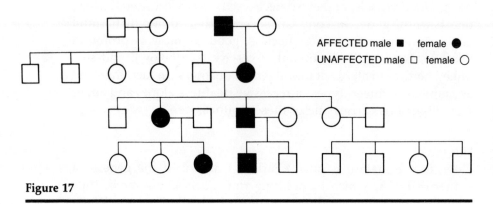

AFFECTED male ■ female ●
UNAFFECTED male □ female ○

Figure 17

A typical family tree involving dominant inheritance — as seen, for example, in Huntington disease, heart disease due to hypercholesterolemia, and hundreds of other disorders

experience of taking a key to a locksmith to procure duplicates, only to find that your new duplicate key does not fit the lock properly, although it was copied from the correct key. On your return to the locksmith, he simply files a portion of the key and you go back to your lock to discover that it now fits perfectly. Of course, this simple mechanical correction is not possible for genes, and the mutation is preserved and is copied identically thereafter. Most mutations in human beings occur without known reason. Some diseases (discussed in chapter 13) seem to occur more frequently as a result of a mutation in an older father. Irradiation,

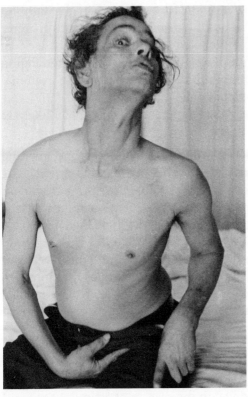

Figure 18

A man with Huntington disease. Dancelike movements of the head, body, and arms interfere with speech, muscle control, balance, coordination, and posture. (Courtesy of Dr. Michael R. Hayden, University of British Columbia, Vancouver, Canada.)

either from X rays, atomic energy, radium, or from the atmosphere, is known to cause mutations in genes. This is why it is unwise to irradiate the testes or the ovaries.

CHARACTERISTICS OF DOMINANT INHERITANCE
Certain general "rules" govern this form of inheritance, in which the gene in question is located on a regular, or *autosomal*, chromosome, rather than on one of the two sex chromosomes.

1. Any child of an affected person has a 50-percent risk of inheriting the gene from that parent.
2. Unaffected family members do not transmit the harmful gene. An apparent exception can occur when the harmful gene is in fact present but the disorder is not at all apparent, even after the most detailed examination. This lack of so-called *penetrance* may explain cases in which a generation was "skipped." More often, however, incomplete examination is the better explanation.
3. Males and females are equally affected and equally likely to pass on the harmful gene.

If a dominant genetic disorder is a common one, mating between two affected individuals will not be infrequent. The consequences of being born with a "double dose" of a dominant disease gene vary from fatal to serious. Two achondroplastic dwarfs, for example, who may prefer their child to be dwarf, have a 25-percent risk of having a child with a "double dose" of that gene, who will invariably die in the womb or in the first weeks or months of life because of severe bone abnormalities that interfere with lung or brain function. The common (1 in 500) familial hypercholesterolemia that can result in early heart disease is also due to a dominant gene. A "double dose" in the offspring of a couple who *both* have this disorder results in heart attacks even before puberty! Huntington disease may be the only known exception, thus far, in which the offspring of a couple who *both* have the dominant gene develop the disease at a similar time and with comparable severity to the parents. If this proves true, understanding the mechanism will be most instructive.

Inheriting Harmful Genes from Both Parents

The position of each gene along the length of a chromosome is generally constant, and hence a certain function can be mapped more or less accurately at a specific point along one specific chromosome. Every gene for a particular structure or function from one parent is matched with a gene with the same function on the corresponding chromosome of the other parent. The actual function controlled or directed by the matching genes is a reflection of their *combined* action. If one of the matching genes inherited from one parent is defective, then the other normal gene acts to provide half the needed function — which is usually enough to keep that person functioning normally.

If you are perfectly healthy and free of unusual disorders, you might wonder why manifestations of the twenty or so harmful genes you carry are not evident. The answer is that your harmful genes are *recessive*. The recessive gene usually has no (or little) obvious effect on the body if it has been paired with a normal gene from the other parent, though the genetic function of this pair of genes (one defective, one normal) will probably be half of what is normally found. For instance, if the particular gene in question is concerned with the formation and function of a certain enzyme, then it might be possible to demonstrate that you do indeed possess quantitatively only half the strength or activity of that enzyme.

CARRIERS

If it is found that the activity of a certain enzyme is half of what it should normally be, you can infer that one of the two genes "controlling" that enzyme is defective. This would mean that you are a carrier of, but not a sufferer from, the disease that is characterized by virtually no activity of that particular enzyme. For example, children with Tay-Sachs disease have the enzyme hexosaminidase A missing from their cells, while those who are merely carriers of the disease can be shown to have only half the normal activity of hexosaminidase A.

Thus, if your one "harmful" recessive gene is paired with one normal gene, you yourself usually have no obvious symptoms, but there is a 50-percent chance that each of your children will inherit your defective gene. If your mate has a normal gene of the particular pair under discussion, each of your children will therefore receive at least one normal gene, and will have a 50-50 chance of inheriting two normal genes (one from each parent) and thus of *not* being a carrier. But, most important, there is *no* likelihood in this situation that any of the children will have two of the abnormal genes. This means they will *not* develop this genetic disease, since both defective genes must be present in an individual for a recessive condition to be expressed.

If, however, you should have children with someone who, like yourself, is a carrier of the same recessive gene, there is a 25-percent risk in each pregnancy that you will have an *affected* child; a 50-percent risk that a child will also be a carrier like yourselves; and, finally, a 25-percent chance that a child will not only be entirely unaffected, but not even be a carrier (see figure 19).

The likelihood that someone who is a carrier of a recessive gene

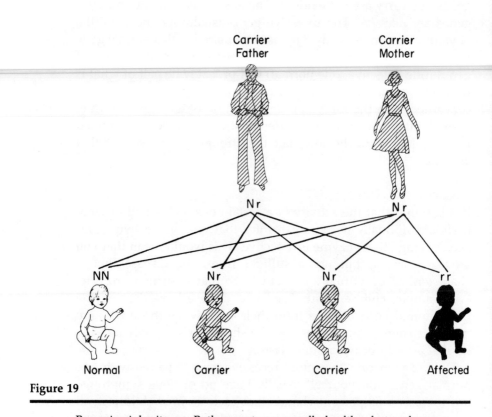

Figure 19

Recessive inheritance. Both parents are usually healthy, but each may carry a defective recessive gene (r), which generally causes no problems when paired with its normal counterpart (N). Disease results when a person receives two of these recessive genes. When both parents are carriers, each child has 25-percent chance of inheriting a "double dose" of the defective gene, a 50-percent chance of being an unaffected carrier, and a 25-percent chance of being neither a carrier nor affected.

will marry an individual carrying a similar defective gene will vary greatly, depending upon the frequency with which the particular gene occurs in the population at large. For example, approximately 4 percent of Ashkenazic Jews are carriers of Tay-Sachs disease. Statistically, it is possible to calculate — and this has indeed been confirmed — that in about 1 in every 900 Jewish couples, both partners will be carriers. In view of the recessive nature of this disease and the 25-percent risk that these carrier couples may have a defective child, it is possible to calculate that (unless preventive steps are taken) 1 infant in every 3,600 live births among Ashkenazic Jews will suffer from Tay-Sachs disease — that is, will appear to develop normally for about the first six months of life, then begin to do poorly, be unable to sit or stand any longer, have seizures, and go blind. Death usually occurs between two and five years of age.

It is obvious that despite there being over 1,400 recognized recessive diseases, their relative rarity makes marriages between two carriers uncommon. It is characteristic of recessive hereditary disease that the genes can pass down through many, many generations without creating any specific family history of the disorder (see figure 20).

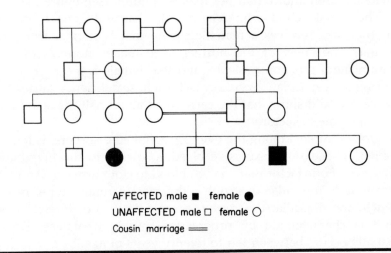

AFFECTED male ■   female ●
UNAFFECTED male □   female ○
Cousin marriage ══

Figure 20

A typical family tree involving recessive inheritance — as seen, for example, in cystic fibrosis. Note that there were no affected individuals in the early generations. It is not unusual to find cousin marriages in the family history of individuals with recessively inherited disorders.

CHARACTERISTICS OF RECESSIVE INHERITANCE

Specific criteria govern recessive inheritance, in which an individual receives the same harmful gene located on an autosomal (nonsex) chromosome from both unaffected parents.

1. Typically, the disorder is not present in the parents or other relatives, but only in brothers and sisters.
2. Males and females are equally likely to be affected.
3. On average, once one child has the disorder in question, the risk of recurrence is 25 percent.
4. The parents of an affected child may be related.

There are also some conditions in which the carrier may have greater genetic "fitness" or advantage than normal individuals. These are discussed in chapter 9.

Harmful Genes from a Mother's X Chromosomes

Harmful genes are sometimes located on one of the two female sex chromosomes. When a female has a harmful gene on one of her X chromosomes, the odds are she will pass it along to half her male children and to half her female children (see figure 21).

The female child who receives the harmful gene from the mother also receives a matching gene that is normal from the father. The effect of the normal gene makes normal (though diminished) function possible, and that female is a carrier. Measuring the particular enzyme or factor whose deficiency causes the disease would show that the carrier had about half the activity of that enzyme or factor.

One example, among the over 250 X-linked disorders, is hemophilia, a disease of excessive bleeding owing to the hereditary absence of one factor that enables blood to clot normally. Another example is muscular dystrophy; the most common type, called Duchenne muscular dystrophy, reveals itself in early childhood and is characterized by progressive muscle weakness. Death usually occurs between ten to twenty years of age.

If a female passes along the hemophilia gene to a male child, he will actually be affected by the disease. Since the defective gene she carries is on only one of her two chromosomes, there is a 50-50 chance that her male child will receive the normal one and not even be a carrier (see figure 21). If, however, the male child

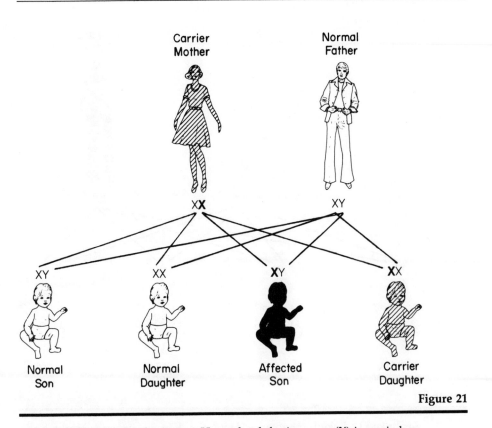

Carrier Mother

Normal Father

X**X**   XY

XY   XX   X**Y**   X**X**

Normal Son   **Normal Daughter**   **Affected Son**   **Carrier Daughter**

Figure 21

Sex-linked (X-linked) inheritance. Here, the defective gene (**X**) is carried on one X chromosome of the mother, who is usually healthy. Disease occurs when the X chromosome containing the defective gene is transmitted to a male. The odds of being affected are 50-50 for each male child, while 50 percent of the daughters will be carriers.

inherits the X chromosome with the gene for hemophilia, muscular dystrophy, color blindness, or some other sex-linked recessive disease, he will have that condition. If he then marries a woman who is a noncarrier and has daughters, all of them will receive the X chromosome with the defective gene and therefore become carriers of the disease (see figure 22). Their brothers, because they receive their father's normal Y chromosome, will neither be carriers nor have the disease. Of course, when an affected male impregnates a female carrier, the probabilities are quite different (see figure 23). The commonest condition transmitted on the X chromosome, by the way, is red-green color blindness: about 8 percent of white males (less in other races) are affected.

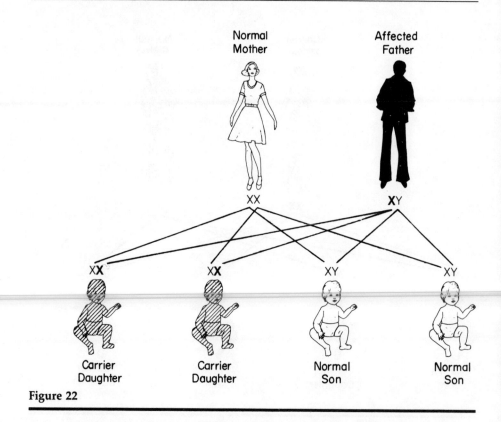

Figure 22

Sex-linked (X-linked) inheritance. Here, the defective gene (**X**) on the X chromosome is on the father's X only. He has the disease (such as hemophilia). He passes his X chromosome to all his daughters, who become carriers, while none of his sons are affected.

Hemophilia is one of the oldest recognized genetic disorders. There are records in the Talmud dating back to before the sixth century C.E. The rabbis at the time exempted from circumcision a male child whose brother had bled heavily following this ritual. These exemptions extended to the sons of sisters of a woman who had had a male who bled excessively after circumcision. In their wisdom, this exemption was *not* allowed for the father's son born to other women.

The British royal family provides a typical family "pedigree" for sex-linked disease (see figure 24). Queen Victoria was a carrier of hemophilia, as was her daughter Princess Beatrice. Two of the princess's sons had hemophilia and one daughter (Queen Ena) was a carrier. Queen Ena in turn had two sons who were also affected. And on and on.

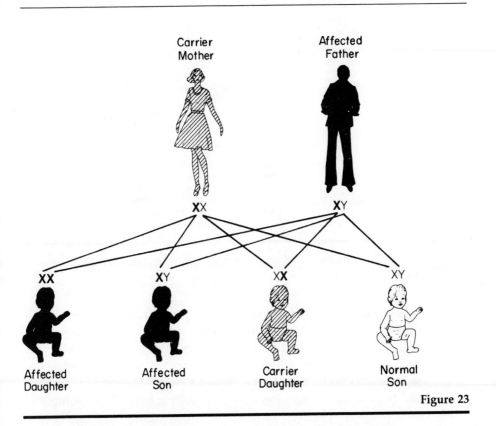

Carrier
Mother

Affected
Father

X X

X Y

X X
Affected
Daughter

X Y
Affected
Son

X X
Carrier
Daughter

X Y
Normal
Son

Figure 23

Sex-linked (X-linked) inheritance. Here, the defective gene (**X**) is present on the X chromosome of the father as well as on one of the X chromosomes of the mother. The father has the disease (such as hemophilia), and the mother is a carrier. There is a 25-percent chance that a daughter will be affected, and a 50-percent chance that a son will have the disease.

MILD SYMPTOMS IN CARRIERS

As mentioned, the woman who is a carrier of the gene for a sex-linked disease does not usually have a similar harmful gene on the *other* X chromosome. But even with this single dose of the "bad" gene, it would not be remarkable for her to show some mild manifestation of the disease in question. If it is muscular dystrophy, she may well have symptoms of weakness when walking that may become increasingly obvious as she tires during the day. Or, she may have weakness walking upstairs or running for the bus. She or others may have noticed that her calves are seemingly well developed, even though she claims that her legs are particularly weak.

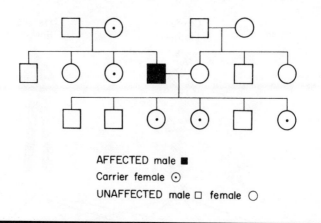

AFFECTED male ■
Carrier female ⊙
UNAFFECTED male □ female ○

Figure 24

A typical family tree involving X-linked disease, such as hemophilia or muscular dystrophy

Likewise, the female carrier of hemophilia may not have entirely normal blood-clotting ability. Mothers who are carriers of a white-blood-cell defect that leads to chronic granulomatous disease, a condition in which the offspring are especially subject to serious infections, can also be shown to have decreased functional ability of their white blood cells to kill bacteria. The mothers of certain albino children may show some pigmentary changes in their eyes.

But it is extremely rare for a female actually to have the sex-linked recessive disease itself. In such cases, the female patient would have to be the daughter of an affected father and a mother who is a carrier, and she would have to inherit an abnormal gene from each of them (see figure 23). How the defective gene might be passed to such a couple's children and grandchildren is shown in figure 25.

CHARACTERISTICS OF X-LINKED RECESSIVE INHERITANCE
Five main criteria make it possible to determine whether or not a disorder is transmitted by a recessive gene on one of the X chromosomes. Even with such criteria, as is the case with dominant and recessive autosomal genes discussed earlier, a certain conclusion cannot always be reached. For the vast majority of family pedigrees, however, these guidelines are valuable:

1. The disorder is passed from an affected man *through all* his daughters to, on average, half *their* sons.

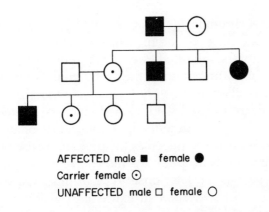

AFFECTED male ■   female ●
Carrier female ☉
UNAFFECTED male □ female ○

Figure 25

A typical family tree showing the sequel of a rare mating between an affected male with an X-linked disease and a female carrier of the same disease.

2. The disorder is never transmitted from father to son.
3. The disorder may be passed through a few carrier females to affected males who are all related to one another through these females.
4. Female carriers may show some signs (mostly mild) of the disorder.

UNEXPECTED INSIGHTS

Techniques of DNA analysis described in chapter 8 have yielded unexpected insights into our understanding of X-linked diseases. The most impressive advances have occurred in Duchenne muscular dystrophy (DMD). Mothers who have had a single son with this disorder have had DNA studies of their blood cells that revealed they were *not* carriers of the muscular-dystrophy gene in the sense described above and in chapter 8. Only after a few of these mothers had *second* sons was it realized that their ovaries must contain egg cells both with *and* without the DMD gene — a phenomenon called *germline mosaicism*. From these studies, a key recommendation has emerged: consider every mother who has only had one son with DMD to be a carrier, and offer her prenatal diagnostic studies (see chapter 23) in future pregnancies.

Another important discovery using DMD gene–analysis techniques has revealed that an *unaffected* male may transmit this lethal gene on his X chromosome to his daughters. Germline mosaicism in these males must also occur, since in families studied thus far not all daughters inherited the DMD gene. The frequency of unaffected males transmitting the DMD gene is still to be determined, but is likely to be very low. Meanwhile, the DMD gene itself has been isolated and cloned and its protein product, dystrophin, has been described. A deficiency of dystrophin in muscle cells is now recognized as the probable cause of DMD. The lower the level of dystrophin, the more severe the DMD, and vice versa. In a similar but later-onset type of muscular dystrophy — Becker type — the molecular structure of dystrophin is abnormal. Measurement of dystrophin in cells obtained by muscle biopsy and DNA analysis of the DMD gene now allow extremely precise diagnosis of Duchenne or Becker muscular dystrophies.

A CURIOUS LACK OF ACTIVITY

One of a female's two X chromosomes becomes genetically inactive within days of conception. Geneticists puzzled for years over this phenomenon that can render inactive either of the X chromosomes contributed by the mother or father. This "decision" within a cell about which X becomes inactive is permanent, but may differ between cells of the same individual. This inactivation mechanism explains a rare but important situation, such as a female with Duchenne muscular dystrophy, a disorder ordinarily confined to males. In such remarkable cases, a female with this muscle-dystrophy gene on one X has an abnormality of the other X chromosome, which is then inactivated. Instead of simply being a carrier with one normal X "neutralizing" the effect of the harmful gene on the other X, the full effect of the gene is expressed and the disease will be overt. Similarly, if one X is missing and the other has the dystrophy gene, you find the remarkable situation of a female with Turner syndrome who has Duchenne muscular dystrophy!

Female carriers of defective genes on the X chromosome may show no features of a disorder or may have mild to moderate signs. One major factor accounting for this variability is the random inactivation of one X chromosome.

Genes from the Male Y Chromosome

Unlike the important and serious diseases that originate from harmful genes on a female X chromosome, no disease is yet known to result from a gene on the Y chromosome. Any such condition would be passed from father to son and subsequent males, just like a surname. The trait for hairy ears is transmitted this way.

X-linked Dominant Inheritance

The main characteristic of X-linked dominant inheritance is that an affected male passes the harmful gene, and hence the resulting disease, to *all* his daughters and to *none* of his sons. The X-linked blood-group system (called Xg) is transmitted in this way. Only a few, very rare diseases are transmitted this way.

A few rare conditions passed by X-linked dominant genes occur only, or almost exclusively, in females; affected males die before birth. The family pedigree in such situations often shows many miscarriages — presumably of males.

CRITERIA FOR X-LINKED DOMINANT INHERITANCE
Specific "rules" also govern this rarer form of inheritance.

1. Affected males have no affected sons and no normal daughters.
2. Affected females pass the disease gene to half their children of either sex.
3. Affected females are usually much less ill than males.

Harmful Extranuclear Genes from Mothers Only

Earlier in this chapter, I discussed genes that exist within all cells but that are located outside the nucleus, in mitochondria. The DNA of mitochondria, which is double-stranded and circular, is suspended in the fluid inside the sperm and does *not* enter the egg at all when fertilization occurs. Only the male nucleus, containing half the individual's 50,000 to 100,000 genes, gains entry. The egg, however, always has the mother's mitochondrial genes and will transmit them to *all* her children, of *both* sexes. Males, therefore, have no role in this mode of transmission, which is called *mitochondrial, cytoplasmic,* or *maternal inheritance.*

A few rare genetic diseases — each involving the brain, the kidney, the heart, and other muscles — have long been suspected as arising from defective mitochondrial genes. Proof has been long in coming, but has now been provided: a point mutation in one mitrochondrial gene has been shown for a specific type of inherited blindness called *Leber hereditary optic-nerve neuropathy*. Other disorders thought likely to fall in this category of hereditary disorders include a specific type of inherited epilepsy (*myoclonic*) and an inherited heart-muscle failure. Specific DNA tests are sure to follow for those affected but still without a definite diagnosis. Therapy, perhaps, might not lag far behind.

Sex-limited and Sex-influenced Disorders

A disorder that is transmitted on a nonsex chromosome but that is seen in only one sex is described as *sex-limited*. The best example is precocious puberty, in which boys reach puberty at about four years of age. They grow rapidly and develop pubic hair, a larger penis, and a deeper voice. Although it is transmitted by a dominant gene, the condition does not manifest itself in females.

Sex-influenced disorders occur in both sexes but vary widely in how frequently they show up in each sex. Baldness, which is vastly more common in males, is a good example. In contrast, 80 percent of those with trisomy 18, a sex-influenced chromosomal disorder, are female.

Mild to Severe Disease

Analysis of the pattern of inheritance from the family pedigree can be most confusing. One reason may be that a "bad" gene may not always be expressed in the offspring to the same degree as in the parent from whom it came. For example, there is a dominant hereditary disorder called neurofibromatosis that affects the brain and nervous system. The affected person may have birthmarks on the skin that look like coffee stains, soft little lumps along the course of nerves, and possibly larger tumors originating from nerves in vital organs such as the brain, kidneys, or lungs. Because the gene for this disorder may vary in its strength of expression, the parent might simply have a few typical marks on the skin, while the affected children (50 percent of the offspring would be affected) might have not only the birthmarks, but also multiple tumors that could, because of their site, even lead to early death.

While variations in gene expression are perceived more often and more easily in the dominant hereditary disorders, it is not unusual to find variability in the recessive disorders as well. Out of several children in the same family with albinism, for instance, some may have complete and others partial involvement. In cystic fibrosis, one sibling may be more severely affected than another, and so on.

Awareness of this possible variation in gene expression is important in interpreting family pedigrees. Occasionally, it might be thought either that there is no family history whatsoever for a particular disorder or that a generation was skipped. It is therefore important always to be certain that the allegedly unaffected relative has indeed no sign of the disease. Such certainty may be possible only after actually examining the parents or close relatives, or performing special tests on them.

In a dominant hereditary disease called tuberous sclerosis, for instance, the affected person may experience an acne-like rash over and around the nose and cheeks, mental retardation, epileptic seizures, tumors of muscle (especially heart), as well as tumors behind the eye. When a child is born with signs of this disease, the parents should automatically be examined because of the dominant mode of inheritance in about 50 percent of these cases (the other 50 percent arise spontaneously due to mutation). Careful physical examination of both parents (using a Wood's fluorescent lamp to detect white skin spots) may reveal no signs of the disease, and the false conclusion of a mutation could be made. However, this kind of conclusion can be safely made only after extensive and careful study, including X rays of the skulls of both parents and CAT scans of their brains, as well as chest X rays, ultrasound testing of their kidneys, and so on. The reason for this very critical examination is that the risk of recurrence will be zero if both parents are genetically normal, but 50 percent if one parent is found to have some sign of the disease.

The Age of Onset

One of the most intriguing aspects of genetic disease relates to the age of onset of a particular disorder. There are many variations. As mentioned earlier, a child born with muscular dystrophy of the male-only, sex-linked variety may appear to be entirely normal during the first one to five years of life. Children born with Tay-Sachs disease appear perfectly normal at birth and for the first five to six months of life. But from the first day of life — indeed, from the ninth week of fetal life (or earlier) — a certain biochemical enzyme deficiency can be demonstrated. Some hereditary disorders, such as Huntington disease, may not become apparent until forty to sixty-five years of age, though rarely there may be clinical signs discernible at age fifteen or beginning after sixty-five.

Many of these observations really make you think about the factors that govern our normal development. Why, for example, does puberty occur when it does? What conditions our growth spurt as adolescents? What "body clock" mechanism causes the menopause? Why do some people become diabetic late in life? Historically, much of the research on rare genetic diseases has shed unexpected light on normal body processes, such as aging, as well as on the function of many genes. A dramatic advance would be for us to understand the mechanism, in place from the time of conception, that delays the appearance of an inherited disorder for months to decades. Imagine if we could prolong this delaying effect for a lifetime!

Genetic Disease Caused by Multiple Factors

There are many elements, not the least being the environment, that interact with minor gene abnormalities to produce serious defects. It is not really understood how environmental factors operate, but the correlation of many common diseases with geography and social condition is well known. In this category we find cleft lip and palate, clubfoot, congenital hip dislocation, spina bifida (open spine) and anencephaly (brain defect), some congenital heart defects, and possibly predisposition to coronary artery disease, certain types of diabetes, hypertension, and allergies — to name but a few.

The environmental causation is linked in some way to a genetic predisposition. We find that seasonal clustering is especially characteristic of some of these conditions, such as anencephaly (which especially affects babies born in autumn and winter), dislocation of the hips, and certain heart defects (including patent ductus arteriosus, coarctation of the aorta, and ventricular septal defect). German measles (rubella) may account for some of these defects, as well as widespread viral infections in a community at a certain time. In fact, virtual epidemics of anencephaly in the newborn have been recorded in New England, strongly suggesting the action of a virus.

Geography must play its part also. In Great Britain, for instance, the incidence of neural-tube defects (spina bifida and anencephaly) falls dramatically as one moves from Northern Ireland to the south of England. In the early 1970s, 7 in 1,000 births in Belfast reduced to about 2 per 1,000 in southern England. (Today, for reasons we do not understand, the incidence has dropped to about 4 per 1,000 and 2 to 3 per 1,000, respectively.) The effect of the original home place on groups of people seems to remain even when they move to a drastically different locale: the Irish in Boston have kept their higher risks of spine and brain defects.

Economic conditions play a still unconfirmed role, though it is well recognized that poor whites suffer a frequency of nervous-system malformations at birth that is higher than whites in upper social classes. Yet blacks in the United States have a low incidence of these malformations, just as they do in West Africa and in Great Britain, in spite of their generally disadvantaged condition. The higher frequency of neural-tube defects in the Sikhs in the Punjab is also reflected in births to Indians and Pakistanis (also mostly from the Punjab) in certain areas of England. Some data suggest that Irish women married to black West Indian husbands still have a higher frequency of births with these brain and nervous-system defects.

Finally, both maternal age and birth order are factors that influence the birth of babies with spina bifida and anencephaly. The firstborn and the fourth and later children in a family are most in danger. Mothers under twenty and over thirty-five seem to have defective children in excess of those in the middle range of their reproductive life.

There are, then, in essence, four different categories of hereditary disease. The first type (discussed in chapter 2) involves abnormal chromosomes: either in number or structure. The second type is related to harmful genes with major effects that we inherit or that occur through mutation; these cause recessive, dominant, or sex-linked diseases. The rare third type is caused by maternal transmission of genes from outside the nucleus. The fourth type are those disorders that reflect the interaction of specific genes with certain as-yet-unrecognized agents in the environment.

Heredity, of course, is not responsible only for devastating diseases. Some less serious disorders and some common inherited traits, such as hair and eye color, as well as some other interesting genetic characteristics or conditions, are detailed in appendix C. Appendix D answers some questions you might have about particular inherited conditions.

Clearly, then, we all have harmful genes, which are inherited from one of our parents or originate at our conception. It is already possible — and it will become more common — to determine which harmful genes we carry. There are some good reasons why you should be concerned and well informed, and they form the substance of the next four chapters.

8

The "New" Genetics: Gene Analysis, Gene Therapy, and Genetic Engineering

We are at the beginning of a new era in medicine. Our knowledge of basic genetics from work over three decades has coalesced to yield opportunities heretofore deemed impossible. Actual diagnosis of inherited diseases through gene analysis is one of the first direct benefits. Potential for actual *cure* of certain genetic diseases is no longer regarded as myth. And the new gene technology has revolutionary implications not only for medicine, but also for industry, farming, and agriculture. Let us broadly consider the "new" genetics — one of the most exciting areas in medicine and biology ever developed.

Diagnosis through Gene Analysis

The defective proteins genes make or the ill effects that result still constitute the main methods used to diagnose genetic diseases. Only recently has it become possible to locate and analyze (albeit mostly indirectly) single genes that cause disease. The first such analysis using new technology facilitated detection of the location of the defective gene that causes Huntington disease. Since enormous progress is anticipated in the next decade — and your family, among many, may benefit — understanding how this innovative technology works is important.

CUTTING GENES
To analyze and manipulate genes, it is first necessary to cut DNA into small fragments. Once again, we have been helped by discoveries made on bacteria. These organisms possess enzymes (which happen to defend them from attack by viruses) that can be

101

used to cut human DNA into fragments. Each of these enzymes, called *restriction endonucleases* (about 300 are known), cuts a strand of human DNA at very specific sites. Even these *cleavage* or *restriction sites*, as they are called, are inherited. Every restriction endonuclease will therefore predictably cut at the same site, yielding the same size fragments of DNA every time for that person. Fragments vary in size from a few hundred to several thousand base pairs long.

SPREADING GENES

The fragments of DNA, once cut, need to be separated for analysis to proceed. The mixed fragments are placed at one end of a slab of a gelatinlike substance (agarose gel) and an electric current is applied across the gel. The DNA fragments migrate from one end of the gel to the other at a rate dependent upon their size, small fragments moving faster than larger ones. This process, called *gel electrophoresis*, yields fragments, visible after special staining, that are arranged in order on the gel by size. Usually, whole genes have been cut into pieces of different sizes and are present in multiple fragments of different sizes. Further steps are therefore required for analysis.

PROBING FOR GENES

The single and cut strands of DNA — representing many bases joined together, including pieces of genes — can now be studied further. The study focuses on a particular pair of chromosomes, one from each of the parents (see figure 26). I just mentioned that restriction enzymes are used to cut DNA and that the sites at which they cut (A, B, and C in figure 26) are inherited, together with the chromosome from that parent. Thus, in figure 26 it can be seen that two fragments (AB and BC) result from one parent, while the same enzyme yields only one fragment (AC) from the other; the latter is seen not to have the cleavage site B. Because a particular gene would normally occupy the same location on each of a chromosome pair, *probes*, which are pieces of DNA tagged radioactively so that their location can be easily detected, are used to home in to identical sites on a matching chromosome pair (figure 26). Probes thus can be used to distinguish one fragment from another. Notice that the probes home in to the same site on both chromosomes, but help recognize two fragments of different lengths (AB and AC). You can best imagine the probe as a set of jigsaw-puzzle pieces seeking their complementary mates (see figure 27), in this case on a fragment such as AB or AC.

Once specific fragments of differing lengths are identified by a complementary DNA probe, the same studies are performed on closely related family members, including those affected by a specific defective gene (see figure 28). The same cutting enzyme is used on DNA from all family members. Results are then compared and the distances are calculated from various sites the probes recognized to the approximate site of the defective gene. Sheerly by repeatedly cutting DNA, probing the fragments, and calculating distances, an increasingly accurate conclusion can be made about the exact location of the gene in question. Clearly, if probes recognize sites on both sides of the gene — sites called *flanking markers* — their use is especially valuable. Since genes in the same

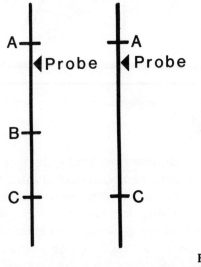

Figure 26

Probing for genes. The vertical lines represent strands of DNA from a pair of chromosomes. The letters represent sites at which DNA is cut by a specific enzyme, yielding fragments of different lengths. DNA probes home in to corresponding sites on matching chromosome strands. Each probe "marks" a fragment, helping to distinguish different lengths — in this case, AB and AC.

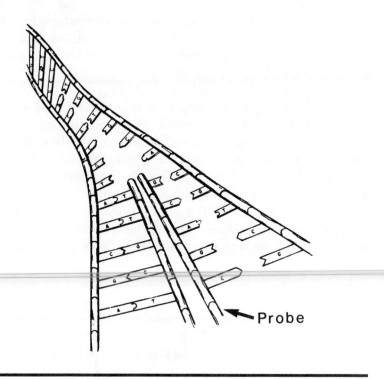

Probe

Figure 27

The "unzipped" single strand of DNA can be studied by using a complementary probe.

region of a chromosome tend to remain together when they pass down through a family, statistical calculations make it possible to estimate the distance between the probe sites and the location of the gene in question. In this way, specialists can *statistically conclude* whether a gene has been inherited or not, and can use this information for actual diagnosis, carrier detection, and prenatal diagnosis.

Even if the *exact* location of the gene is uncertain, results of 95 to 99 percent certainty are achievable. This would also apply even if the basic mechanism of the disease is still a mystery and even if the disease, albeit inherited, has not yet appeared. The reason why a 100-percent certainty is not provided reflects the fact that a gene may actually swap places (during cell division and chromosome splitting) with its opposite number on its chromosome mate. This extremely important phenomenon is called *recombination*. Only cutting sites reasonably "close" to the gene are used for diagnostic purposes and in such cases the chances of a swap happening is usually less than 5 percent. The closer the cleavage site is to the suspect gene, the lower the risk of recombination and the more accurate the diagnosis can be.

Once the defective gene is *precisely* located on a particular chromosome, its exact structure can be determined (*sequenced*) and its function studied.

BLOTTING FRAGMENTS
The DNA fragments simply diffuse into the gel as soon as the electric charge is stopped. An essential piece of the technology

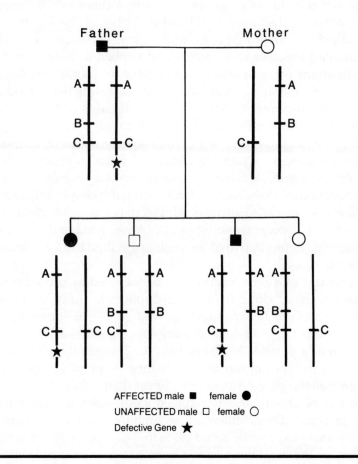

Figure 28

A family pedigree of a couple and their four children. Note how a defective gene at a specific site on complementary strands of DNA can be passed to the next generation.

was devised to recover these fragments immediately and hold them in their final positions. For further analysis, the double-stranded DNA fragments are separated by soaking them in alkali — a process that separates the double strands of DNA into single strands. Then, by a key process invented by Dr. Edwin Southern of Edinburgh, Scotland, a special filter paper is used to blot the gel, and the fragments lift off the gel onto the filter paper.

GENE MAPS

This technique and others have been used to construct gene maps — that is, to assign genes to specific locations on a chromosome. Mapping has been facilitated by virtue of the fact that, as mentioned, neighboring genes on the same chromosome tend to be inherited together (they are thus described as *linked*). In fact, rare situations occur in which two or three (or more) neighboring genes may be missing — a deletion — thus causing the individual, if he or she survives, to have all of the diseases coded by these genes. These are collectively referred to as a *contiguous gene disorder*. One example on the X chromosome involves concomitant Duchenne muscular dystrophy, impaired function of white blood cells (chronic granulomatous disease), mental retardation, deficient function of the adrenal gland (adrenal hypoplasia), and an enzyme disease called glycerol kinase deficiency. (A child who had all these disorders, and whose DNA was used to discover the specific gene defect in two of these diseases, died in an automobile accident!)

Knowledge of a gene location can be followed by procedures to isolate the gene, clone it (make a duplicate), study its protein products, analyze the malfunction, and set about correcting the disease. Thus far, over 4,000 disease-causing genes are known to exist, out of a possible 50,000 to 100,000. There is therefore much discussion in Washington about whether to spend $2 billion and sequence all the genes! Meanwhile, fewer than 100 disease-causing genes out of about 4,000 recognized single-gene disorders have been mapped. Straightforward techniques exist (for example, enzyme analysis of white blood cells) to diagnose or detect carriers of many of the conditions once a gene has been mapped. Similar diagnostic methods are available for prenatal diagnosis in these disorders. For the vast majority of single-gene diseases, however, none of these tests have been available.

DIAGNOSIS BY ANALYZING DNA

As you know by now, we all harbor at least twenty harmful genes. For the most part, we rarely get to know the actual identity of any of them. Now, with gene technology advancing so rapidly, this state of usually blissful ignorance will eventually end — if we so desire. The secrets of our genes will steadily yield to combinations of family-pedigree analysis, ancillary tests on family members, and use of natural and synthetic DNA markers.

Dramatic new advances in detecting genetic diseases using the DNA technology just described are increasing the number of disorders in which precise prenatal diagnosis, postpartum diagnosis, and/or carrier detection can now be achieved. In many cases, a diagnosis can be made long before a disease becomes apparent. A selected list of genetic disorders in which DNA technology is proving valuable is provided in table 6, together with the chromosome location of the disease-causing gene. Some of these examples are so new that only one laboratory has the available probes for diagnosis, and such a lab may still regard its efforts as being in the research arena. Nevertheless, the published findings herald the availability of new tests, for the future, if not for immediate use. Note that although the basic defect is not understood for most of the diseases listed in table 6, it is still possible to obtain valuable diagnostic information for families. Should you recognize any of these diseases in your family, consult with a clinical geneticist in a medical school–affiliated hospital to determine whether your family could benefit from these new tests. Given the rapid pace of progress in this field, any family with a serious genetic disease should remain in touch with a clinical geneticist.

One unexpected spin-off of DNA technology is its use in the resolution of disputed paternity and, less often, maternity. Dr. Alec Jeffreys and his colleagues at Leicester University in England devised a method of recognizing genetic markers that are as unique to an individual as that person's fingerprints. They used the basic technique of cutting or cleaving DNA into fragments and tracking these variable-sized fragments through families. Their technique, described as "DNA fingerprinting," depends upon the actual lengths of highly repetitive pieces of DNA that are *between* genes — the introns (discussed earlier). This enables them to monitor twenty different DNA cleavage sites simultaneously and makes this the best genetic technique ever found to analyze our genes.

**Table 6. Selected Genetic Diseases in which DNA Technology Is
Available**

Genetic Disorder	Chromosome on Which Gene Is Located
Adrenoleukodystrophy	X
Agammaglobulinemia	X
Albinism (ocular form)	X
Alpha-1-antitrypsin deficiency	14
Alzheimer disease	21
Aniridia	11
Charcot-Marie-Tooth disease	X
Chronic granulomatous disease	X
Congenital adrenal hyperplasia	6
Cystic fibrosis	7
Familial amyloid neuropathy	18
Familial hypercholesterolemia	19
Familial polyposis of the colon	5
Fragile-X syndrome	X
Growth hormone deficiency	17 and 20
Hemochromatosis	6
Hemophilia A	X
Huntington disease	4
Incontinentia pigmenti	X
Lymphoproliferative syndrome	X
Manic depression (bipolar type)	11
Muscular dystrophy (Becker type)	X
Muscular dystrophy (Duchenne type)	X
Muscular dystrophy (myotonic type)	19
Neurofibromatosis	17
Ornithine transcarbamylase deficiency	X
Osteogenesis imperfecta	17
Phenylketonuria	12
Polycystic kidney disease (adult type)	16
Retinitis pigmentosa	X
Retinoblastoma	13
Sickle-cell anemia	11
Spinal muscular atrophy	X

Table 6 (continued)

Genetic Disorder	Chromosome on Which Gene Is Located
Thalassemias	11 and 16
Tuberous sclerosis	9
von Willebrand disease	12
Wilms' tumor	11
Wiskott-Aldrich syndrome	X

NOTE: If you recognize one of these disorders in a near relative, contact a clinical geneticist in a teaching hospital to determine if you need genetic counseling and carrier or other tests.

The Leicester workers observed that there is such great variation between individuals that the system could be used for the first time ever for positive proof of paternity or identity, not just exclusion of paternity or identity, which has been the rule with all other genetic tests used thus far. British researchers calculated that the probability of unrelated individuals sharing the same DNA fingerprint is, for all practical purposes, zero! Even among siblings, the chance of sharing the same pattern of markers was found to be 1 in 100 million!

Consequently, this technique has been characterized as revolutionizing police work. Jeffreys and his colleagues have already successfully demonstrated DNA fingerprints from four-year-old bloodstains and from sperm separated from other cells in the vagina following rape. These same techniques have been used successfully in North Carolina and Florida to convict a murderer and several rapists, including a father accused of raping his daughter, who bore his child. (Chapter 9 describes the use of this technique in an immigration case in which the authorities disputed *maternity*.)

DIRECT GENE ANALYSIS

Isolating a gene by the techniques described might suggest that direct analysis of its structure would best satisfy the goal of precise diagnosis. For sickle-cell disease and its causal point mutation, which virtually never varies, this method would be fine, but simpler accurate tests are available. For most other genetic diseases, however, different mutations are likely to exist along the length of a gene; many families with the same disease have a slightly different structure of the defective gene. Thus, for many years to come, analyzing the gene's structure — sequencing it — in a family will not be practical.

However, analysis of viral genes and those of other micro-organisms has already begun to play an important role in the more precise diagnosis of infectious diseases.

INFORMATIVE FAMILIES

Use of DNA analyses requires comparative study of the genes of key family members, since the location of most defective genes is unknown. This indirect approach is important, because some families will not be able to benefit from these techniques — at least until the gene is cloned and its protein products studied.

In the earlier discussion on probing for genes, I emphasized that the sites where restriction enzymes cut are inherited. Notice in figure 28 that each chromosome strand with its restriction (cleavage) site is transmitted, and that each one passes to half the children. The restriction site itself is actually a genetic signpost or marker that makes it possible, for example, to distinguish one person's number-4 chromosome from another's. If, when using a probe, fragments are identified that are of the same size, it will not be possible to distinguish which fragment came from which parent. In such instances, in regard to that particular probe, the family is described as *uninformative*. Individuals and families in whom different-size fragments can be identified when using the same enzyme are regarded as *informative*. The key to success in determining the presence of a mutated gene is for at least one parent to have different restriction (cutting) sites. Variations in the sequences of the bases at the cutting sites are extremely frequent and occur randomly about once in every 100 or so base pairs. Hence, the different-size fragments again simply reflect our unique genetic individuality.

To benefit from these techniques, the genes of a closely related affected family member are required for comparative study. In practice, this usually means obtaining a blood sample for DNA analysis from the affected relative. You might think this is a fairly simple matter, and certainly no obstacle to a couple concerned about *avoiding conception* of a defective child. One couple's recent experience, which is not unusual, will serve to enlighten.

Esme married Sam, who knew that her mother's one sister had two boys with Duchenne muscular dystrophy. The boys were in their teens, wheelchair-bound and dying slowly, when Esme and Sam came to me to discuss their risks and options vis-à-vis a child of theirs being affected by Duchenne muscular dystrophy. I explained their good fortune: that there was a high probability that the new gene technology could make it possible to determine whether Esme carried this gene, and even whether their fetus in early pregnancy was affected by muscular dystrophy.

The awkward silence in my office spoke reams. Preferring to stay away from her "sick cousins," Esme had had little contact with her aunt for over ten years. But the couple left the consultation intent on asking Esme's aunt the favor of getting a blood sample from one of her sons.

I heard later that the aunt resisted but did promise to ask the boys themselves. Neither of the boys could remember seeing Esme. Both refused to have blood drawn to help their cousin. Esme and Sam, who still plan to conceive, will have to take their chances without the benefit of these latest advances.

BANKING GENES

Banking genes to help others in the close and extended family is a new dimension that has received insufficient attention. Because of further expected rapid technical advances in gene analysis, you should be aware of any family member with a serious or fatal genetic disease. In the coming years, having their genes available to analyze in conjunction with yours (and possibly those of your fetus) may prove incredibly important to you. Every effort should therefore be made to obtain a blood sample from the affected relative after you have arranged with a geneticist to bank this gene source. (Our Center for Human Genetics at Boston University School of Medicine provides such services.)

IMMORTAL DEPOSITS

Once a blood sample is obtained, one portion can be used to extract the genes, freeze them, and put them away for later analysis. Such samples may not be usable indefinitely. Moreover, they may not contain sufficient DNA if repeated analyses are required because a number of family members wish to benefit from genetic testing of the sample. Therefore, the white blood cells are taken from the other portion of the blood sample and are infected with the infectious mononucleosis virus. This virus is incorporated into the nucleus of these cells and literally immortalizes them; that is, they become able to grow continuously, without ever aging. These immortal cell lines can be deposited in sterile ampoules, frozen in liquid nitrogen at minus 196 degrees Centigrade, thawed out, grown again, and refrozen! We have hundreds of such cell lines frozen away in liquid nitrogen, primarily for research purposes, but some also for concerned families planning ahead.

ETHICAL DILEMMAS

Would you want to know that you are destined to be stricken with a fatal genetic disease? The new DNA technology now allows such fateful predictions for diseases such as Huntington's as early as eleven weeks of pregnancy. Given all the personal and family implications, I have serious concerns about the use of this test, say, for an untreatable fatal disease such as Huntington's. While every child of an affected parent has a 50-percent risk of this disease, it may literally take decades to manifest itself. The suicide rate among such persons at risk is about seven to eight times higher than average; and the half who undergo genetic testing and hear bad news have an even greater burden to bear. There is reason for optimism that some effective treatment will emerge in the next few decades, if not sooner. People at risk who choose not to know will have to live on in uncertainty. It is, of course, an individual choice. A questionnaire study of 155 persons at 50-percent risk for Huntington's, done *since* the new DNA tests have become available, indicated that 63 percent of this high-risk group would have a predictive test even if no treatment were available. However, only 4 of 24 respondents (17 percent) planning to have children said that a positive test result would deter them from doing so. So far as prenatal diagnosis was concerned, 76 of the 155 respondents (49 percent) said they would use such a test, and 32 of these 76 (42 percent) stated they would abort a pregnancy if the fetus had the Huntington-disease gene. Notwithstanding the conclusions of this survey, only time will tell how many will actually *have* the test. I suspect it will be far fewer than the number in the survey who responded in the affirmative.

Other ethical and legal issues abound. For example, if a geneticist injudiciously advises a presymptomatic test for Huntington's, the person tests positive, and promptly commits suicide, is the doctor legally vulnerable? Should children be tested? At what age?

Can insurance companies demand test results as they normally do, and can they request that such tests be done before health or life insurance is provided? After all, such demands are now being made for AIDS.

Much professional and public discussion is necessary to develop sound public policy on these issues. Huntington disease is simply the first of many fatal or crippling diseases we will be able to predict with certainty long before a disease appears. In the absence of an effective treatment or avoidance regimen, who among us wants to know where the "genetic land mines" lie? Clearly, this technology must be used to help rather than harm people.

Gene Therapy

Introducing normal genes into the cells of at least one body organ to remedy or even cure a single-gene defect is a gene therapy that is foreseeable. Laboratory experiments have been successful in introducing functioning genes into the cells of mice, rabbits, pigs, and sheep. A distinction must be made between inserting genes into regular body cells as opposed to putting them into an egg, sperm, or single fertilized egg. Experiments on germ cells of humans in which any gene can be inserted and thereafter transmitted to offspring are presently strictly forbidden. Such efforts in mice, however, have provided valuable lessons.

In one such experiment with mice, researchers used the *rat* gene that codes for growth hormone and injected copies of this gene into the fertilized eggs. In the few mice in which the procedure worked, giant offspring resulted. In another experiment, the fertilized eggs of mice with thalassemia (the same hemolytic anemia seen in humans) were used. Copies of the normal *human* gene coding for the necessary protein (globin) were injected into fertilized eggs, which were then implanted in surrogate mothers. The offspring were born free of this disease.

Despite these successes, considerable difficulty has been encountered. Once introduced into a cell, genes may not function at all, or may do so poorly or at too slow a rate. Moreover, viruses that are frequently used to "carry" genes into cells have themselves become incorporated into the genes of the recipient animal. We already know that insertion of viral genes into normal genes can cause new genetic disorders (such as limb defects), and it would be no surprise if cancer was another possible consequence.

Another approach to learning about gene function was taken by Harvard University researchers who have bred a strain of mice that carry a specific cancer-causing gene. In 1988 these scientists were awarded the first United States patent on a living animal!

These techniques have also been successful in introducing normal functioning human genes into bone marrow of mice and monkeys, with an eye to remedying diseases such as thalassemia or a fatal immune-system disorder called combined immunodeficiency. A major obstacle is represented by the many genetic diseases in which mental retardation is a major feature. Even when it is clear that an inherited enzyme deficiency can be corrected *in a test tube* by transferring a normal gene into the cells of an affected person, getting such normal genes safely and effectively into the brain cells *in the body* remains an unachieved goal. Restrictions on the use of viruses to carry genes into cells, with or without a specific target organ, continue to delay human applications. Moreover, inserting a single gene into human DNA, even if the gene could be precisely targeted to the correct chromosome, may still upset neighboring genes. Clearly, gene therapy remains on hold for the present.

Genetic Engineering

The introduction of DNA technology has been compared to the discovery of fire in the context of benefits to mankind. This is no exaggeration when you consider the vast implications, first for health, but stretching to encompass farming, agriculture, entomology, and the drug and chemical industries. Realistic expectations that will revolutionize agriculture include new forms of food crops, such as those that grow in salt water, those that will not require traditional fertilizers for nitrogen, and those that are bred to resist environmental hazards — frost, disease, or insects, for example. This new molecular knowledge will facilitate the breeding of larger and more fecund food animals. Newly created but safe microorganisms will assume roles hitherto undreamed of. Already such organisms have been developed that feed on oil spills!

The transfer of plant genes has already become a routine procedure. Genetically engineered plants are being produced to improve their quality or quantity, or both, and to confer resistance to herbicides or infections. Mixing genes of different fruits and vegetables is likely to result in novel combinations, such as topatoes (a tomato/potato vegetable) or, depending upon the expression of the gene used, pomatoes! In 1988 the first patent on plants was granted; it involves a technique for increasing the protein content of alfalfa and other forage crops. The possibilities for agriculture are obviously limitless.

Health implications of genetic engineering include prolonged life expectancy through earlier and more precise diagnoses, prediction of susceptibility and avoidance of hazardous exposures and occupations, new and more effective treatments, new types of drugs, the likely cure of many cancers, and effective treatments for heart disease, diabetes, and various genetic disorders.

The essentials of this technology are not difficult to understand. The objective is to obtain large amounts of a specific gene's pure product, such as insulin, growth hormone, and so forth. The first step involves cutting DNA into fragments by using the enzymes known as restriction endonucleases (discussed earlier). Then, employing the natural or synthetic probes mentioned previously, specific genes can be isolated.

The next step in the process involves bacteria. In some bacteria there are circular bits of DNA called *plasmids*. They can be removed, and the same restriction enzyme that was employed to cut the human DNA can then be used to cut, and hence open up, the circular DNA of the bacterial plasmid. Since the cut DNA fragments have sticky ends, the human DNA (or gene) can be spliced into the now open circular plasmid DNA and its circular shape reconstituted. These plasmids containing human genes are then introduced into specially modified bacteria that cannot survive outside artificial laboratory conditions. These bacteria are then cultivated and can be used to make gene libraries for further research or grown into huge numbers, with the gene product finally derived for human use. Other techniques with the same goal involve bacterial viruses called *phages*.

Gene products already made this way include insulin, growth hormone, fertility hormones, Factor VIII (the deficient-clotting factor in hemophilia), erythropoietin (which stimulates red-blood-cell formation, and may prove valuable in treating sickle-cell anemia), and hepatitis-B vaccine. Novel medicines, new vaccines (such as against malaria), replacement proteins for certain genetic diseases, new drug delivery techniques, more precise diagnostic tests, and on and on — the gene products of the "new" genetics will provide infinite advances for the benefit of mankind.

9

Genes, Ethnic Origin, and Blood Groups

Each of us is unique, not only by virtue of our parents' genes but because of how our own genes function together and in concert with varying environmental factors. The first major difference among us is our racial origin. Three distinct groups are easily distinguishable: African, Caucasian, and Oriental. Representatives of all three groups today inhabit every continent, and, over the centuries, very significant mixing has occurred. It is easy to distinguish a person's race when the color differences are really striking, but mixing has tended to make it harder and harder to determine the origins of some people. Study of the differences between gene effects has been useful in this regard. It has become clear, however, that genetic differences between races appear small in comparison to those between the ethnic groups within races.

A few genetic characteristics, such as certain blood groups, even used alone, may indicate an individual's ethnic group. Africans virtually all have the Duffy blood group (named for the person in whom it was first recognized), in contrast to only 3 percent of whites with this genetic marker. Rh-negative blood (explained below) occurs rarely among Africans, and is most frequent in whites; and there are other genetic markers that give us clues that allow differentiation of racial groups.

Genetic Adaptations

Many years ago, researchers noticed that in areas where malaria was rife, there was also a high frequency of sickle-cell anemia — a progressive, hereditary, ultimately fatal hemolytic anemia that mostly affects blacks, Greeks, and Asiatic Indians, causing about 100,000 deaths yearly. Ultimately, it was clearly shown that those individuals carrying the gene for sickle-cell anemia were much more resistant to malaria. You will recall that sickle-cell anemia is a recessive disorder, and the gene is carried equally by both parents. While the individual affected by this anemia is ill, the carriers are not. There is, however, an increased risk of sudden death among sickle-cell carriers who indulge in vigorous and prolonged physical exercise. Among black recruits undergoing extremely rigorous basic military training, 1 in 3,200 died suddenly. Individual blacks should therefore know their carrier status (a simple blood test makes this possible), and carriers should avoid severe, prolonged exercise in heat and especially at high altitudes.

As mentioned, carriers of the sickle-cell gene turn out to have an advantage, or be "fitter," in that they are able to resist malaria much better than those of us without the gene. Other genetic adaptations to disease are known to confer advantages on the bearer of the specific gene, too. Thalassemia, for example, the fatal, hereditary anemia similar to sickle-cell disease, involves changes in bones and skin, as well as spleen enlargement. It is found mainly in people who live in or originate from the Mediterranean regions — especially Italians and Greeks — as well as in those from Africa and India. Again, carriers are more resistant to malaria. Female carriers of the sex-linked enzyme disorder called glucose-6-phosphate dehydrogenase are more resistant to malaria; this disorder also occurs in those who live in the malaria belts of the world.

One highly speculative and unproved selective advantage has been suggested for Tay-Sachs disease. The supposition is that carriers are more resistant to tuberculosis, a disease that was rife in the overcrowded ghettos of Eastern Europe from which these carriers originated.

GENES AND CLIMATE

It may have occurred to you that skin color, which is determined by your genes, is closely related to the amount of sunshine in the region of ethnic origin. The closer to the equator, the stronger the sunlight, and the more likely it is that the skin of the people who originate there will be dark. Hair and eye color — both gene-determined — have similar distributions to skin color.

Dark-colored skin, eyes, and hair are well adapted in hot, bright climates.

The so-called peppercorn hair of Africans may prevent rapid evaporation of sweat, possibly protecting their heads from excessive heat and sunstroke. The absence of facial hair on Orientals who originated in northern and eastern climes has been considered an advantage in preventing frostbite. The narrow eyes and the eye fold near the nose in Orientals may protect the eye from the elements. Since body heat is retained better by larger bodies, it should come as no surprise to you, at this stage of the discussion, to discover that body size increases with latitude! So much for gene advantages. Let us move on to what is more serious: disadvantageous genes.

GENES IN ISOLATION

Many populations in the world have among them groups of people who isolate themselves from the rest on religious or other grounds. Such a group may remain as an "isolate" for many generations, inbreeding being the expected and actual consequence, with more and more members sharing the same harmful genes they have inherited. The children born in any isolated group show the manifestations of inbreeding by virtue of an increased frequency or unusual nature of birth defects or genetic disorders.

In some isolated alpine villages, the frequency of albinism is high. Other villages have high rates of deaf-mutism, blindness, or mental retardation. On one Pacific Ocean atoll, Pingelap, about 5 percent of the population experience total color blindness (achromatopsia), a condition that is associated with severe nearsightedness and other serious eye problems. In one community of the Amish in Pennsylvania, a particular type of dwarfism with extra digits has been recognized. The Amish — a religious and social isolate — intermarry, and few genes from the "outside" are ever added to the community.

Ethnic groups, to an extent, represent isolates because of inbreeding. Let us explore the consequences of our ethnic origins.

Hereditary Disease and Ethnic Group

The risks that you or a child of yours will suffer from or carry a hereditary disorder may in large part depend on your ethnic origins. Tables 7, 8, and 9 list some of the possibilities. The high frequency of some diseases, such as Tay-Sachs disease and familial dysautonomia, that are found among Ashkenazic Jews is not seen in Sephardic Jews. While familial Mediterranean fever may be common to Sephardic Jews, it is rarely found in Ashkenazic Jews. Blacks in the United States who carry the sickle-cell gene inherited it from ancestors who brought it with them from Africa. About 1 in 12 black Americans now carries this disorder, and a child with sickle-cell anemia occurs in about 1 in 650 black births.

Cystic fibrosis is the most common genetic killer of white children. About 1 in every 2,500 are born with it, and about 1 in every 25 whites is a carrier. Cystic fibrosis in blacks or Orientals is positively rare. When it does occur, it is almost invariably related to white admixture somewhere in the family history. Phenylketonuria (PKU), a hereditary biochemical disease that causes mental retardation when untreated, similarly is mostly found in whites and is rare in blacks and Orientals.

It should be made clear that any of these diseases could occur, through intermarriage or mutation, in any ethnic group. Tay-Sachs disease is found in children who are not Jewish, even though it is at least 100 times less frequent. (One remarkable exception: French-Canadians, who carry the gene for Tay-Sachs disease about as often as Ashkenazic Jews. Roughly 1 in 27 French-Canadians are carriers; hence, these people, most of whom are Catholics, are advised to have a carrier test for Tay-Sachs disease *before* conception, in order to know their risks and exercise their options.)

Questions about an individual's country of origin or ethnic group, therefore, can be crucial in alerting the physician to the possibility of a rare and unsuspected disease. A recent example illustrates this point.

Pietrus was a twenty-four-year-old, married graduate student who came to the emergency ward complaining of severe abdominal pain and vomiting. His symptoms had begun on the very day that he came to the hospital. He said that up to that day he had been in excellent health except for some constipation, though he admitted to one previous similar episode that had also landed him in the hospital and resulted in the removal of his appendix. This had been some five years ago, and he reported that the doctor had said after the operation that the appendix had, surprisingly, looked perfectly

Table 7. Common Genetic Disorders in Various Ethnic Groups

Ethnic Group	Genetic Disorder
Africans (blacks)	Sickle-cell disease and other disorders of hemoglobin
	Alpha- and beta-thalassemia
	Glucose-6-phosphate dehydrogenase deficiency
	African-type adult lactase deficiency
	Benign familial leukopenia
	High blood pressure (in females)
Afrikaners (white South Africans)	Variegate porphyria
	Fanconi anemia
American Indians (of British Columbia)	Cleft lip or palate (or both)
Armenians	Familial Mediterranean fever
Ashkenazic Jews	A-beta-lipoproteinemia
	Bloom syndrome
	Congenital adrenal hyperplasia
	Dystonia musculorum deformans
	Familial dysautonomia
	Factor XI (PTA) deficiency
	Gaucher disease (adult form)
	Iminoglycinuria
	Meckel syndrome
	Niemann-Pick disease
	Pentosuria
	Spongy degeneration of the brain
	Stub thumbs
	Tay-Sachs disease
Chinese	Thalassemia (alpha)
	Glucose-6-phosphate dehydrogenase deficiency (Chinese-type)
	Adult lactase deficiency
Eskimos	E_1 pseudocholinesterase deficiency
	Congenital adrenal hyperplasia
Finns	Congenital nephrosis
	Aspartylglucosaminuria
French-Canadians	Tay-Sachs disease
	Neural-tube defects
Irish	Neural-tube defects
	Phenylketonuria
	Schizophrenia
Italians (northern)	Fucosidosis

Table 7 (continued)

Ethnic Group	Genetic Disorder
Japanese and Koreans	Acatalasia Oguchi disease Dyschromatosis universalis hereditaria
Maori (Polynesians)	Clubfoot
Mediterranean peoples (Italians, Greeks, Sephardic Jews, Armenians, Turks, Spaniards, Cypriots)	Thalassemia (mainly beta) Glucose-6-phosphate dehydrogenase deficiency (Mediterranean-type) Familial Mediterranean fever Glycogen-storage disease (type III)
Norwegians	Cholestasis-lymphedema Phenylketonuria
Yugoslavs (of the Istrian peninsula)	Schizophrenia

NOTE: This table simply provides some perspective on the extent of hereditary ethnic disorders. Any specific concern is best discussed with your doctor.

normal. No family history was available, since he had been adopted.

General physical examination revealed a well-built young man with only slight fever, persistent vomiting, dehydration, and complaints of severe abdominal colic with only some abdominal tenderness. A series of blood tests followed by X rays still did not reveal the cause of his abdominal pain, and questions began to arise about the possibility of poisoning. An alert young physician noted that Pietrus had been born in South Africa, which prompted the doctor to suggest a urine test to check for a condition with acute attacks that has a mortality rate of about 24 percent. This condition, correctly considered and ultimately diagnosed in Pietrus by the young resident, is acute intermittent porphyria, a complex, hereditary biochemical disease of proteins that affects the nervous system and other organs — a disorder that indeed is common in South Africans of Dutch descent.

Pietrus probably inherited the gene for this dominant genetic disease from one of his parents. (Recognizing that he carried it was important. It not only saved him from additional surgery, but also alerted him never to allow himself to be given barbiturates, since such drugs often precipitate severe attacks in porphyria patients and may prove fatal. Moreover, he then received careful genetic counseling, during which it was explained that he and his wife had a 50-percent risk in every pregnancy of having an affected child.

Table 8. Chances of Genetic Disorders for Various Ethnic Groups

If You Are	The Chance Is About	That
black	1 in 12	You are a carrier of sickle-cell anemia
	7 in 10	You will have milk intolerance as an adult (for example, develop diarrhea)
black and male	1 in 10	You have a hereditary predisposition to develop hemolytic anemia after taking sulfa or other drugs
black and female	1 in 50	
	1 in 4	You have or will develop high blood pressure
white	1 in 25	You are a carrier of cystic fibrosis
	1 in 80	You are a carrier of phenylketonuria
Jewish (Ashkenazic)	1 in 30	You are a carrier of Tay-Sachs disease
	1 in 100	You are a carrier of familial dysautonomia
Italian-American or Greek-American	1 in 10	You are a carrier of thalassemia
Armenian or Jewish (Sephardic)	1 in 45	You are a carrier of familial Mediterranean fever
Afrikaner (white South African)	1 in 330	You have porphyria
Oriental	100%	You will have milk intolerance as an adult

Pietrus elected to have a vasectomy and the couple adopted first one and then a second child.

Extensive and laborious studies by Dr. G. Dean, in South Africa, established that porphyria was introduced in that country by a pair of immigrants from Holland who married in 1688. Today, about 1 in 330 white South Africans possess the gene for porphyria and are actually affected by this disease. Because anesthesia, barbiturates, and other drugs may prove fatal to affected individuals, many hospitals in South Africa routinely test every patient for porphyria on admission.

The Genes and History

Why, you may have wondered, should a disorder such as Tay-Sachs disease be carried by almost 4 percent of Ashkenazic Jews regardless of where they live? The same type of question could be asked about various other diseases that especially affect certain ethnic groups. The most likely and obvious explanation is that these individuals have a common ancestry and therefore may share the same "bad" genes. Historical events serve to explain why Ashkenazic Jews are afflicted by genetic diseases that are quite different from those of Sephardic Jews. It seems that Ashkenazic Jews descended from those who fled to northeastern Europe after the sacking of Jerusalem by the Romans in 70 C.E. The Sephardim (Oriental or Spanish Jews) are probably descended from those who fled to Babylon after the destruction of their first temple in 586 B.C.E. The Jews in Iran and many of those now in Israel are the likely descendants of those exiles, as are those who fled westward from the Romans in 70 C.E. to North Africa and

Table 9. Chances of Genetic Disorders in a Child for Various Ethnic Groups

If You Are	The Chance That Your Child Will Have	Is About
black	Sickle-cell anemia	1 in 650
	Thalassemia	8 in 1,000
white	Cystic fibrosis	1 in 2,500
	Phenylketonuria	1 in 25,000
Jewish (Ashkenazic)	Tay-Sachs disease	1 in 3,600
	Familial dysautonomia	1 in 10,000 to 20,000
Italian-American or Greek-American	Thalassemia major	1 in 400
Armenian or Jewish (Sephardic)	Familial Mediterranean fever	1 in 8,000

NOTE: These approximate risk figures apply only if both you and your spouse are of the same ethnic group.

Spain. Subsequent changes in the genes from mutation and inbreeding probably largely explain the occurrence of different diseases in these two groups of Jews, as well as the frequency with which they occur.

Through blood-group studies, it has been possible to come to some conclusions about the movement of the Celts, who include peoples of Cornish, Welsh, Irish, and Scottish descent, and the Vikings of Norway (Norsemen). The Vikings invaded Ireland and Britain and temporarily settled there until their final eviction in the late tenth and early eleventh centuries C.E. Some Norsemen, after leaving Ireland and Britain, settled in Iceland, taking with them Celtic wives and slaves. Genetic evidence shows that most of the early settlers in Iceland were Celts rather than Norsemen. The genetic evidence is based on the peculiar frequency of specific and different blood-group systems in peoples of those areas today. From similar studies, it has been deduced that there is a common ancestry for inhabitants of the west coasts of Norway, Scotland, and Ireland. The most likely explanation appears to be the importation of Celtic women as wives and slaves of Norsemen to Norway in the Viking period.

Scholars have long pondered the origin of the European Gypsies. The majority of these people once lived in Hungary, as isolated groups in which there was much inbreeding. It has been noted that a remarkable difference exists between the frequencies of the ABO blood-group system in Gypsies as compared to Hungarians. The blood-group systems in Gypsies, however, are remarkably similar to those in Asiatic Indians, and it is thus probable that the Gypsies have Indian origins.

Blood Groups and Disease

Each of us possesses different genes that control the development of protein substances that determine the different blood groups. The ABO blood-group system was discovered by Dr. Karl Landsteiner in 1900. Your blood group is determined as A, B, AB, or O — and there are many other blood groups recognized by virtue of specific proteins that coat the surface of red blood cells. Since 1900, some 250 different proteins on and in the red blood cells alone have been recognized. Correctly matching blood is therefore critical before blood transfusions are given.

Not unexpectedly, ethnic differences occur in the frequency of blood groups. The B gene, for example, is three times more frequent in Orientals than in whites, whereas in virtually all American Indians, the B gene is almost entirely absent.

The Rh blood groups were discovered in 1939, and are thus named because the blood of rhesus monkeys was used in the experiments. Persons who have the specific Rh protein, or *antigen*, on their red blood cells are designated "Rh positive"; they constitute about 85 percent of all Caucasians; the other 15 percent do not have this protein and are called "Rh negative." Orientals are rarely Rh negative, and only about 1 percent of blacks are Rh negative. (Actually, there are eight Rh antigens from the Rh gene complex, but for practical purposes I will discuss only the main and most important one.)

If an Rh-negative woman becomes pregnant by an Rh-positive man — and this really is the main hazard — then the fetus may be Rh positive. In such a case, fetal red blood cells with the Rh antigen cross over into the mother and are recognized by the mother's immune system as "foreign invaders." The Rh-negative mother's body mounts an attack by producing a protein *antibody* to counter the entering foreign protein. Forever after, the mother's body makes the anti-Rh antibody, which will cross over into the fetus in *subsequent* pregnancies. Every time that a fetus is Rh positive (after the first pregnancy, which in most cases will be fine), the Rh-negative mother's antibody will attack and destroy the fetal red blood cells. Fetal death, stillbirth, or severely ill newborns with hemolytic anemia and jaundice may be the result. Blood transfusion of the fetus in the womb or exchange transfusion after birth is usually lifesaving, and will also prevent later mental retardation, deafness, or cerebral palsy. An Rh-negative mother may have become sensitized from a previous incompatible blood transfusion, which would affect the very first pregnancy. Similarly, sensitization of the mother may occur from fetal red blood cells crossing over during an unrecognized miscarriage very early in a first pregnancy. Then, in the second pregnancy, which the mother thinks is her first, the fetus may be affected. Sensitization may also occur immediately after an amniocentesis if not prevented.

If you plan to have a child, it is obviously enormously important not only to know your genes but to find out your Rh group. The reason is that it is possible to prevent an Rh-negative woman from becoming sensitized simply by giving her an injection of a specific protein, immunoglobulin, upon the birth of her first baby. Because of this immunoglobulin treatment, hemolytic jaundice in the newborn is steadily being wiped out. Remember, however, to find out your Rh group (and that of your mate) before pregnancy; and ensure that the first visit to the doctor occurs early in pregnancy, rather than waiting some two to four months. Rh immunoglobulin should be given immediately (and certainly within seventy-two hours) in situations in which fetal red blood cells may enter the Rh-negative mother's circulation, such as following birth, miscarriage, and amniocentesis.

Incompatibility between mother and baby in the ABO blood groups only occasionally causes problems (anemia, jaundice, or both) after birth. Curiously, ABO incompatibility between mother and child is strongly protective against Rh sensitization. For example, if the fetal cells are A or B and the mother's cells are not the same, the mother's antibodies destroy them so rapidly that there is not sufficient time for the slower process of Rh sensitization to occur.

The search for associations between ABO blood groups and disease has been relatively unrewarding. Perhaps the most firmly established association is between the blood group O and duodenal ulcer. Even then, Type O occurs only about 1.4 times more often in people with duodenal ulcer. Pernicious anemia and stomach cancer are slightly more common in those with Type A blood, while stomach (as opposed to duodenal) ulcers are a little more common in Type O individuals. Type A is associated with a rare birth defect called the nail-patella syndrome, which is characterized by totally absent or abnormal nails, and absent or defective kneecaps. A dominantly inherited muscle disease, myotonic muscular dystrophy, is associated with another blood-group system called the Secretor system. Young women who take oral contraceptives have a higher incidence of blood-clotting complications if they have Type A, B, or AB blood rather than Type O.

THE HLA COMPLEX

In addition to those having to do with red blood cells, other inherited-blood-group systems have been found that vary among individuals, including M, N, Xg, Secretor, Kell, Duffy, and Lutheran. Over thirty different antigens can be detected on white blood cells. This group constitutes what is called the *HLA system* — one that is crucial to the careful matching necessary for organ transplantation.

Coating the surfaces of virtually all our cells are certain proteins — *glycoprotein antigens* — that are produced by our genes and hence unique to each of us. Initially called HLA — for *human leukocyte antigen* — because these glycoproteins were first found on leukocytes, or white blood cells, we now know how ubiquitous they are: they occupy the surface of all cells except red blood cells that are devoid of nuclei. The particular set of cell-surface antigens our genes make is not only unique to us but critically important if it becomes necessary to transplant a vital organ such as the kidney or heart. The donor's organ must match with these antigens to avoid the potential disaster of tissue rejection.

The set of HLA antigens each person possesses also functions in the killing and disposal of virus-infected cells, as well as stymieing any mechanism that might lead the body to make antibodies against its own proteins — a malfunction that occurs in so-called *autoimmune diseases,* such as thyroid disease and diabetes.

The HLA cell-surface glycoproteins are made by a remarkable major complex of genes located on chromosome number 6. This gene cluster — called the *major histocompatibility complex,* or *MHC* — is made up of genes at five sites; these sites are designated A, B, C, D, and DR. Alternative genes, or *alleles,* could occupy each of these locations, and these genes are given numbers — 1, 2, 3, and so on — for identification. Where some uncertainty exists about an alternative gene, a number is preceded by W. Many alternative genes occur. For example, twenty are known for the A gene site, or *locus;* over forty are known for HLA (*locus*) B; and so on. The set of five HLA genes aligned along one number-6 chromosome is called a *haplotype,* and since we each have a pair of number-6 chromosomes, we each have two haplotypes (see figure 29). Hence, if on one chromosome at the A gene-locus, a number-10 alternative gene is present, you are designated as HLA group A10. If at the B locus there is a number-7 alternative gene, you are HLA-B7; and similarly for all five gene locations on each chromosome. In figure 29 you can see how the HLA group is inherited, and how an individual might transmit his or her set of HLA genes. As it happens, these HLA genes are so tightly linked that (except for rare exceptions) they travel in a bunch tightly knitted together. There is a 1 in 4 chance that two siblings will have the identical HLA-gene set. Hence, when a transplant is needed, the search begins among the brothers and sisters, since the parents are very unlikely to have matching HLA groups.

We have known since 1974 that individuals with certain HLA

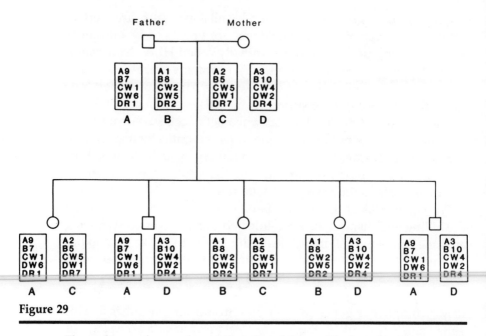

Figure 29

The pedigree of a couple and their five children. The boxed pairs represent segments of DNA in a number-6 chromosome pair. The letters and numbers within indicate specific HLA haplotypes. Note how they "travel" as a unit from each parent.

types are at greater risk of developing specific diseases. For example, a Caucasian (white) with HLA-B27 has a 69-times-greater risk of developing ankylosing spondylitis, a form of rheumatoid arthritis of the spine. While HLA-B27 occurs in about 1 in 12 whites, the incidence of ankylosing spondylitis is 1 in 2,000. On the other hand, 89 percent of all whites affected with this painful disorder have HLA-B27. Therefore, if you are a close relative of someone with the disorder but do not have this HLA type, it is virtually certain you will not develop this condition. Fortunately, many with HLA-B27 do not get ankylosing spondylitis, which suggests that some other factor, such as infection, is necessary to precipitate it in a genetically susceptible person.

A surprisingly long list exists of diseases associated with particular HLA types and the relative risks people have of developing such conditions (table 10). The vast majority of these disorders tend to be familial but not due to one gene. Virtually all seem to involve the immune system.

Table 10. Selected Significant HLA Types and Disease Associations

Disease [a]	Nature of Disease	HLA Antigen	Race [b]	Percentage of Affected Persons with This HLA	Percentage of Unaffected Persons with This HLA	Relative Risk [c]
Narcolepsy	Recurrent episodes of irresistible sleep in the daytime, often associated with collapse due to loss of muscle tone and an associated abnormal electro-encephalogram	DR2 DR2	O C	100 100	34 22	358.1 129.8
Ankylosing spondylitis	A chronic rheumatoid-like arthritis of the spine	B27 B27 B27	O C B	85 89 58	15 9 4	207.9 69.1 54.4
Reiter disease	A syndrome involving inflammation of joints, eyes, urethra, and/or cervix	B27	C	80	9	37.1
Dermatitis herpetiformis	A blistering, intensely itchy disease of the skin that may be associated with cancer	DR3 B8	C C	82 75	20 22	17.3 9.8
Pemphigus vulgaris	A serious/fatal blistering disease of the skin and mucous membranes of the mouth and respiratory tract	DR4 A26 B38	C, J C, J C, J	91 60 59	32 20 21	14.6 4.8 4.6
Goodpasture syndrome	A disorder with hemorrhage from the lung associated with kidney disease	DR2	C	88	27	13.8
Sjögren syndrome	An immune disorder with progressive destruction of mucous-producing glands, resulting in a dry mouth, dry eyes with serious implications, and other organ involvement	DW3 B8	C C	64 50	24 24	5.7 3.3
Pernicious anemia	An autoimmune disease resulting in anemia and vitamin B_{12} deficiency	DR5	—	—	—	5.4
Behçet's disease	A syndrome with recurrent ulcers of the mouth, eyes, and genitals, often leading to blindness	B5 B5	O C	68 31	33 12	4.5 3.8

Continued on next page

Table 10 (continued)

Disease[a]	Nature of Disease	HLA Antigen	Race[b]	Percentage of Affected Persons with This HLA	Percentage of Unaffected Persons with This HLA	Relative Risk[c]
Hyper-thyroidism	An autoimmune disease that makes the thyroid gland overactive	B35	O	42	14	4.4
		DR3	C	56	25	3.7
		B8	C	43	23	2.5
Paralytic poliomyelitis	A viral infection involving peripheral nerves and the spine that may result in paralysis and occasionally death	BW 16	—	—	—	4.3
Juvenile rheumatoid arthritis	Chronic arthritis involving joints and other systems with onset in childhood	B27	C	25	9	3.9
		DR5	C	34	15	3.3
Ulcerative colitis	A chronic inflammatory bowel disease of unknown cause	B5	—	—	—	3.8
Acute anterior uveitis	Inflammation of the iris and closely related parts of the eye	B27	C	47	10	8.2
Chronic active hepatitis	An ongoing inflam-mation of the liver from varying causes that may result in liver failure, cirrhosis, and death	DR3	—	—	—	6.8
Juvenile diabetes mellitus	Sugar diabetes with onset most often in childhood	DR4	B	46	11	6.7
		DR3	O	38	14	4.8
		DR4	C	51	25	3.6
		DR3	C	46	22	3.3
		DR3	B	57	28	3.2
		DR4	O	49	25	2.6
		B8	C	40	21	2.5
Idiopathic hemo-chromatosis	A disorder in which excess iron is stored in many organs, involving liver, pancreas, heart, and pituitary, and causing serious tissue damage	A3	C	72	28	6.7
		B7	C	48	26	2.9
Idiopathic membranous glomerulo-nephritis	An immune disease of the kidneys that often progresses to kidney failure	DR3	—	—	—	12.0

Table 10 (continued)

Disease [a]	Nature of Disease	HLA Antigen	Race [b]	Percentage of Affected Persons with This HLA	Percentage of Unaffected Persons with This HLA	Relative Risk [c]
Celiac disease	A disorder of food malabsorption associated with intolerance to a protein found in wheat	DR3 DR7 B8	C C C	79 60 68	22 15 22	11.6 7.7 7.6
Psoriatic arthritis	A chronic arthritis of unknown cause (but with a hereditary predisposition) associated with psoriasis	B27 BW38	— —	— —	— —	10.7 9.1
Adrenal insufficiency	Serious lack of adrenal gland function resulting in weakness, low blood pressure, increasing pigmentation of the skin, and eventually collapse if untreated	DW3	—	—	—	10.5
Psoriasis vulgaris	A scaling, red palpable skin disorder involving the scalp and outer parts of the arms and legs	CW6 B37 DR7 CW6	O O O C	27 20 10 56	4 2 1 15	8.5 8.4 7.6 7.5
Rheumatoid arthritis	A chronic, multisystem disease of uncertain cause involving the immune system that leads to variably severe arthritis	DR4 DR4	C O	68 66	25 39	3.8 2.8
Myasthenia gravis	An autoimmune disorder that interferes with nerve-muscle transmission, resulting in abnormal weakness and fatigue on exertion	B8	C	44	19	3.3
Systemic lupus erythematosus	An autoimmune disease of unknown cause that affects all systems of the body and may be fatal	B8 DR3	C C	40 42	20 21	2.7 2.6
Multiple sclerosis	A chronic disease of the nervous system that results in partial loss of nerve coverings and associated damage, which in turn leads to intermittent and recurrent attacks that interfere with normal function of the nervous system	DR2 B7	C C	51 37	27 24	2.7 1.8

Continued on next page

Table 10 (continued)

Disease[a]	Nature of Disease	HLA Antigen	Race[b]	Percentage of Affected Persons with This HLA	Percentage of Unaffected Persons with This HLA	Relative Risk[c]
Polycystic kidney disease	A dominantly inherited disease of the kidneys that results in the formation of cysts, in progressive destruction of the kidney, and in death if untreated	B5	—	—	—	2.6

[a]Diseases are listed in desending order according to how frequently they are associated with an HLA type.
[b]B = black; C = Caucasian; J = Jewish; O = Oriental.
[c]Relative risk = the risk of developing the disease for those with the HLA type divided by the risk of developing the disease for those without the HLA type.

Knowledge that the HLA genes travel in a tightly linked bunch has helped in an unexpected way. The gene for the not-infrequent condition congenital adrenal hyperplasia (discussed elsewhere) is closely linked to HLA-BW47 and this has enabled prenatal diagnosis of the disorder.

The incredible variation in the possible sets of HLA genes and their product antigens is so broad that we can use them to distinguish one person from another. In a Los Angeles case of disputed paternity, HLA-typing proved spectacular. Two men were involved with the same woman, who became pregnant and delivered nonidentical twins. HLA studies showed that each man was the father of one twin! She had ovulated twice and had been impregnated by each male within hours of each other.

Consanguinity and Incest

For the most part, the harmful genes we all carry have little effect on our health or that of our offspring. It has been estimated that about 1 individual in 3 carries a gene for severe mental defect. As you have seen, when we marry and procreate with someone of the same ethnic group (a Celt with Celt, a Jew with Jew, and so on), then the chance that we have the same harmful gene in common rises significantly. Taking this one step further, it should be clear that if close relatives such as cousins marry and have children, they are even more likely to have the harmful genes in common, and therefore would have an increased risk of having genetically defective offspring.

The importance of *consanguinity* — a blood relationship between mates — will depend on how closely the couple is related. Because first cousins have two grandparents in common, they might share ⅛ of their genes. Their risk of having a child with a birth defect would approximate 6 to 8 percent in each pregnancy. This would be in contrast to the known 3- to 4-percent birth incidence of serious birth defects or mental retardation in the population at large. First cousins thus have a risk about double that of the average couple. While it is known that first-cousin marriages are noted more frequently among the parents of mentally retarded individuals, first cousins nevertheless do have a better than 90-percent chance of having normal offspring. Second cousins have only ¹⁄₃₂ of their genes in common, and they *probably* have only a slight (if any) increased risk of having a genetically defective child.

Many countries and at least half the United States prohibit marriage between uncle and niece or aunt and nephew, as well as between first cousins. In contrast, the latter is actually encouraged in some societies, as in certain areas of Japan, where 10 percent of marriages occur between first cousins. In one state of India, Andhra Pradesh, uncle-niece marriages are encouraged, and such unions may constitute about 10 percent of all marriages in the area. In uncle-niece or aunt-nephew marriages, a threefold increase in birth defects would be expected.

Based on the above calculations and remarkably little good data, most countries prohibit marriages between relatives as close as first cousins. In the United States, fewer than 1 in 1,000 marriages are between first cousins.

Though the highest reported rate of marriages between relatives is found in Japan (4 in 1,000 marriages), in certain isolated groups around the world the rate of marriage between close relatives may be as high as 25 percent! As far as the genes go, such marriages in these lonely communities reduce the number of carriers but increase the number of offspring with genetic disease.

Intercourse between brothers and sisters, fathers and daughters, or mothers and sons constitutes incest and is not only illegal but considered taboo in most societies. The reported studies of incest all note a devastating increase in the occurrence of serious birth defects, mental retardation, or both, when offspring result. Such unions would be expected to result in about a fivefold increase in the birth of defective offspring. A 1982 Canadian report on 29 children born of incestuous unions showed that most had severe abnormalities, low birth weight, mental retardation, or medical problems. Four (13.8 percent) had specific recessively inherited defects.

Rape and Murder

The uniqueness of an individual can be established biochemically by measuring not only the blood group, but also by assessing exactly other genetically transmitted proteins found on the surface of and within the red and white blood cells, in the bloodstream (serum), and in various body secretions such as saliva or semen. There are literally hundreds upon hundreds of different proteins that can be used when there is a specific need to identify a particular person. For the red blood cell alone, more than 250 proteins have been recognized. The HLA blood groups are even more specific. Hence, in cases of murder or rape, examination of a tiny speck of blood (and remember, you cannot even see a single red blood cell with the naked eye) or some semen can rapidly assist in the identification of a suspect. The fruits of genetic research thus have become important in the solution of such crimes of violence. The accurate distinction that now can be made between human beings on the basis of various genetic blood tests has made identification more certain than fingerprinting! Indeed, the unique characteristics of specially stained individual chromosomes, of ABO, Rh, and other blood groups, and of serum proteins and HLA groups, together with the latest DNA analyses, make it possible to identify virtually every individual — and often to provide conclusive verification of a murderer's or rapist's identity.

And Who Is the Father?

Court proceedings involving disputed paternity are common, since the United States government nowadays forces absent fathers to pay for their children's upbringing. The tests noted above are valuable for settling cases of disputed paternity — especially the new DNA tests discussed in chapter 8. (The courts, however, have been slow in using the latest and best tests.) In some circumstances, a man can be excluded as the possible father if both he and the mother lack the specific blood group that the child has. For instance, a child with Type A blood cannot be produced if both parents are Type O. The possible blood groups that could be found in children of individuals with various known ABO blood groups are listed in table 11.

Our uniqueness is reflected by the remarkable diversity of our genes and their products. Human ingenuity in applying the spectacular technology just discussed should surprise no one. The Immigration Service in England, for instance, was faced with the demand for permanent reentry advanced by a particular Ghanian male who claimed to be the son of a settled immigrant (his mother). To make the most accurate determination possible, officials arranged for DNA analysis of blood samples collected from the boy, his mother, and maternal relatives. The father was not available. This DNA "fingerprinting" method, detailed in chapter 8, showed that the boy truly was the son of his mother, as claimed.

Table 11. Possible ABO Blood Groups in Parents and Their Children

Parents' Blood Group			Children's Blood Group	
Partner 1	×	Partner 2	Possible	Impossible
O	×	O	O	A, B, AB
O	×	A	O, A	B, AB
O	×	B	O, B	A, AB
O	×	AB	A, B	O, AB
A	×	A	O, A	B, AB
A	×	B	O, A, B, AB	None
A	×	AB	A, B, AB	O
B	×	B	O, B	A, AB
B	×	AB	A, B, AB	O
AB	×	AB	A, B, AB	O

Sensory Genes

The way we recognize our environment is through our sensory nerves, which bring messages to our brains about sounds, odors, tastes, sights, and things we touch — everything we encounter moment to moment. Whether it is the ability to hear the chirping of crickets (which are genetically controlled sounds themselves) or the inability of some people to taste certain substances, our senses are governed by genes. Those who are tune deaf — sometimes incorrectly called "tone deaf" — are genetically unable to recognize certain tunes, and have poor musical ability. The opposite situation undoubtedly exists and could not simply be explained by environment in certain musical families.

Our body odors are also influenced by our inner chemistry — governed, of course, by our genes. One aspect of our uniqueness is the odor we emit. For example, police dogs are said to be unable to differentiate the scents of identical twins. Mice, it appears, prefer to mate with partners that have a different HLA group (H_2 in the mouse), who probably have a different scent. Some people cannot detect the characteristic odor certain individuals' urine takes on after they eat asparagus. The gene that controls our ability to smell this substance is dominantly inherited. Another dominant gene probably dictates whether or not we actually manufacture and excrete the smelly substance (methanethiol).

Even our body cycles, called *circadian rhythms*, are probably genetically influenced, such as the time of puberty and menopause. We know little about our own inner biological clocks, but we have already learned valuable lessons from fruit flies, bees, and other insects. One gene mutation in fruit flies seems to knock out their "clocks" altogether. Sleep-wake cycles are lost and they appear to be insomniacs. Male fruit flies normally sing a mating song by beating their wings in a sixty-second interval pattern. This mutation allows the song to continue, but it has no rhythm.

Our unique individuality is conferred by our genes and yields the rich diversity of the human species. It is that very uniqueness that can be corraled and used for specific genetic tests. It also explains certain aspects of intelligence and behavior, the topics I will concentrate on next.

10

Heredity, Intelligence, Learning Disorders, and Behavior

Through personal experience, you may be aware of certain families rife with feeblemindedness for generations. You may also have been puzzled to see remarkably brilliant individuals emerge from average or mediocre families. Is there really any way to correlate intelligence with heredity?

Definitions of intelligence are quite vague and refer, in essence, to "the ability to carry out abstract thinking." As early as 1905, IQ tests, as we know them today, were established, based mainly on the work of the French psychologist Alfred Binet. The tests he devised on behalf of the French Ministry of Education were specifically for measuring potential achievement in school. The American revisions of the Binet test were prepared initially at Stanford University; hence the name Stanford-Binet IQ test.

Actual measurement of IQ is beset with a variety of problems. As currently applied, IQ tests are administered as a specific test in a specific way under specific conditions. They do attempt to measure a general intelligence factor, but fail in a number of ways. While they may indeed provide a fair measure of verbal, numerical, and inductive-reasoning abilities, they fail to distinguish many other specific abilities (perhaps 120 in all!).

Many years ago, before revision, the Stanford-Binet test showed higher average IQs for boys than for girls. While there are those who might still believe that males have superior intellects, the facts are that careful revision of the test questions over many years now gives almost identical IQ distributions for males and females. Although the test has been used as a measure of scholastic potential or success, it has fallen into disrepute because of its inability to account for the varying cultural, racial, or socioeconomic backgrounds of children in a mixed society. Hence, perfectly intelligent children from different cultural backgrounds may provide so-called wrong answers. A variety of tests have been developed to take care of these sociocultural biases, but none has yet fully achieved this goal. As far as whites are concerned, IQ tests do correlate very well with ultimate scholastic achievement. For other cultural backgrounds or ethnic groups, IQ tests must still be viewed with considerable skepticism.

Some studies have concluded that the IQ remains fairly constant between two and eighteen years of age. Other studies, in contrast, have shown changes of all sorts during the school years. Environmental factors that showed some effects were extreme parental permissiveness or extreme strictness, both being apparently responsible for decreases in IQ with increasing age.

Heredity and IQ

In 1906, well before the widespread use of any kind of IQ test, some researchers were concluding that 80 to 90 percent of intelligence is determined by heredity. Professor R. J. Herrnstein at Harvard came to a similar conclusion in 1971, as far as whites were concerned, after reviewing the history of intelligence testing. Professor Leon J. Kamin of Princeton University, on the other hand, wrote a perceptive analysis concerning the inheritability of IQ and concluded that virtually no good data exist to prove that IQ scores are in any degree inheritable. In the light of such contrasting views, what does the best evidence available show?

TWIN STUDIES

I refer many times in this book to the use of identical versus nonidentical twins in the assessment of inheritable characteristics. Since identical twins have identical genes, they provide a unique opportunity to study IQ. But since they develop together in the womb, and are usually reared together, the importance of environment may make interpretation of any study quite difficult. Hence, a number of different researchers elected to study identical twins who had been separated and reared apart. The expectations were that if heredity was of consequence in the development of IQ, the scores achieved by parted identical twins should be extremely similar. As expected, some of the most extensive studies showed highly significant correlations between identical twins reared apart, suggesting a high degree of inheritability of IQ.

Professor Kamin and others, however, have pointed to the fundamental fallacies of these studies. Besides the critically important and valid technical criticisms that he leveled at past studies of identical twins reared apart, he drew attention to the very similar environments in which these children were reared. In many instances in these studies, infants were placed in the homes of relatives or family friends. Hence, the effects of similar environment could have been confused with hereditary effects. He also suggested that the theoretical expectations of the person performing the test unconsciously bias the actual IQ measurement of twins. Furthermore, Kamin noted another possible source of bias: the researchers depended upon the twins themselves to provide detailed information about the extent of their separation. Kamin's conclusion from previous separated identical twins studies is that there is not a simple inheritance pattern for intelligence.

In another, more recent study at the University of Minnesota, identical twins reared apart were again shown to have remarkably similar IQs. Forty pairs of identical twins who were separated at less than four months of age and who remained apart for an average of nearly thirty years were compared to 20 nonidentical twin pairs who were separated at about one year of age and reunited after nearly forty years. While less than perfect, this study provides further evidence that heredity plays a significant role in the development of intelligence, through genetic mechanisms still unknown.

IQ STUDIES OF ADOPTED CHILDREN
You may have thought that the practice of adoption would provide unique opportunities to determine the degree of hereditary influence on the development of intelligence. The adopted child, with genes from his or her biological parents, grows up in a new environment. If IQ is largely genetically determined, the adopted child's IQ should correlate closely with that of his or her true parents. So much for theory. In one large study, the average IQ of adopted children, 117, was compared to the average IQ of their true mothers: it was 86! This large difference probably corresponds to significant differences in the socioeconomic status of the adopting families. Obviously, adopted children are very often placed in homes that are more advantageous than those of

the biological parents. The environmental effects on those living in a "selected" family are not negligible.

In another study, the correlation of IQ of the adopting parents with their biological children was even slightly lower than was the correlation with their adopted children. This would suggest no effect of genes at all! The opposite has also been found. Certainly expectations that studies of adopted children would clarify the importance of hereditary influences have not as yet been realized. The innumerable factors that impinge upon the development of intelligence make it difficult to come to solid conclusions about the relative effects of environment and heredity.

Factors That Influence the Development and Measurement of IQ

FAMILY FACTORS

Studies in Scotland, France, the Netherlands, and the United States on children between six and nineteen years of age all indicate that the intellectual level generally declines as family size increases. While it is known that family size differs across the strata of society, these studies revealed that IQ declined with increased family size independently of status, though it goes without saying that socioeconomic conditions affect IQ scores.

American and Dutch studies furthermore show that IQ also declines with birth order. That is to say, the IQs of children born toward the end in a large family are generally lower than those of the first few children in the family. One further dimension to this question of birth order is the spacing of children within families. There appears to be a relationship between the general decline of IQ with birth order when the intervals between children are especially short.

Single-parent homes are more likely to constitute inferior intellectual environments, which could be reflected in generally lower intellectual abilities in children from such homes. Early loss of one parent would be expected, similarly, to produce a more intellectually suboptimal environment than would a parental loss at a later age. Studies have indeed come to these conclusions. One study, for example, showed that fatherless students scored significantly lower on an American college entrance examination compared with children from intact homes. Indeed, differences in intellectual performance found between children from fatherless homes and from intact homes have been found to be greater the earlier the father's death or departure and the younger the child when parental absence began. Death, divorce, or separation is also invariably associated with emotional chaos or stress in the home, which again, as expected, is associated with diminished intellectual performance. Such intellectual deficits have been found to occur even when the father's absence is temporary and where there is no significant stressful situation because of his absence. The return of the missing adult has clearly been shown to be beneficial. For example, the remarriage of the remaining parent of a young child has been shown to result in the improved intellectual performance of that child.

Besides the peculiarities of the administered IQ tests in reflecting differences between the sexes, other hidden factors may also be affecting these assessments. For example, intervals following male births appear to be somewhat longer than those following female births (possibly because of parental preference for male children). Females are born later in the family more frequently than are males. Finally, it has been known for many years that more fetal and newborn deaths occur among males than among females. Fetal deaths also occur with a higher frequency in older mothers. All these factors could therefore combine to produce sex differences as reflected by IQ tests.

TWINS AND IQ

Twins, whether identical or nonidentical, are known to score consistently and substantially lower on IQ tests and other tests of intellectual performance than other children. Not unexpectedly, triplets (in general) score lower than twins. Probably the most important factor explaining these differences is the birth weight of twins, which is in most cases lower than in nontwins. The higher frequency of prematurity and the associated complications due to lack of oxygen, hypoglycemia (low blood sugar), jaundice, and other complications constitute adverse factors in the development of intellectual function. Although that sounds sufficient as an explanation, it is not the whole story.

You would expect that the intellectual performance of twins who were separated early in life would be higher than that of twins reared together (because of more individual attention and care per hour per child). The importance of environmental effects on the development of intellectual growth of twins is reflected in a report showing that twins whose co-twins were stillborn or died within four weeks achieved nearly the same IQ as nontwins! Low birth weight and its associated complications in twins are clearly not the only explanations for why twins usually do less well than others.

RACE AND IQ

You are undoubtedly aware of the extensive public discussion and controversy that have centered in recent years on the relationship of IQ to race. That discussion has mainly sought to explain why the IQs of blacks, in particular, have generally been found to be lower than whites. There are those very vocal scientists who have tried unconvincingly to maintain that blacks are genetically inferior to whites. Perhaps the absurdity of any such claim is already obvious to you in the light of the foregoing discussion. A brief summary of the issues that impinge on the IQ-and-race question could be useful.

The difference in IQ scores between blacks and whites in the United States has been found to be of the order of 15 IQ points. You recall that I referred to the fact that there is, at present, no truly culture-free IQ test; available tests can hardly be considered an assessment of innate ability, simply because they depend on development having taken place in a specific cultural background. Those scholars claiming a genetic basis to explain the differences in IQs between blacks and whites have essentially failed to consider fully the extent of environmental influences on the development of IQ. I have referred above to most of the factors that would adversely affect the development of IQ in blacks. Their generally lower socioeconomic status, poorer nutrition, more frequent parental absence, higher frequency of low birth weight and prematurity, larger family sizes, shorter spacing between children, associated sociocultural deprivation, decreased educational opportunities, and so many other factors all unequivocally make more blacks than whites socially disadvantaged. The consequences for IQ development should be obvious. Moreover, being black may also carry with it a social burden not suffered by whites. The role of chronic though low-grade lead poisoning — a plight of poor urban black (and white) communities in the United States — cannot be ignored either.

Of additional interest is the finding that the IQs of children born to older mothers are consistently higher than those born to younger mothers. Hence, it may be significant that white mothers in the United States are on the average nearly three years older at the birth of their first child than are black mothers.

If, indeed, there were genetic differences between blacks and whites as far as IQ is concerned, such an argument should be supported by differences observed between black and white children raised in the same environment. A study in England compared black, white, and mixed-parentage children and revealed no significant differences between the groups, although white children tended to have the lowest average scores of three tests. United States studies of transracial adoptions have repeatedly shown that black adopted infants reared by whites in stable homes do as well intellectually as their white peers both in and out of their homes.

In another study, an attempt was made to correlate the IQ of American blacks with their probable degree of white ancestry as judged by blood-group similarities. The data tended to show that the IQ for blacks seemed to be lowered by white ancestry, but the results were not statistically significant.

Just as with stature, it is likely that the quality of intelligence is inherited to some extent. That any individual's final abilities reflect interaction between genetic endowment and environment should be obvious to all. Equally obvious, perhaps, is that it is counterproductive and incorrect to suggest the inferiority of one race compared to another simply on the basis of a questionably valid IQ test.

HEREDITARY DISORDERS AND SUPERINTELLIGENCE
Superior intelligence has been reported in association with six heritable disorders: torsion dystonia, retinoblastoma, phenylketonuria, infantile autism, gout, and myopia (nearsightedness). In torsion dystonia, the affected person has dreadful muscle spasms, is unable to walk or eat without aid, and is generally totally incapacitated, yet exhibits great intelligence. People affected by the hereditary eye tumor retinoblastoma have also been observed to have high IQs, although some experience mental retardation. Some individuals with gout whose blood uric acid is high exhibit higher IQ levels; questions have arisen about whether this may be due to the uric acid stimulating the brain. Brothers and sisters of patients with phenylketonuria have also been found to have above-average intelligence. A similar observation has been made

about the parents of autistic children. A recent report takes issue with claimed associations of high IQ and torsion dystonia, retinoblastoma, and relatives of phenylketonurics. More studies are needed for final clarification.

Meanwhile, studies on myopic students have shown an association with high IQ even for those tested ten years before they knew they were nearsighted. The results showed that at age seventeen to eighteen those youngsters who were nearsighted scored higher on IQ tests than their peers. The frequency of nearsightedness was highest among those students who had the highest IQs. So the bespectacled individual may not only look bright, but often is.

Learning Disorders

Brain damage or dysfunction may lead to learning diabilities. The first obvious type is when the brain has been damaged during birth — for example, by lack of oxygen or by hemorrhage. This can result in varying degrees of brain dysfunction, and is associated with other signs of brain damage, such as spastic limbs or inability to speak. A second, though not so obvious, type of brain dysfunction is suspected to be the result of damage to certain areas of the brain in the fetus that cause disabilities in the development of language, speech, and hearing, as well as abnormal body movement, as in the hyperkinetic child. This syndrome is frequently referred to as attention-deficit disorder. In actuality, no characteristic brain damage has been discovered in these affected children.

The third type of brain dysfunction probably results from hereditary influences that interfere with the development of perceptual function, including the ability to read and the acquisition of language. This is the group in which we recognize dyslexia — which, by the way, is derived from Greek and simply means reading disorder. Of all the known learning disabilities, the best information about hereditary aspects, though incomplete, refers to dyslexia, to which we will confine our considerations. The term *learning disability* refers to those disorders involved in understanding or using language, spoken or written, which may manifest themselves as an imperfect ability to listen, think, speak, read, write, spell, or do math.

DYSLEXIA

The term *dyslexia* was first used 100 years ago. We now know that it is an extremely common condition that affects about 15 percent of the population in Western countries. Dyslexia is regarded as a disorder manifested by difficulty in learning to read despite proper teaching and adequate intelligence and sociocultural opportunity. The traditional concept of dyslexia has implied a clearly inherited condition, frequently associated with a family history of the condition and not linked to any preceding brain damage. Typically, dyslexics have difficulties with language and words — especially their use, meaning, pronunciation, and spelling. The reading disability often persists into adult life and may be associated with handwriting problems, including reversals and rotations of letters, as well as additions, deletions, or substitutions of them. IQ tests do not accurately reflect the abilities of dyslexic persons.

Recent work has focused on efforts to determine which part of the brain is malfunctioning in dyslexics. Analysis of the brains of male dyslexics who died from accidental and other causes has revealed a consistent abnormality in the "architectural" arrangements of nerve cells in language-related areas of the left side of the brain. Sophisticated electrical studies of brain function have also confirmed abnormalities in the left brain hemisphere. In 1987, a study from the Massachusetts Institute of Technology concluded that dyslexics have a visual disability in *interpreting what is seen* — that they retained more accurate letter identification at the periphery of their field of vision than toward the central area, where letters were harder for them to interpret.

Parents and doctors alike know that children with learning disorders frequently have behavioral or emotional difficulties. On occasion, it is asked whether these difficulties are a cause or a consequence of the learning disorder. It is also well recognized that children with learning disorders not infrequently come from chaotic home environments or from families with serious problems; therefore, it is sometimes difficult to distinguish environment from heredity as the prime cause. Obviously, it is important to distinguish the hereditary aspects, since if a problem can be anticipated, early remedies may successfully be sought. Under favorable conditions, most children with dyslexia can be taught to read and spell so as to be fully capable of pursuing any career of their choice.

Hereditary Aspects of Dyslexia: As long ago as 1905, evidence suggested that dyslexia tended to occur within families. One of the most careful and important studies on its genetic aspects was done by Dr. B. Hallgren on children attending the Stockholm

Child Guidance Clinic in Sweden. Some 276 patients were carefully evaluated, all but six of the children being personally examined by Dr. Hallgren. He observed that 88 percent had a family history of one or more members with reading problems. From a genetic point of view, the conclusion was that dyslexia was transmitted as a dominant disorder. That is, an affected parent would pass along the gene for dyslexia to 50 percent of his or her children. Further Scandinavian studies at the same center focused on twins with dyslexia. In each case studied, twins who were identical both had dyslexia, which in genetic terminology is described as "100-percent concordance." For the nonidentical twins, only 1 in 3 pairs was found in which both members had the disorder. Other studies have come to similar conclusions.

Dyslexia affects males much more often than females. Various estimates suggest that there are between 4 and 10 males with dyslexia for each female so afflicted. In the Scandinavian studies, marriage between first cousins or other close relatives was not found to be a consequential factor in the genetic aspects of this disorder.

It would seem highly likely that there is more than one hereditary type of dyslexia. You could therefore expect that in certain families one of the parents may have the disorder in an extremely mild form — perhaps barely detectable. Many dominant diseases are the result of mutation, and this would explain isolated cases within a family (a single-gene mutation would thus be responsible). In such cases, with both parents normal, the risk of having another dyslexic child would be zero, in contrast to the situation in which one parent is clearly affected, with a 50-percent risk of passing it on to each subsequent child. It is also possible with dominant inheritance for the disorder to involve one sex especially. An exciting finding by researchers in Boulder, Colorado, indicates that a gene for dyslexia may be located on chromosome 15. Confirmatory studies are still necessary.

Another likely form of inheritance is via the X chromosome of the mothers, as in hemophilia and muscular dystrophy. The mother might be the carrier of the dyslexia gene, with a risk of 50 percent of her sons being affected and 50 percent of her daughters being carriers.

More still needs to be done on the genetics of dyslexia. Analysis and evaluation of hereditary patterns have been confounded because many individuals have different combinations of disabilities, such as speech problems together with dyslexia.

Handedness, Dyslexia, and Allergies: About 85 to 90 percent of adults are right-handed, and in 90 to 95 percent of adults the left side of the brain governs speech and language and word control. Left-handers are not all identical in this respect; some use the right, others the left brain for verbal processing, while still others use both sides. Not every left-hander performs all tasks with the left hand, and some may kick or throw a ball with their right.

For at least sixty years, we have known that left-handedness occurs more often than usual among children with learning disabilities. Some studies suggest that left-handers forced to write with their right hand may account for at least 50 percent of stutterers. Left-handers are found in 20 percent of all the retarded and in 28 percent of the severely and profoundly retarded. On the other hand, left-handers comprise 20 percent of the members of Mensa, an organization for individuals who score at or above the 95th percentile on IQ tests. A study of gifted mathematics students showed almost identical findings; in addition, 60 percent had allergies or asthma, while 70 percent were nearsighted.

Most studies have concluded that the majority of left-handers are no different from right-handers. Nevertheless, the late Dr. Norman Geschwind, a neurologist at Harvard Medical School, proposed that left-handedness and immune-system disorders such as allergies or asthma might occur together and frequently be linked to dyslexia, stuttering, autism, or even genius — especially artistic, musical, or mathematical genius. Although girls do better in language and at language-related tasks, mathematically gifted students are almost always boys.

Dr. Geschwind theorized that an excess of the male hormone testosterone during fetal development can influence brain cells so that the *right* side becomes dominant for language-related abilities and the person becomes left-handed. Sensitivity to testosterone and the activity of the immune system are probably genetically linked.

Further study will be needed to explore this theory. Meanwhile, it seems clear that there are at least three groups of left-handed writers — naturals, learned, and those with some brain injury or disorder. Genetic influences on handedness as evidenced by identical twins being the same in 80 to 100 percent of cases has been challenged and the matter remains unresolved. Much more research is necessary to unravel the inheritance of handedness and to determine why more males are left-handed, have immune system disorders, are dyslexic, have autism, stutter, and are math geniuses.

Heredity and Behavior

By this point in the discussion, you may have developed a healthy skepticism about conclusions made from twin studies on matters of intelligence. You should, then, not be surprised that abundant reports on temperament and personality show that identical twins are similar very much more often than nonidentical twins. One famous German study, begun at a camp for twins in 1937, was continued for over thirty-one years. Two major features — capacity to absorb information and capability for abstract thought — remained concordant for identical twins. Mental responsiveness remained concordant till about twenty-three years of age. Once again, there is evidence for a heritable component — even for behavior traits.

HOMOSEXUALITY

Homosexual behavior has not traditionally been regarded as genetic in origin. Nevertheless, a 1953 study of 44 identical male twin pairs and 51 nonidentical male twin pairs showed extremely high concordance rates for those identical twins with a high degree of homosexual behavior. Overall, for all degrees of homosexual behavior, identical twins were alike 100 percent of the time while nonidentical twins were concordant in only 25 percent. Subsequent studies have continued to reflect higher concordance for identical than nonidentical twins, but at a lower rate. Moreover, identical twins have similar active or passive roles in homosexual relationships.

The first obvious explanation for such high concordance is that one twin may have seduced the other. This, however, was studied and not found to be the case. What other "external" (in the home) or "internal" (in the womb) environmental factors are involved remains obscure. Earlier, in the dyslexia discussion, I discussed a theory having to do with hormone effects on the developing fetal brain — a possibility that could apply to homosexuality. There are at least two other examples of this phenomenon. Females exposed in the womb to masculinizing hormones exhibit tomboyish behavior much more often than their unexposed peers. Similarly, girls who evidence an adrenogenital syndrome with excess male hormone effects also show a similar high rate of tomboyish behavior. Experiments with pregnant chimpanzees and mice also confirm that subsequent behavior of offspring exposed in the womb can be affected by hormone exposure. The idea that some brain-influencing effect occurs in the womb in those who become homosexual remains conjectural.

ALCOHOLISM — GENETICS OR ENVIRONMENT?

Over 90 percent of adults in most Western countries drink alcohol at some time during their lives. In the United States, at least 1 in 3 men will experience a temporary alcohol-related problem. The lifetime risk of alcoholism is about 10 percent for men and 3 to 5 percent for women.

The familial nature of alcoholism is well known; there is a three- to fourfold increased risk for this disorder in the children of alcoholics. This realization does not, however, clarify whether the family environment or genes are responsible. Twin studies should help, since both types of twins would be exposed to the same family environment. In fact, identical twins are concordant for alcoholism twice as often as nonidentical pairs. Taking the question of heredity a step further, studies of the adopted children of alcoholics were found to have a three- to fourfold higher risk of alcoholism for their sons and an increased risk for their daughters.

The mechanisms that confer the predisposition to alcoholism are not known exactly. Enzymes that degrade alcohol in the body are under genetic control and could be involved. This might explain why Japanese individuals, 90 percent of whom have a slightly different, faster-acting enzyme, show the phenomenon of flushing after drinking only a little alcohol, which causes a red face, increased pulse rate, and a feeling of illness.

In a search to find other inherited markers of susceptibility to alcohol, the recorded brain waves, or electroencephalograms (EEGs), have been examined at rest and in response to alcohol. Thus far, two typical EEG patterns have emerged in the sons of alcoholics; and identical twins appear to have genetically determined differences in how their EEGs react to alcohol. The current consensus is that alcoholism is a multifactorial, or *polygenic,* disorder that results from the interaction of both environmental and genetic factors. The evidence obtained through tests and studies of family history is already clear enough to warn the sons and daughters of alcoholics to completely stay away from alcohol.

11

The Gene Screen

Most of us do not know which harmful genes we carry. If we have a child with a hereditary disease, we may know — at least for one gene. If a close relative is affected, we should be alerted to have special tests. The first clue may, however, be too late to avert tragedy. Advances in medical technology enable all newborns to be tested for certain genetic disorders that cause mental retardation. Programs have also been developed to determine carriers of specific genetic disorders in various ethnic groups at risk, thereby alerting couples to other options prior to the conception or birth of a defective child. It is possible, then, to screen for genetic disorders or for carriers of specific inherited conditions.

Screening for Genetic Disorders

The screening of newborns for *disease* has been adopted worldwide for obvious reasons. Early detection has allowed for the initiation of early treatment and the prevention of mental retardation in a few of the hereditary disorders, such as phenylketonuria and, more recently, hypothyroidism — deficient thyroid-hormone production (usually not genetic), which results in mental retardation if not treated soon after birth. There have been few objections to such programs, except perhaps to their overall cost. Some screening programs have focused on diseases without cure and for which early treatment is said to delay complications and prolong survival. Sickle-cell disease and cystic fibrosis are the best examples. For sickle-cell disease, screening of newborns is useful, because affected infants in whom the disease is undetected may succumb anytime later from bacterial infection (especially pneu-moccocal infection) for which vaccines are now available.

While diagnostic screening is certain for sickle-cell disease, screening tests for cystic fibrosis (which aim to detect a pancreatic-enzyme deficiency in babies' stools) have both false positive and false negative results. Screening tests for Duchenne muscular dystrophy (which use a blood test for the muscle enzyme creatine phosphokinase) may have some false positives too, but early diagnosis, at present, will not help the medical condition.

Early detection of all these genetic disorders yields one critically important potential benefit: timely genetic counseling to alert couples to their high risks — 25 percent — of having another affected child and their options for prenatal diagnosis.

Screening for Carriers

Screening programs to detect *carriers* of harmful genes aim to inform individuals of their personal risks and future options. In the 1970s, efforts were made to screen discrete ethnic groups for carriers of disorders that were much more common among them than among the general population. Blacks were screened for sickle-cell disease and Ashkenazic Jews for Tay-Sachs disease in community-wide programs. Tay-Sachs-disease carrier testing was reasonably successful and many couples were identified who unknowingly had a 25-percent risk of having a child with the catastrophic disease. Unfortunately, these programs have now largely dwindled, and to a large extent Jewish couples must rely upon the rabbi who officiates at their wedding to advise them about the availability of a carrier test. Not all rabbis will or do provide this important advice, however, and some Jewish couples have a civil ceremony. Tay-Sachs disease in a child is an unmiti-gated tragedy and every Ashkenazic Jew should know about the blood-carrier test and available prenatal diagnosis.

Sickle-cell-carrier screening programs fared poorly from the start and broke down completely for lack of proper educational efforts, together with problems of privacy and confidentiality. The high rate of illegitimacy among blacks also made carrier testing at the time of marriage almost superfluous. The chronic suffering of those afflicted by sickle-cell disease should be sufficient to alert all blacks to check their carrier status *before* having children.

Thalassemia carrier testing of people of Mediterranean extraction has not been successful in community-wide programs in the United States. Concern about stigmatization, antipathy to abortion, and lack of knowledge about the disease probably rank as the main reasons. In contrast, such programs in London have been successful among Cypriots, but less so among certain Asian groups. Given the availability of a blood carrier test and prenatal diagnosis, couples in which *both* partners have ancestral origins in the Mediterranean should be tested. Carrier testing, to succeed, involves a number of critical, though minimal, prerequisites, including:

1. definite knowledge about the pattern of inheritance for the particular disease (for example, if the disease to be tested is recessive, both parents must be carriers);
2. an available, well-defined ethnic community with a higher risk for the particular disease that is being screened;
3. an accurate, reliable, rapid, simple, automated, and inexpensive method of detecting individuals who carry these disorders; and
4. the possibility of diagnosing the actual disease in the fetus so that the couples who are both found to be carriers can elect to prevent serious or fatal genetic disease through prenatal diagnosis, early treatment (when possible), or pregnancy termination.

At this time, sickle-cell disease, Tay-Sachs disease, and thalassemia meet all these criteria, with cystic fibrosis rapidly coming closer.

Screening a population for carriers makes a lot of sense. The goal is to prevent the occurrence of serious hereditary disorders by providing carriers in their reproductive period with as many options as possible. They may, for example, be influenced in the choice of a mate, or elect to have no children, to have artificial insemination by donor, to choose in vitro fertilization, to contract with a surrogate mother, to plan on adoption, or to undergo sterilization (the man may opt to have a vasectomy, or the woman may have her tubes ligated). Still others may decide to have their own children, reassured by the availability of prenatal diagnosis. Certainly, it is of the greatest importance to you to determine before you start to have children whether or not you are a carrier of a disorder that can be diagnosed in the fetus in early pregnancy.

It is obviously too late for this approach when you have already had a child with a genetic defect. The ideal time for many couples to determine whether they are carriers of certain harmful genes is when they apply for their marriage license, or at least before the union is consummated. Sadly, the logic of that theme, which I originally advocated in 1973, has only recently begun to prevail. Only in this way can couples be certain of being spared the tragedy of having deformed or defective children.

Currently, the law simply requires a blood test for venereal disease. Few recognize that genes and germs can be communicated in the same way — and that the consequences of harmful genes are often much more catastrophic! During the early screening programs, unforeseen difficulties were encountered. Because many of the programs were begun in haste and without careful preparation, a number of mistakes were made. One of the most critical problems that arose was the confusion engendered in those who were screened. In the sickle-cell-anemia testing program for carriers, for example, many carriers who were screened thought that they themselves had the disease and thus needed special dietary or other care. This, of course, was incorrect. Equally disturbing was what happened during a similar type of screening program in Greece, where young women discovered to be carriers found themselves ineligible for their prearranged marriages. In the United States, some carriers of sickle-cell anemia were unreasonably asked to pay higher insurance premiums and were even excluded from some branches of the armed forces. Finally, since both parents are carriers of sickle-cell anemia if an affected child is born to them, the screening program deeply embarrassed couples when it turned out that one partner was in fact *not* a carrier. The unmasking of infidelity on the part of the mother naturally caused grave problems in a family already struck with the tragedy of a defective child.

Screening and the Future

The future of screening seems clear. Patients and physicians alike will continue to seek and determine inborn weaknesses, predispositions, or susceptibilities. Screening for hereditary diseases will be seen as a very significant effort in the prevention of disease, and the number of genetic diseases that can be diagnosed accurately, as well as the number of disorders in which the carrier state can be determined, will continue to increase steadily. It is not difficult to imagine that a time could come when each of us has a genetic-identity card. Such a card would assist us not only in the selection of mates and in the prevention of genetic disease, but also in the prevention of diseases that we may be susceptible to at any point in our lifetime. Certain occupations may be hazardous to your health if you are genetically susceptible to some chemical or other agent engendered by such work. A good example is the inherited (autosomal-recessive) disorder called *alpha$_1$-antitrypsin deficiency*. Chronic lung problems with breathing difficulties starting in the third and fourth decades are typical. Tobacco smoking or polluted air aggravates dangerous lung infections and hastens deterioration in health. Remarkably, even those who are gene *carriers* of this condition — they are easily detected by a blood test — are advised not to smoke and to avoid work in polluted atmospheres. Estimates are that refraining from smoking can add fifteen years to the lives of those with alpha$_1$-antitrypsin deficiency.

We still need drastic changes in the way we think about disease. To seek medical attention only when you are sick is a hazardous way of life. It is especially hard to convince young adults of the truth expressed in the time-honored adage "An ounce of prevention is worth a pound of cure." In the cases of hereditary disease, ten dollars' worth of prevention should be measured against a lifetime of sorrow, thousands of dollars for alleviation, and no real cure.

Though screening millions of infants and adults for genetic *diseases* is now routine, it will take a change in the public's attitude to achieve large-scale efforts to detect *carriers* of serious common disorders. It is crucial for you to research your family tree (see appendix A) and recognize your own genetic susceptibilities.

Drugs and Other Perils

Each time you take a drug — even an aspirin — you consider the dosage and are at least aware of possible side effects. But have you ever thought that you might have an inborn adverse reaction to a particular drug? Has it ever crossed your mind that a dosage of some other drug may be too much or too little for you because of your unique biochemical makeup? Some people who have seizures may need much more antiseizure medication than others because of the different ways drugs are bound and transported by blood proteins. These differences are inherited. Similarly, an individual may have a violent reaction to a single injection of penicillin that could be fatal. Again, this reaction, or *anaphylaxis*, probably is related to the transmission through the family of an inherited allergic predisposition.

Inborn Mechanisms

As we all differ in appearance and personality, so do basic chemicals in the body differ slightly from one person to the next. These biochemical differences are important to understand since every individual body handles drugs in a unique way. The evidence is clear that this pharmacological individuality is a reflection of hereditary variability. As an example, many drugs we take have to be processed through the bowel and via the liver to the bloodstream. The drug has to be broken down by chemical reactions in the body and it may have to be transferred around the body via a particular protein, to which it may be hooked in various ways. The drug, or the products resulting from its breakdown, then has to pass through the cell wall to get into the cell, where it can have its effect.

The actual sequence of events is even more complicated than the simplified description I have just given, but it should be sufficiently clear that heredity makes a difference in the handling of a drug by the body at each step. Inherited deficiencies of those enzymes necessary to break down a drug or to help in its excretion by the body may lead to accumulation of the drug and finally to toxic effects on the heart, the eyes, the liver, and so on. Alternatively, because of a unique, inherited metabolic system drugs may be so rapidly broken down by a person with a different genetic makeup that the usual dosages of drugs taken have little or no effect. In fact, about half the population of the United States and Canada breaks down isoniazid (used in the treatment of tuberculosis) slowly. In contrast, only 5 percent of Canadian Eskimos do so, while 83 percent of the Egyptians are able to break it down this way. The implication, again, is that if a person breaks down a drug such as isoniazid slowly, much more is given than is necessary, and it may rapidly accumulate and cause such complications as nerve damage.

A Little Is Too Much

Quite a number of other drugs, including phenobarbital, are handled differently by different individuals. A potentially cat-astrophic situation may arise in a person who lacks an enzyme whose "job" is to take care of particular drugs. The two most important such drugs are succinylcholine and suxamethonium, which are usually given as muscle relaxants as part of general anesthesia. The "sensitive," susceptible individual, unable to break down a drug efficiently, accumulates high concentrations of the drug in the bloodstream. Such an individual may suffer from prolonged paralysis following anesthesia and die if the situation goes unrecognized. The sensitivity to these drugs is inherited equally from parents who are both carriers. Those who carry this predisposition do not themselves show any abnormal sensitivity to these drugs. The situation is, however, *not* rare, since about 1 in 2,000 individuals and about 2 out of every 100 whites might carry this disorder! Curiously, Alaskan Eskimos have a remarkably high incidence of this sensitivity.

A Lot Is Too Little

The opposite situation has also been noted. In other words, an individual may have a much more efficient mechanism to break down the drug, and therefore may need, for example, almost three times more succinylcholine to achieve the same degree of relaxation under anesthesia as does a normal individual. Also, a number of persons have been identified for whom the dose of an anticoagulant drug necessary to achieve its action was about twenty-five times more than that expected for a normal individual. This resistance to an anticoagulant was shown clearly to be inherited.

Combinations May Be Hazardous

I am sure you are already aware that one drug may interfere with another drug taken simultaneously. For example, a person taking isoniazid for the treatment of tuberculosis may also be taking Dilantin for epilepsy. In such a situation, the isoniazid may interfere with the handling of Dilantin, which could then accumu-late in the body, yielding toxic concentrations. Such a consequence would then lead to the complications of Dilantin toxicity, including jerky eye movements, loss of balance and unsteadiness in walking, drowsiness, and so on. Some of the drug-to-drug interactions probably result from inherited biochemical mechanisms.

Hereditary Disorders Associated with Adverse Drug Reactions

In all countries, there are many people born with hereditary conditions that by themselves cause few if any problems. In some of these disorders, however, exposure to specific drugs may lead to serious complications. A large number of blacks and Orientals may have an inherited deficiency of a particular enzyme within the red blood cells called glucose-6-phosphate dehydrogenase — a condition known as G-6-PD deficiency. If they take sulfa drugs, drugs to reduce fever, certain antimalarial drugs, and a host of others, including aspirin, they may develop a hemolytic anemia. Between 5 and 40 percent of Italians, Greeks, and other Mediterranean and Middle Eastern peoples also have G-6-PD deficiency. The defect is especially important in Greeks and Chinese, since it is commonly associated with serious jaundice in the newborn, which if not promptly treated may cause brain damage. Another complication occurs when the fetus has G-6-PD deficiency and the mother takes certain medications, such as sulfa drugs. The drugs cross over into the susceptible fetus and may cause severe anemia and even fetal death. A person born with a variation in hemoglobin, one of the proteins in the red blood cells, may also suffer from anemia following the taking of similar drugs.

A whole host of other hereditary disorders are known that may cause particular drugs to produce unusual or undesirable effects. Patients with inherited glaucoma — high pressure in the eye — develop even higher pressures within the eye if they are inadvertently given atropine or other drugs to dilate the pupils; the condition may also be aggravated by some types of cortisone. Hence, it is important to find out through a simple test used by all eye specialists whether you have glaucoma or not.

Patients with Down syndrome are known to be extremely sensitive to atropine. This drug, usually given immediately before an operation, may even be fatal in these patients. Patients with familial dysautonomia, a disorder of the involuntary nervous system that controls blushing, blood pressure, and so on, may show a marked increase in blood pressure when given the drug noradrenalin. It also appears that some of the symptoms can be temporarily relieved by treatment with the drug Urecholine.

One day in 1932, a scientist complained that the powder his colleague was working with in the same room caused a very bitter taste. The substance, called phenylthiourea — or phenylthiocarbamide (PTC) — did *not* taste bitter to the man actually handling it. Quite by accident, these two had stumbled onto what is now well recognized: some people are born with the inability to taste PTC. Their inability to taste this bitter compound is inherited as a recessive gene from both parents, while the capacity to taste it is transmitted by a dominant gene from one parent to half the children.

The curious thing about nontasters is that they have a much higher frequency of lumps or tumors of the thyroid gland. Tasters, in contrast, more often develop overactive thyroid glands, or hyperthyroidism.

There seems to be no end to the remarkable diversity of the genetic mechanisms of disorders that plague us. Consider yet another typical story:

> J.T. was twenty-two years of age, and decided that she had had enough of recurrent severe tonsillitis and ear infections. Her physician, in desperation, had recommended a tonsillectomy, and J.T. finally decided to proceed.
>
> During the operation, her temperature suddenly skyrocketed to 108 degrees Fahrenheit, her pulse raced at 200 per minute, and her heartbeat became grossly irregular. Her body developed extreme muscular rigidity and she became as stiff as a board. Fortunately, her quick-thinking and alert anesthesiologist knew immediately what to do: stop the anesthesia, give oxygen, force respiration, cool the body rapidly with ice, treat biochemical consequences, and so on.

J.T. survived unharmed, but nowadays about 60 percent of individuals still die in this situation. Her inherited defect is called *malignant hyperthermia,* and it can be precipitated by a variety of anesthetic agents, resulting in catastrophe. Tragedy may strike during the first, second, or subsequent operations under anesthesia. About two-thirds of the victims are affected during their first experience with general anesthesia. Such individuals may appear to be perfectly healthy, but a very careful examination may reveal one or more of the following features: droopy eyelids, crossed eyes, curvature of the spine, loose-jointedness, large muscle bulk (although the patient may complain of muscle weakness or cramps), and difficulties with temperature control. Until recently, most people never knew they had malignant hyperthermia until it evidenced itself, and this lack of awareness often killed them. The triggering factor appears to be general anesthesia, during which the patient develops sudden and tremendously high temperatures. One woman is known to have reached a temperature of 112 degrees Fahrenheit and to have survived without apparent brain damage — probably a world record.

Hyperthermia is transmitted as an autosomal dominant trait, so an affected person takes a 50-percent risk in each pregnancy of passing it on. About half the cases arise spontaneously from a gene mutation. There are so few clues on physical examination that the first identification of the disorder may be during surgery under general anesthesia. And that may be too late! It is possible, and very important, to test anyone with a family history of hyperthermia. Muscle biopsy enables a diagnosis to be made before an operation in about half the cases. Electrical studies of muscles, called electromyograms, show a typical pattern in about half the cases. Unfortunately, a reliable diagnostic test is still not available. Raised levels of the muscle enzyme creatine phosphokinase are common in those affected. Currently, the most effective emergency drug treatment is with the muscle relaxant dantrolene. It is also used *before* surgery begins as a possible protection in susceptible individuals..

Counterparts of many human diseases have been found in a variety of different animals, including cats, dogs, cows, pigs, mice, and hamsters. This provides opportunities to study the particular disease in a model system, and recent observations have shown that the pig can have a disease similar to malignant hyperthermia. Not only are these animal models extremely valuable in research for discovering the basic mechanisms of a human disease, but they also provide opportunities to try out different kinds of treatment.

Helpful Considerations If You Have a Hereditary Disorder

While it is not yet possible to cure an inborn hereditary disorder, it is possible to do a number of constructive things. Specific treatments are available that may allow a person with a genetic disease to live without undue pain, discomfort, or even ill health.

The kinds of treatment available for hereditary diseases are discussed in some detail in chapter 33. But I really cannot emphasize enough the need to be continually on the lookout for the specific complications associated with a hereditary disorder you may have.

Being aware is one thing; anticipating and preventing harm is another. A person who suffers, for instance, from the dominant hereditary disorder called pheochromocytoma, which is characterized by tumors that occur along the autonomic nervous system, has a high risk of developing cancer of the thyroid gland. Here, watchfulness is all important. Blood-pressure checks, examination of the thyroid gland, and blood and urine tests would all help to anticipate possible complications and catch them in time to prevent a serious outcome. In another rare condition, vitamin D resistant rickets, the disease is transmitted mainly from affected males to all female offspring. Recognizing this disorder and knowing its mode of transmission will make it possible to insist that all daughters are given high doses of vitamin D, plus additional phosphorus in the diet. If a female transmits the disease, then half the sons and half the daughters will require the same treatment.

Another example is that of adult polycystic kidney disease, which an affected parent will pass along to half of his or her children. (See the discussion about new diagnostic tests in chapter 8.) Since high blood pressure is a serious hazard here, it is important to recognize one special condition that is being associated rather frequently with polycystic kidney disease. It is a small bulge (called a berry aneurysm) in the wall of a blood vessel inside the skull that can burst when the blood pressure goes too high. Hypertension therefore cannot be ignored in this situation, and kidney transplantation may have to be considered.

Hereditary disorders may not be fatal, though some of their complications may cause eventual death. It is therefore vital for the affected person to remain very alert about the possibilities. Take, for example, the condition called familial polyposis of the colon. In this condition, large numbers of polyps occur in the colon and are often premalignant. Hence, whenever these polyps are found, they should be surgically removed. Such an affected person is advised to have a sigmoidoscopy — examination of the colon under direct vision through a tube passed through the rectum — at least once or twice a year for life. Some have even chosen surgical removal of the entire colon to prevent any such further possible complication.

In the case of the more common ailment of sugar diabetes, or diabetes mellitus, many physicians feel that all close relatives of diabetics should be tested for it once a year. It is recommended, too, that first-degree relatives of those afflicted by glaucoma, also a hereditary condition, should have yearly examinations.

There are a number of other diseases in this category with long names that would not be meaningful unless you yourself had one of them. The point really is that you should know what you have and should be aware of what possible complications may ensue. Anticipatory intervention or early diagnosis may save your life or the life of your loved ones. Therefore, consult your doctor and explore these possibilities with this approach in mind.

To sum up:

You may discover which harmful genes you carry if

1. you yourself have a genetic disorder;
2. you have a child with a specific hereditary disease;
3. you had yourself tested because of a family history of a particular disease;
4. you are *automatically* a carrier (for instance, the daughter of a father with hemophilia); or
5. you had yourself tested because you belong to a certain ethnic group (such as blacks, Italians, or Jews of Ashkenazic origin).

Ultimately, screening of large segments of the population may become more widespread as further research leads to progress in medical knowledge and technology. Meanwhile, if you find yourself identifying with any of the five groups just described or have concerns about symptoms I have discussed, then I advise you to consult your doctor. If the answers provided leave you uncertain, then a consultation should be sought with a medical geneticist in a large medical center associated with a medical school.

My ardent hope is that you will consider your genetic status and that of your mate before you marry, before you have children. You owe it to yourself and to your children at least to have proceeded with the best available knowledge of your genetic endowment. You have a right to know — in fact, a distinct personal obligation to know.

12

Twins

Twins, even in today's matter-of-fact world, still invariably evoke comment, whether it be in the hospital at the news of their birth, on the street, in school, or elsewhere. At least nowadays the birth of twins is not regarded as a threat to society, as was the case not so many years ago. Various taboos arose from fear and ignorance of what was then considered an uncanny event.

During medieval times and earlier, the mother and her twins were reviled in many countries, since she was thought to have been unfaithful and each twin was assumed to have been sired by a different father (as indeed sometimes happens) or by an evil spirit. Mythology is replete with horrors in which both children were killed, or at least one, and frequently the mother as well. (One idea is that the revulsion stemmed from identification of twins with the multiple births of animals.) You may recall that the twins Romulus and Remus, born in ancient Rome, were destined to die with their mother.

In our scientific age, interest has focused on the mechanism and causes of twinning and the diseases that occur more often in both identical and nonidentical twins than in single-birth individuals.

Types of Twins

The distinction between identical and nonidentical twins was first made in 1874. *Identical twins* arise from a single fertilized egg that divides into two separate embryos within fourteen days after fertilization. *Fraternal* (or *nonidentical*) *twins* originate from the fertilization of two different eggs, usually at about the same time. Overall, about one-third of twins are identical and two-thirds are nonidentical. Rarely, blood cells from one nonidentical twin may cross into the other, seed into the bone marrow permanently, and result in an individual with two sets of genetically different cells — even male and female in one person. Such a person, referred to earlier, is called a *chimera*.

How Frequently Do Multiple Births Occur?

Identical twins are born at a fairly constant rate of about 1 in 250 births worldwide. In contrast, birth rates for nonidentical twins vary enormously. Rates are high in Africa, low in the Far East, and intermediate in the United States, Europe, and India. In the United States, Britain, and Europe, about 1 in 89 pregnancies in whites concludes with twins of either type; the rate is somewhat higher in blacks. The highest rates reported are from tribes in western Nigeria, where close to 1 in 25 births are twins. Various drugs used to stimulate release of an egg during infertility treatment often result in multiple ovulation. The incidence of multiple fetuses following stimulation ranges between 8 and 40 percent!

Triplets occur somewhere between 1 in 5,000 and 1 in 10,000 births. Until fairly recently, quadruplets were thought to occur about once in 500,000 births, and quintuplets once in 50 million births. The use of the new fertility drugs has clearly increased the frequency of these multiple births in the last few years. The first quintuplets who lived to attain adult age were born in Canada in 1934 — the Dionne sisters. Many other sets have now been reported, all over the world. One elegantly studied case involved quintuplets born in Danzig, Poland. Their blood groups showed that they each arose from a different fertilized egg.

The Causes of Twinning

A variety of possible influences may induce twinning. The matter is complex and can possibly best be considered under two familiar headings: environmental and genetic.

ENVIRONMENTAL INFLUENCES
Identical twins are born at about the same rate all over the world, and factors such as race, the age of the mother, and nutrition do not appear to influence the rate. The exact cause that leads a fertilized egg to split into two embryos is unknown. It seems to be

a random phenomenon, but may be related to certain environmental factors or a transient lack of oxygen to the embryo.

In contrast, a variety of environmental factors seem to be operating in nonidentical twinning. In Finland, for example, the frequency of fraternal-twin births is highest in July and lowest in January. This observation prompted the suggestion that the long exposure to sunlight in the Finnish summer stimulates hormone production in the mother, which results in the release of more than one egg at a time. A study in the state of New York, however, failed to show any seasonal difference in the rate of twinning.

The likelihood of twins appears to be greater than normal when women become pregnant within the first three months after marriage. Interestingly, stress or impaired nutrition may decrease the likelihood of nonidentical, though not identical, twin births, as was noted during the Second World War.

It is well recognized that an increased frequency — as much as fivefold — of nonidentical twins occurs with increasing age of the mother, reaching a peak between thirty-five and thirty-nine years, at which point the frequency drops off rapidly. The likely reason for this age effect is that the level of a hormone that stimulates formation of an egg — follicle-stimulating hormone, or FSH — increases with maternal age. High FSH levels are also seen in the Nigerian tribes noted above. Naturally, the more children a woman has, the more likely she is to have nonidentical twins. Some workers have also noted that the chances of having twins increase with the height of the mother; the taller the mother, the greater the chances. Also, obese women appear to have a greater likelihood of having nonidentical twins than thin women. Finally, and wholly unexplained, is the observation that twins occur more often in illegitimate pregnancies compared with the rate in the general population.

GENETIC INFLUENCES

Throughout this book, I refer repeatedly to studies of twins for different diseases. The idea has been that identical twins share identical genes and should therefore be subject to the same diseases — if hereditary — in contrast to nonidentical twins. A genetic predisposition to have nonidentical twins is, however, a different matter. It should already be clear to you that having identical twins does not appear to be influenced by genetic mechanisms or family history.

Most of us have heard of families in which nonidentical-twin births have appeared in successive generations. There certainly are some remarkable examples of fraternal-twin births reported in the medical literature. There is the reputed case of a Dr. Mary Austin, who, during thirty-three years of marriage, apparently had 44 children, including 13 pairs of twins and 6 sets of triplets. It was said, in 1896, that one of her sisters had given birth to 26 children, and another sister to 41 children. An unusual case is reported of an American fisherman, whose wife gave birth to their first pair of twins in October 1945. The couple subsequently had a second pair of twins in October of the following year, and a third pair of twins again in October of the year after that. This gave them a total of six children in three years. In 1938, a woman was noted to have given birth to six pairs of nonidentical twins, between whom were interspersed some single births. This woman's father had a set of triplets with his second wife. In another remarkable case, reported in 1918, a woman had triplets in her first pregnancy and delivered a second set of triplets in her second pregnancy nine months later. She had six sons in one year. (The Internal Revenue Service should have paid her that year!)

In certain families, the father has been attributed as being important in transmitting the tendency. A celebrated case turned up in 1914 — a Russian peasant who married twice. It was said that his first wife gave birth to 4 sets of quadruplets, 7 sets of triplets, and 16 pairs of twins. Clearly inexhaustible, he married again and with his second wife had 2 sets of triplets and 6 pairs of twins. This gentleman, therefore, had a grand total of 87 children, 84 of whom survived!

Again supporting possible hereditary factors in the male was the reported case of a man who, himself a twin, had 9 pairs of twins with his first wife. His wife remarried and subsequently gave birth to 6 single sons.

Except for these very unusual families, the father generally does not appear to have a significant role in the frequency of twinning. If anything, the best available data imply that the role of the mother is much more important. Studies based on the excellent family records of the Mormon Church of Salt Lake City, Utah, showed that women who were themselves nonidentical twins and their sisters had twins much more often than did the general population. Men who were themselves nonidentical twins and their brothers did not have twins more often than others in the community. Other studies have borne out these conclusions. More recent evidence shows that mothers who have identical twins do not have a higher likelihood of repeating such births; but mothers of nonidentical twins appear to have about a four-times-higher-than-average chance of having another twin birth. Her sisters have a similar increased chance.

Birth Defects in Twins

A higher rate of birth defects has been noted in identical twins. While this may imply hereditary mechanisms, some surprising evidence from animal experiments suggests that identical twinning itself may be influenced by environmental factors.

In general, birth defects occur two to three times more often in twin than in single births. The United States Collaborative Perinatal Project — an intensive study of all births over a fixed period — obtained information on 1,195 twins. Some 219, about 18 percent of them, had minor and major birth infirmities, with more than one defect present in a vast majority of these infants. It appeared, however, that this large figure was caused for the most part by identical rather than nonidentical twins. Also, birth defects occurred more frequently in black than in white twins, and more among male than female twins.

Both members of an identical pair are frequently subject to the same defects — which suggests, of course, a hereditary influence. Curiously, while heart defects are more common in identical twins than either nonidentical twins or singletons, they usually affect only one of the pair. This serves to emphasize that identical twinning may be an error of development itself: whichever factor that causes splitting of the fertilized egg may also result in maldevelopment of the heart. In contrast, it is exceedingly rare to find both twins, either nonidentical or identical, affected by anencephaly, the severe brain and skull defect discussed in other chapters. This observation further suggests the important role of unknown environmental factors as the cause of this defect.

Understanding the occurrence of chromosomal defects in twins is even more difficult. Without going into a complicated explanation, suffice it to say that chromosomal defects may occur in both members of a pair, in either identical or nonidentical twins. One author studying the frequency of Down syndrome in twins found that 159 sets had been reported. In all of these sets of twins, one twin only was affected in 155 of the sets; in the remaining 4 sets, both twins had Down syndrome. Another chromosomal defect that affects males — Klinefelter syndrome, which is characterized by an extra X chromosome — is more likely to affect both members of a twin pair. Indeed, a higher frequency of twinning has been noted in the brothers and sisters of people affected with Klinefelter syndrome.

FUSED (SIAMESE) TWINS

A morbid fascination will probably always attend the birth of twins joined together at some part of their bodies. They are called Siamese because of a pair of twins, Chang and Eng Bunker, born in 1811 in Siam (now Thailand), who were joined from the breastbone almost to the navel. They married two sisters. Despite being joined, Eng managed to have twelve children and Chang had ten. Initially, they appeared as curiosities in Barnum's circus in the United States. Later, they lived in the Carolinas until Chang had a stroke, dying some two years later. Dr. F. Vogel and Dr. A. G. Motulsky, who wrote briefly about these twins, indicated that Eng, healthy up to that time, died two hours later. At autopsy, it was found that the two were also connected by their livers.

Such twins occur about once in 33,000 to 165,000 births and most commonly are joined together at the chest. All such conjoined twins have the same sex and almost invariably have associated birth defects. About 70 percent of them are female. (Curious and totally unexplained is the fact that twins in chickens are invariably female.) The most likely reason for Siamese twins is the incomplete separation of the dividing cells following fertilization of a single egg. Most experts now believe that unknown but definite environmental influences affect this incomplete fission. Certain drugs may cause the same phenomenon of conjoined twins in hamsters, and a diminished oxygen concentration or simply raised water temperatures may be the cause in zebra fish. There does not, however, appear to be any clear hereditary factor recognized thus far that would predispose an individual couple to have fused twins.

On occasion, during the development of joined twins, the lack of separation may be quite bizarre. Incredible freaks may now and then be born as "monsters" with two heads and one body, or one head and two bodies, or one body and four arms and four legs, and so forth. These "monsters" are not usually born alive, but if so, rarely live more than a few hours.

Death in Twins

An increased rate of death in twins at the time of birth or soon thereafter is well recognized and relates mainly to the higher rate of premature delivery and low birth weight. The occurrence of death in the first month of life is seven times higher for twins than for single births, and the likelihood of death while still in the womb is three times greater. It turns out that identical twins have a much greater risk of dying than nonidentical twins during the first month of life. As could be expected, most commonly the second delivered twin is at the greater disadvantage in terms of survival and development. The use of instruments for delivery, the length of anesthesia, and the lack of oxygen associated with the delay all adversely affect the outcome for the second twin. Sometimes, in about 1 in 1,000 deliveries, twins actually interlock during delivery — a complication that leads to death of both babies in about one-third of such cases.

Pregnancy with Twins

Family history aside, it is not really possible to predict who will have twins. It has been noticed, however, that on the average, women who weigh more in relation to their height before pregnancy are more apt to have twins than their slimmer peers.

A woman or her doctor is usually alerted to a twin pregnancy by the larger womb size, faster and bigger weight gain, and the actual discernment of multiple fetal parts during an abdominal examination. Early diagnosis of a multiple pregnancy is important and is often accomplished through maternal serum alpha-fetoprotein screening and ultrasound (both described in chapter 23). Mothers carrying twins need greater care since virtually all the common complications of pregnancy occur more frequently and with more devastating consequences in multiple pregnancy. Twins who are relatively heavy in the womb are diagnosed earlier, and weigh more when born, which is good; the smaller, late-diagnosed twins are more wont to be born prematurely, which is not good. Failure to diagnose twins at all before labor poses greater hazards to the second one delivered, for various reasons, including harm caused by drugs used to contract the uterus after delivery.

The Vanishing Twin

Twins are conceived in 1 in 19 to 1 in 30 pregnancies, but in only 1 in 89 are twins actually delivered! The phenomenon of the "vanishing twin" has been realized only since the routine use of ultrasound in early pregnancy. A Pennsylvania study of 1,000 early pregnancies showed that in 21.2 percent of twin pregnancies, one twin died and was reabsorbed by the mother's body, often leaving no trace. Such events were sometimes signaled by transient vaginal bleeding. While the reason for loss of one twin is uncertain, the rate of loss is very similar to the miscarriage rate of a single fetus.

The Study of Twins

Identical twins have identical genes. Nonidentical twins are like siblings and have on average one half of their genes in common. A genetic disease is therefore likely to strike *both* twins much more often in identical twins than in nonidentical twins. Twins are said to be concordant if *both* show the same disease and discordant if only one has the disease. The higher the concordance rate, the more likely is it that a condition is inherited. Diseases caused by a single abnormal gene (for example, cystic fibrosis) will affect both identical twins 100 percent of the time, whereas the concordance rate in nonidentical twins will be much less and equal to the rate found in single-birth siblings.

Twin studies have therefore been used extensively to investigate whether a disease is totally or partly inherited. Such studies have helped to determine that there are genetic components in all of the selected disorders listed in table 12. For example, as indicated, when manic depression occurs in one identical twin, it will also affect the other 70 percent of the time, whereas if it affects one nonidentical twin, the co-twin will develop this disorder in 15 percent of cases.

For many diseases that result from environmental factors interacting with one or more genes (and almost all of the disorders listed in table 12 are in this category), exposure would have been similar in all twins, and therefore such studies would not be

Table 12. Likelihood of Both Twins Having Certain Disorders

Disorder	Identical Twins	Nonidentical Twins
Manic depression	70%	15%
Schizophrenia	60	10
Psoriasis	61	13
Diabetes mellitus	55	10
Congenital dislocation of the hips	41	3
Mental retardation (IQ less than 50)	60	3
Allergies	50	4
Hyperthyroidism (overactive thyroid)	47	3
High blood pressure	30	10
Clubfoot	23	2
Gallstones	27	6

helpful. However, superb studies have been done on twins reared apart since early infancy. As we have seen, criticisms of such studies are possible, since the adopting homes are likely to have been similar; furthermore, the twins still had about nine months together in womb sharing environmental exposures! Notwithstanding these reservations, twin studies have yielded extremely important clues to understanding how various diseases are caused.

Birth Defects

Genetic Counseling

Newer Ways
of Making Babies

13

Hereditary Birth Defects and Mental Retardation

Superstition and Birth Defects

Throughout the ages, men and women have reacted to the birth of malformed offspring with awe, fear, admiration, or with a foreboding of evil and imminent disaster. Artifacts depicting remarkable malformations such as double-headed single-bodied "monsters" or other species of conjoined twins occasionally served as prototypes of gods or demigods with magic powers. It seems that in very ancient times, grossly deformed infants were regarded as divine and probably worshiped. Our fascination with major birth defects has been expressed in art for thousands of years. A sculpture of a double-headed-twin goddess was discovered in southern Turkey that dates back to Stone Age civilizations.

From time immemorial, some men and women have believed that a major birth defect was God's punishment for sins committed or a sign of punishment to come. It is not unusual, even today, to encounter parents of a child with a birth defect who confide their innermost belief that God is punishing them for acts or sins they have committed.

One common belief is that a shock or stress or excessive worry may produce birth defects in a child. No solid evidence has accumulated to confirm such a view. During the Second World War, the bombing of London and other major English cities did not lead to an increase in the frequency of birth defects, despite overwhelming stress.

Another age-old idea, that the thoughts of the pregnant mother may affect the development of the fetus, has also not been substantiated. The power of these mental impressions, as they are called, was believed to act from the moment of conception. Children conceived out of wedlock, allegedly in great passion, were thought to be artistically gifted. Many centuries ago in Greece, expectant mothers were encouraged to look at beautiful statues and pictures in order to make their children strong and beautiful; indeed, the Spartans made laws to that effect. Mothers have also blamed their own negative thoughts, moods, and emotions during pregnancy for children born with harelip or cleft lip.

So-called irrational beliefs or ideas of the past may turn out to have some basis in fact. For example, a sudden shock *could* release the adrenal hormone cortisone, which has been proved to cause cleft palate in some strains of mice. Also, chickens exposed to excessive noise have produced offspring with a higher frequency of birth defects. Whether the mechanism is through the release of cortisone because of stress or some other metabolic pathway is unknown, but we cannot and should not extrapolate too literally from the animal to the human condition.

Hereditary Birth Defects

The term *birth defect* generally implies that a child has been born with some kind of physical sign that is abnormal, though invisible biochemical disorders are also correctly dubbed birth defects. The obvious congenital defects may vary in intensity from a single birthmark on the skin to a hole in the heart, grotesque facial abnormalities, or even two-headed "monsters." A group of birth defects that consistently occur together are referred to as a *syndrome*. Congenital defects, as well as mental retardation, may be due to hereditary factors, even in the absence of a family history of them. On the other hand, they may occur together or separately due to known factors that affected the fetus during development in the womb, but were not hereditary; these are known as *acquired* (or *environmentally caused*) *birth defects*.

Hereditary birth defects can result from the effects of single harmful genes — dominant, recessive, or sex-linked — as discussed in chapter 7. Over 4,000 have been identified, and together they account for about 7.5 percent of all hereditary birth defects. Interactions between multiple genes and environmental factors are important causes, but account for a significant but unknown number of birth defects.

Environmentally caused defects result from a host of recognized factors, including infections, chemical toxins, X rays, fever, drugs, alcohol, diseases of the mother, and actual problems within the womb or the fetus itself. These factors, considered in the next three chapters, collectively are extremely important and account for at least 10 percent of all birth defects.

It may not be possible to discover the cause of either a specific birth defect or of mental retardation in over 60 percent of cases. Difficulty may also be experienced in distinguishing hereditary from acquired causes, as in cleft lip and palate, which are most often hereditary, but may also be caused by drugs taken during pregnancy.

Very careful analysis of the family history may point to a particular genetic disease. Photographs of affected family members living elsewhere may lead to recognition of an unusual disorder. Hospital records and autopsy reports on a previously affected child or relative are, again, of great importance in diagnosis. Finally, actual physical examination of the parents of an affected child may provide some diagnostic leads. Such a scrutiny of the parents may even include X rays of the skull. Tuberous sclerosis, as I said in chapter 7, may be inherited directly from one parent who looks entirely normal, though X rays may reveal characteristic deposits of calcium or other features in the brain, implying that that parent actually has the disease but shows no outward manifestations.

Obviously, every effort should be made to find out the cause of a birth defect or mental retardation, since recognition of the cause is really the primary step to treatment and prevention of its recurrence.

HOW FREQUENTLY DO BIRTH DEFECTS OCCUR?

Major birth defects with or without serious mental retardation occur in 3 to 4 percent of all births. By seven years of age, after disorders not obvious at birth have been diagnosed, these numbers have doubled. In the United States alone, estimates suggest, about 160,000 babies each year are born with serious birth defects, mental retardation, or both. Between 2 and 3 percent of the U.S. population are mentally retarded (mostly mildly), while 0.25 percent of the population are severely retarded. On a cumulative basis, this implies that there are over 6 million individuals in our country who are mentally retarded! This is a very serious public-health concern.

As I mentioned earlier, in the United States and other Western countries, 25 to 30 percent of all major children's hospital admissions stem from birth defects, genetic disease, or mental retardation. Minor birth defects occur in up to 40 percent of all births (see table 13). These and other minor signs can generally be ignored per se, though when three or more are present, there is about a 20-percent risk that they signal the presence of something more serious, such as a heart or brain defect.

Gross structural birth defects are a different matter. They constitute the second-most-common cause of infant deaths, exceeding 9,000 annually in the United States.

Hereditary Causes of Birth Defects

DISORDERS OF THE CHROMOSOMES

I discussed in earlier chapters how a person may be born with too many chromosomes, too few chromosomes, or abnormalities in the structure of particular chromosomes. About 1 in 156 children are born with chromosome abnormalities that may or may not be associated with mental retardation.

Chromosome defects are the cause of about 5 percent of all birth defects. In the United States alone, each year about 24,000 children are born with chromosomal abnormalities, over 16,000 of which are of serious significance.

Confronted by a child with an unusual appearance or mental retardation, it is a relatively simple matter to obtain a blood sample to document the normality or abnormality of the chromosomes. The commonest disorder in this group of conditions is, of course, Down syndrome (see chapter 2).

DISORDERS OF SINGLE GENES

Some 4,000 genetic disorders due to the effects of single harmful genes have been cataloged. These genes wreak havoc when they are transmitted from one usually affected parent (dominant inheritance) or after arising anew (mutation), from both usually unaffected parents (recessive inheritance), or from the usually unaffected mother (sex-linked inheritance). Some disorders, such as microcephaly and some types of muscular dystrophy, may result from either one or two or even three of these modes of inheritance. Fortunately, although there are many disorders in the single-gene category, most are rare and many are neither fatal nor serious.

Table 13. Examples of Minor Defects or Variations of Normal

Head and Neck

Skin tag in front of ears (especially among blacks)
Dimple or tiny hole (sinus) in front of ears (especially among blacks)
Droopy eyelids (ptosis)
Comma-shaped fold of skin covering inner corner of the eyes
Flat bridge of nose
Prominent bridge of nose
Eyebrows that meet
Speckled (concentric) iris (Brushfield's spots)
Upturned nostrils
Scalp defect (skin)
A third soft spot (fontanel) on baby's head
Prominent or flat back of head (occiput)
Bossed forehead
Unusual hair whorls
Tiny hole (sinus) in the neck or ear
Ear shape, size, and location abnormalities (many types)
Cleft lip or gum
Cleft uvula
Excess skin folds in the neck
Small chin
Webbed neck

Chest, Abdomen, and Genitals

Hernia of navel (umbilicus; especially among blacks)
Single umbilical artery
Dimple over shoulder blade, lower spine, or other bone
Extra nipples
Displaced opening of penis (hypospadias)
Vaginal skin tag

Limbs

Extra digits (polydactyly; especially among blacks)
Single transverse crease (simian crease) across palms
Sole crease
Incurved fifth finger

Continued on next page

Table 13 (continued)

Limbs (cont.)

Fifth finger with a single or extra crease
Wide space between first and second toes
Fused or bunched second and third toes (syndactyly)
Overlapping fingers or toes
Recessed fourth or fifth toes
Prominent heels
Fingernails underdeveloped or excessively curved
Tapered fingers

Since I have prevention in mind, I will concentrate on those that shorten life or cause serious disability.

Birth Defects Transmitted from One Parent: There are many different recognized birth defects that can be transmitted from one affected parent with each child having a 50-percent chance of being affected. Some of these dominantly inherited disorders are minor and include conditions such as extra digits or webbed toes (especially the second and third). Some are more severe, as when one parent has a hand or hands resembling a lobster claw. Each new pregnancy in such a family carries a 50-percent risk that the offspring will have a similar deformity.

A parent may have markedly underdeveloped cheekbones but no other defects. An affected child (the risk is 50 percent) may be born missing an ear, and the underdeveloped cheekbones may also affect the shape and appearance of the eyes; these defects form the Treacher Collins syndrome.

A white forelock of hair in a parent may signal that half the children may be born deaf, or have disabilities of less consequence. The disorder is known as Waardenburg syndrome.

Extra digits, or *polydactyly*, is seen in infants quite frequently. Close to 1 in every 100 black newborns has an extra finger on one or both hands, an extra toe on one or both feet, or extra toes as well as extra fingers. Polydactyly may also be simply dominantly inherited from one parent. Occasionally, parents will vehemently deny any family history of this minor defect. On being asked to look at their hands, some of them find a tiny scar at the base of their fifth fingers, which they had not known about. The scar indicates the site of the extra tiny finger that had simply been tied off at birth, and which then fell off shortly thereafter.

Much has been written about birth defects as related to the increasing age of the mother. There are, however, a few defects — albeit rare — that are associated with the increased age of the father. Achondroplasia (chapter 7) is one example of such a defect. In about 7 out of 8 born with achondroplasia, a gene mutation is responsible. The father is frequently over forty or fifty years old in these cases. Once a person is born affected, the disorder is transmitted as a dominant condition. A few other such conditions — ones in which correlation with older fathers has been shown — involve bones, muscles, connective tissues, or joints; these are called Marfan syndrome, Apert syndrome, and myositis ossificans progressiva. One other rare disorder in this group is a defect or absence of the iris (the colored part of the eye), which may signal a tumor of the kidney. Yearly or twice-yearly physical examinations of the child are necessary to ensure that no genetically predisposed defect — for example, cancer of the kidney — develops without being recognized in time.

Birth Defects When Both Parents Are Unaffected But Are Carriers: Disorders in this category result from the corresponding harmful recessive gene being transmitted from *each* parent, causing a disorder in the child. Many of the disorders involve biochemical disturbances in the child that usually lead to mental retardation, but without deformation of the limbs or face. Over 100 of these biochemical disorders of metabolism are diagnosable prenatally, and the chance that they may occur in each pregnancy is 25 percent (for example, Tay-Sachs disease). Earlier, I mentioned birth defects and mental retardation resulting from the marriage of close relatives, from marriage within the same ethnic group, or as a consequence of incest. After acquired diseases such as ear infections have been excluded, the most common cause of congenital deafness turns out to be recessive genes from both parents, usually without any family history of hearing problems but often with ancestors from the same country. Similarly, after excluding acquired conditions such as diabetes, infections, and injuries, blindness will often be inherited through hidden recessive genes.

Sometimes disorders in this group cause visible physical abnormalities, as in certain types of microcephaly, or small-headedness, associated with mental retardation. Some syndromes may be transmitted from both parents through recessive genes, with neither parent actually affected. In each pregnancy, the risk of recurrence is 25 percent. The otopalato-digital syndrome is one example; the affected child will have unusual eyebrows and nose, a "long" head, cleft palate, markedly enlarged tips of fingers and toes, extremely short big toes, and slow physical and mental development.

One particularly important disorder in this group may be confused with serious defects of the brain, such as anencephaly, or spinal-cord defects, such as spina bifida, which are discussed in detail below. Recognition of this condition — Meckel syndrome, which is invariably associated with fatal malformations of the brain, extra digits, and kidney abnormalities — is very important, because the risk of recurrence in each pregnancy is 25 percent, as opposed to about 2 percent for the anencephaly group.

Birth Defects with Mothers as Carriers and Only Sons Affected: It has been known for about fifty years that more males than females with mental retardation are found both in institutions for the mentally retarded and in the general population. This is most pronounced among those who are severely retarded. While it is also known that, for some strange reason, males more often suffer birth injury, this is not sufficient reason to account for the disparity between the sexes in subsequent mental retardation. Careful studies of the family histories in these cases have drawn attention to the fact that in a remarkable number of instances the mother is the carrier. Such mothers have a 50-percent chance that each of their sons could have mental retardation. These boys (if they do not have the fragile-X syndrome) may have smaller or larger heads than normal and occasionally other signs. Many will have no other associated birth defects. Nor is it possible yet to detect the carrier mother of the non-fragile-X-syndrome cases. One case I vividly recall illustrates the point well.

I saw Mr. and Mrs. C in consultation, along with their two sons, aged four and eight. Both boys were moderately retarded and no cause had been found by the referring doctor. Mrs. C's pregnancy had been perfectly normal. She had no infections, no fevers, had taken no drugs, and suffered no injuries. Both she and her husband were perfectly normal.

Physical examination of both boys showed no abnormalities

other than the mental retardation. Furthermore, a whole battery of special tests to try to determine the cause of their mental retardation were all unproductive.

The answer lay in their family history. On Mr. C's side of the family, there were no instances of mental retardation, birth defects, or genetic disease. However, Mrs. C had one brother who was mentally retarded (from no known cause) and four sisters. Two of the sisters had each had one son with mental retardation, and the third sister had had three sons afflicted with it. The fourth sister was unmarried and childless. I looked further back into Mrs. C's family. On her mother's side there were two uncles who were retarded as well. No retarded females were noted in the entire family pedigree.

This was a clear case of a sex-linked disorder: retardation affecting half the males born to female carriers. The family was reassured that they could safely have girls. This they did by using prenatal sex determination. They achieved the birth of two girls in consecutive pregnancies — something they might not have done without the reassurance provided by prenatal diagnosis.

Some experts have calculated that as many as 20 percent of severely mentally retarded boys — those with IQs under 50 — may be the victims of sex-linked inheritance. Clearly, therefore, careful analysis of the family history is crucial in these cases — since prenatal sex determination is now possible, and parents may avoid the risk of having a retarded child by choosing to have girls. Specific chromosome studies should be performed to exclude the fragile-X syndrome (see chapter 6) since it is the most common serious disease in this category.

About 250 birth defects, with or without mental retardation, are transmitted by the mother in the same way. Hemophilia and Duchenne muscular dystrophy are the most common serious diseases in this category. Types of hydrocephalus (large heads with increased fluid in the brain, often associated with mental retardation), a failure of the intestine to open at the anus (imperforate anus), a small, deformed, blind eye (microphthalmia), and certain other biochemical disorders of metabolism are but a few other examples.

DISORDERS FROM INTERACTING HEREDITARY AND
ENVIRONMENTAL FACTORS

This form of multifactorial, or polygenic, inheritance reflects the combined effects of several minor gene abnormalities acting in concert with environmental factors and results in many different disorders. Two of the most serious, both previously discussed neural-tube defects, occur most commonly in children of Irish ancestry. The brain defect, anencephaly, is incompatible with survival, while children with the spinal defect, spina bifida, may have major health problems and a markedly decreased life expectancy. A child affected with anencephaly is likely to be stillborn or die within minutes to hours of birth, or sometimes may linger on for two or three months before dying. The spinal defect, in contrast, is not fatal, but the open spine usually results in paralysis of the legs (paraplegia) and, frequently, lack of control of bladder and, possibly, bowel functions. The child may be mentally normal but confined to a wheelchair, and may leak urine, and sometimes stool, constantly. The effect on the child and family can well be imagined. Occasionally, the spinal lesion is sufficiently small to be repaired without any significant disability remaining, and the person is able to walk and function normally. (Prenatal diagnosis of these defects and their risk of occurrence and recurrence are discussed in chapter 23).

Environmental factors seem especially strong in the cases of both anencephaly and spina bifida. As I mentioned before, in Northern Ireland, the frequency of these conditions is about 3 in every 1,000 births, and this high rate drops gradually as one moves to the south of England, where the frequency is about 2 per 1,000 births. Strange as it seems — and totally unexplained — these anencephaly and spina bifida conditions occur with the same high frequency in the Punjab in India and in Alexandria, Egypt, as in Northern Ireland. The Welsh, the French-Canadians of Quebec, and those who live in the mountains of Tennessee also have higher rates of these neural-tube defects. Besides geographical effects, there are recognizable seasonal effects, in that the defects show up more frequently in winter and fall births than in spring and summer ones. Some relationship between poverty, as well as illegitimacy, among whites has been shown for these defects. Most recently folic acid (an essential nutrient) and possibly vitamin-A deficiency have been implicated. While definitive studies are still in progress, evidence has emerged in England of the beneficial effect of folic acid taken prior to conception and through the first three months of pregnancy. Many fewer neural-tube defects than expected recurred in supplemented pregnancies that followed the birth of an affected child. The matter is, however, very complex; certain confounding factors may explain these good results. For example, there was a striking drop in the frequency of these defects in England during the period of these studies. Moreover, explanations other than folic-acid deficiency (for instance, zinc deficiency) may explain some cases. Recommendations about which, if any, supplements should be taken in pregnancy will depend upon conclusions of studies still to be completed.

There are other birth defects besides the anencephaly and spina bifida group that are caused by the interaction of genetic and environmental factors. These include cleft lip, cleft palate, clubfoot, pyloric stenosis (a congenital blockage between the stomach and the intestine), dislocated hips, and certain congenital heart defects (for example, ventricular or atrial septal defects) — to mention only the most common physically obvious defects.

The late professor Cedric O. Carter of the Hospital for Sick Children, Great Ormond Street, London, studied how frequently these defects recur in families. Since these conditions are quite common, it might be helpful if I gave some guiding figures for most of the disorders mentioned. (For the nervous-system disorders, note that parents may have a child with anencephaly and then later, in another pregnancy, have a child with spina bifida.)

Risks for all the defects mentioned above except pyloric stenosis are as follows:

If only *one* parent is affected:
• the risk of having an affected child is about 3 to 5 percent

If *both* parents are unaffected and they have had *one* affected child:
• the risk of having another affected child is about 2 to 5 percent

If *both* parents are unaffected and they have had *three* affected children:
• the risk of having a fourth affected child is about 25 percent

The general risks that first cousins, nephews, and nieces of affected individuals will have for all these disorders are thought to be higher than in the population at large. The first cousins of spina bifida patients have about an eightfold increase in risk relative to the general population.

Sex differences are often quite prominent for this group of disorders. Males are much more often affected by pyloric stenosis, cleft lip, cleft palate, and clubfoot. In contrast, many more females are born with anencephaly and congenital dislocation of the hips.

The recurrence figures for pyloric stenosis are somewhat different from the general guidelines expressed above:

If the *mother* is affected:
- the risk of having an affected *son* is 16 to 20 percent
- the risk of having an affected *daughter* is about 7 percent

If the *father* is affected:
- the risk of having an affected *son* is about 5 percent
- the risk of having an affected *daughter* is about 2.5 percent

If a *girl* is affected:
- the risk that any future *brother* will be affected is 10 percent
- the risk that any future *sister* will be affected is 4 percent

If a *boy* is affected:
- the risk that any future *brother* will be affected is 4 percent
- the risk that any future *sister* will be affected is 2.5 percent

Table 14. Minnesota Study: Risks of Mental Retardation

Parent Union	Risk That Next Child Will Be Retarded
For Persons Who Have Not Had a Mentally Retarded Child	
Normal person with one retarded sibling × normal person with all normal siblings	1.8%
Normal person with two or more retarded siblings × normal person with all normal siblings	3.6
Normal person with one retarded sibling × retarded person	23.8
Normal person with all normal siblings × normal person with all normal siblings	0.5
For Persons Who Have Had One or More Mentally Retarded Children	
Normal person, one of whose parents had one or more retarded siblings × normal person with all normal aunts and uncles	12.9
Normal person with all normal aunts and uncles × normal person with all normal aunts and uncles	5.7
Normal (or unknown) person × retarded person	42.1

Many couples who are planning a pregnancy discover that they have a mentally retarded relative. In cases where a definite diagnosis has been made, genetic counseling about their risks and options is usually straightforward. When no diagnosis has been possible after complete evaluation, information about the couple's risks can still be provided on the basis of knowledge gained from pooled previous data. Some of these risk figures are given here for guidance (see tables 14 and 15). They are based on two studies: a Minnesota study of 80,000 persons (including all individuals with IQs of less than 70), and an Australian study of 2,000 consecutive families who attended a clinic for the mentally retarded (individuals with Down syndrome and those families in which information was incomplete were excluded).

Progress in the understanding of genetic causes of mental retardation, especially the fragile-X syndrome, has increased the options available to couples planning to start a family. Those with a family history of mental retardation not due to acquired defects of cerebral palsy, brain infection, or injury are advised to seek genetic counseling *before* having children.

In this chapter I have covered the hereditary causes — chromosomal, single-gene, and multifactorial (involving gene/environment interaction) — of birth defects. Our options for treatment or prevention are still seen to be limited as far as genetic disorders are concerned. In contrast, birth defects that arise from environmental agents such as viruses, X rays, toxins, medications, and fever can largely be prevented. Clear chances to prevent tragedies are uncommon in life. The next three chapters represent just such opportunities that you simply must know about. Remember, *not knowing your genes limits or removes your choices!*

Table 15. Australian Study: Risks of Mental Retardation

Severity of Retardation in One Child	Future Risk of Having a Retarded Son	Future Risk of Having a Retarded Daughter
Profoundly retarded (IQ below 20)	5.6%	4.4%
Severely retarded (IQ 20–35)	2.1	3.6
Moderately retarded (IQ 36–51)	9.1	4.2
Mildly retarded (IQ 52–67)	7.7	3.9
Borderline retarded (IQ 68–85)	4.4	1.5

NOTE: All risks listed are for couples who have had one child with mental retardation of unknown cause.

14

The Unseen Enemy:
Infection, X Rays, and Toxins

Mental retardation or birth defects may be caused by a variety of agents found in the environment. The important point is that, in contrast to those caused by harmful genes, environmentally caused defects can be prevented before conception (though infection may, of course, occur any time after birth and lead to mental retardation). In this and the next two chapters, my focus is on the many factors that affect the fetus in the womb.

Infection

Infection contracted by the mother during pregnancy is an important cause of imperfect development of the fetus. Estimates from major U.S. government studies suggest that as many as 10 percent of all cases of mental retardation may be attributed to infectious diseases. While German measles in pregnancy is well known as a cause of birth defects, other viral or protozoan infections may cause equally devastating results. A characteristic of viral infections that attack the fetus is their long-lived nature. Such viruses may continue to be shed in the secretions and excretions of the baby for protracted periods, threatening those who come in contact. One important problem is that viral infections are often hidden and not clinically apparent. The possibility of hidden infection has worried me all too often, since I know, for example, that about half of all mothers who actually have German measles may simply feel out of sorts, have no fever, but nevertheless be in a position to infect the fetus seriously!

TIMING OF THE INSULT
The stages of development from the fertilized egg, to the embryo, to the fetus are well recognized. In the first three months, when all

systems and organs are developing in the fetus, there are critical periods when the eyes are developing, the heart chambers are closing, the limbs are beginning to form, and so on. Sometimes, looking back, it may be possible to estimate fairly accurately at what stage in the first three months of pregnancy a particular insult or infection to the embryo occurred. The matter is complicated by the fact that viruses or drugs, acting on the developing fetus early in pregnancy, may have very different effects from case to case. Even German measles contracted during the first three months of pregnancy does not invariably cause birth defects.

GERMAN MEASLES (RUBELLA)

An Australian ophthalmologist first observed in 1941 that German measles in the mother during early pregnancy could cause what is now recognized to be a pattern of birth defects called the *congenital rubella syndrome*. Those who have had German measles subsequently develop a natural immunity. However, about 10 to 15 percent of all women of childbearing age today are still susceptible to German measles, having had neither the disease nor the recommended immunization. In spite of immunization efforts, massive epidemics of rubella occur worldwide in cycles. During the last major one in 1964, in the United States alone over 20,000 children were born with serious birth defects.

Women exposed to rubella during the first eight weeks of pregnancy have up to an 85-percent risk that their child will be left with some defect identifiable within the first four years of life. Infection between the ninth and twelfth weeks of pregnancy yields about a 52-percent risk, and between the thirteenth and twentieth weeks the figure declines to 16 percent. The earlier in pregnancy the infection occurs, the more severe and multiple are the subsequent effects.

Major abnormalities affecting multiple organs are typical of the congenital rubella syndrome. Over half of all affected infants have cataracts, sometimes with associated defects of the cornea, the retina, and the eye itself. Cardiac problems largely involve the great vessels, resulting in either a narrowing of the pulmonary artery or its branches or a continued opening of the channel that connects the pulmonary artery and aorta. This channel is easily tied off surgically, but the pulmonary arterial tree represents an untreatable strain on the heart.

Consequences of fetal brain infection are evident in about 25 percent of infected infants. Mental retardation, microcephaly (a small head), deafness, delayed language, delayed muscle development, and even autistic behavior are all characteristic. An ominous aspect is continuing subacute though progressive infection of the brain in some of these children, which eventually results in further degeneration of all vital functions. A host of other defects involve skin, blood, liver, spleen, lungs, and bones. Failure to grow and to thrive is common. Probably the most common nonhereditary cause of deafness in a child with no other abnormalities is exposure to rubella in pregnancy. One further potentially salutary consideration is that *even* after the first three months of pregnancy, rubella can cause mental retardation, learning disorders, deafness, and other disabilities.

Twins exposed to viral infection transmitted by the mother during pregnancy are usually both affected. However, there have been exceptions, even with German measles, when only one twin has been affected. In the case of rubella, there is another extremely important hazard to consider: the newborn baby, infected by rubella virus while a fetus in the womb, is usually still infectious and can therefore infect those who come in contact. Some years ago, I was called in consultation to look at a three-pound-one-ounce baby born six weeks prematurely, who, I found, had eye defects and a major heart defect. At that time, there were no other associated problems and in my consultation note I cautioned that rubella could not be excluded as the cause of the heart defect. I said it would be judicious, therefore, to isolate the baby until the diagnosis of rubella was excluded. For some reason this advice was not followed. Four weeks later, I learned that one of the nurses caring for the baby had come down with German measles herself. Worse still, she was eight weeks pregnant at the time of exposure. Only after she had been diagnosed as having German measles was the child I saw tested. Sadly, but not unexpectedly, that child was found to be excreting rubella virus. The nurse and her husband decided on religious grounds not to terminate her pregnancy. They subsequently had an infant also afflicted with the rubella syndrome, including mental retardation.

How can you prevent infection from German measles? There is only one way: Be sure that the mother-to-be is immunized *before* pregnancy. (Some states have made rubella immunization mandatory for both sexes before entering school or marrying.) Since live virus is used in immunization, it should not be given during pregnancy. After immunization, a couple should wait two to three months before conceiving. If rubella vaccination is given inadvertently without a woman knowing she is pregnant, there might only be a slightly increased risk that the fetus will be infected, since a weakened virus is used for immunization. An additional safeguard is to provide a blood sample prior to pregnancy and have the serum studied and also frozen and stored for the level of antibodies to rubella or other viruses. If natural immunity has been established by previous infection, as reflected by a good level of antibodies, then rubella immunization is not needed. Unfortunately, women previously immunized against rubella infection may still, though rarely, have affected offspring, in contrast to those who have had a case of German measles itself. To reiterate, it is wise for all women to determine if they are susceptible to rubella and to be immunized before pregnancy.

CYTOMEGALOVIRUS (CMV) INFECTION

Next in importance to rubella as a cause of birth defects, mental retardation, or both is infection by a large, common virus known as *cytomegalovirus*, or CMV. This virus may affect as many as 6 in every 100 pregnant women; and a remarkable 1 to 2 percent of liveborn infants harbor the virus. Fortunately, 90 percent of these infants, although infected in the womb, show no effects at birth. A few show signs of viral illness — rash, jaundice, and so forth — and nearly 7 percent suffer the grave effects of brain involvement. These are the ones who develop mental retardation, microcephaly, blindness, and hydrocephalus (enlarged, fluid-filled inner spaces of the brain). Actual birth defects from a failure of organs to form properly are rare in CMV. Brain defects, abnormal teeth, and groin hernias, among other abnormalities, have occasionally been reported. The ominous aspect of this disorder is that there are usually no symptoms in pregnancy, or occasionally only a mild influenza like illness. Certainly, when rubella is not epidemic, this virus is probably the most common infectious cause of mental retardation known.

As in rubella, affected newborn infants may be infectious, and should be strictly isolated in the nursery. Pregnant women should, of course, not take care of such infected infants. Luckily, a significant number of infants with cytomegalovirus simply excrete the virus and have no actual defects.

Blood tests are available to determine if a woman is susceptible to cytomegalovirus. Again, while no vaccine is yet available, a stored serum sample may be useful for comparison with a new sample should there be a suspicious infection during pregnancy. This would allow the doctor to detect a brisk increase in antibodies to the virus, which would indicate an active infection. Though apparently rare, it is possible for a woman to have a second infected child, even if she has antibodies in her blood.

TOXOPLASMOSIS

Another infectious agent, a protozoan called *Toxoplasma*, may also affect the fetus in pregnancy. In the United States, about 1 in 200 women have this infection during pregnancy, which results in the birth of more than 3,300 children with congenital infection each year — close to 1 in 1,000 of all babies born. As in the case of rubella and cytomegalovirus infection, this organism may infect the pregnant mother without causing symptoms. When symptoms do occur, they include mild fever, malaise, and muscle pains, and may be associated with enlarged glands and spleen, as well as a rash. Fortunately, fewer than 20 percent of infants infected in the womb actually show any signs of this disease. Those babies who do have signs of toxoplasmosis at birth can be expected to exhibit severe damage or even a fatal outcome up to 40 percent of the time! There may be acute illness with fever, rash, enlarged glands, and convulsions; or signs may only surface months or years after birth and possibly include mental retardation, hydrocephalus, or microcephaly, blindness or other eye abnormalities, hearing impairment, seizures, or learning disorders.

The severity of toxoplasmosis is directly related to the stage of pregnancy when infection began. Of those who contract toxoplasmosis in the first three months, 15 percent will transmit it; in the second three months, the rate rises to 25 percent; and later in pregnancy, it goes up to as high as 60 percent. The earlier in pregnancy it occurs, the more dangerous this infection is to the fetus.

A French study found that approximately 16 percent of pregnant women were susceptible to infection by *Toxoplasma*. The risks observed seemed greater for younger females. Large differences exist in the incidence of this condition; 80 percent of Parisian women showed evidence of past infection compared to fewer than 40 percent of American women.

It is not widely known that domestic pets such as cats and birds may harbor *Toxoplasma* infections. Unlike rubella, cytomegalovirus, syphilis, and herpesvirus infections, toxoplasmosis is not communicated from person to person. The toxoplasma organisms are present in the soil, often as a result of animal excretions, and may be spread by flies, cockroaches, rodents, and birds, as well as cats. They are also found in raw meat, especially lamb and pork, but can be destroyed by cooking. Obviously, therefore, women who are pregnant or are planning to be sometime soon are best advised to avoid contact with birds and cats (including contact with bird cages and litter pans) and not to eat undercooked meat. Special care should be taken in preparing raw meat for cooking, and fruits and vegetables should be thoroughly washed before eating. Gloves as well as shoes should always be worn when gardening.

Treatment of toxoplasmosis during pregnancy requires the help of a specialist in infectious disease. Unfortunately, there is no simple, safe, and effective treatment; prevention clearly is best.

It may be possible through blood tests to determine before pregnancy whether a woman has already had toxoplasmosis, cytomegalovirus, or some other infection to which she would thus be immune. On the other hand, she may be found to be susceptible — that is, to have no antibodies in her blood against these infections. Unfortunately, no vaccine is yet available. As I have said in connection with other infections, one solution is to store a frozen serum sample; if an infection is suspected during early pregnancy, the sample can be thawed and tested. A high level of antibodies in the pregnancy sample compared to the original one would imply an active recent infection and open the option of pregnancy termination. It is again fortunate that about 90 percent of *Toxoplasma*-infected pregnancies end with normal children. Mothers with toxoplasmosis antibodies in their blood are not usually considered in any danger of having an affected child.

In an important study done in New York City, researchers observed that more affluent white women developed toxoplasmosis in pregnancy than those in lower socioeconomic brackets. They concluded that these women had greater access to all kinds of meat and were probably eating undercooked beef or pork. Animals (whose meat we eat) become infected, just as we do, by contact with animal excretions, such as cat feces.

CHICKENPOX

If contracted early in pregnancy, chickenpox virus infection can also lead to birth defects. The risks for the fetus are, however, very much smaller than those of rubella, possibly less than 5 percent. Damaging effects may come from high fever instead of, or together with, the infection itself.

Chickenpox in adults is usually very much more severe than in children and during pregnancy may even be fatal! Since about 4.5 percent of women in the childbearing age may be susceptible, those who know they have never had chickenpox should be very careful to avoid this infection, especially during pregnancy. Susceptible teachers, child-care workers, nurses, and mothers with young children need to be especially careful. Brain damage, cataracts and other eye defects, and skin scarring are the main signs in an infant infected in the womb. New vaccines now being developed may soon offer protection against chickenpox.

AIDS IN CHILDREN

Acquired immunodeficiency syndrome (AIDS) has become a modern plague, about 40,000 having already died from it in the United States alone. AIDS is caused by a virus — known as *human immunodeficiency virus*, or HIV — that is transmitted by blood, contaminated needles, sexual intercourse, and, around the time of or during birth, from mother to child. Over 500 children have been reported to have AIDS in the United States, and many many more are expected. Cesarean-section delivery does not seem to protect the child from AIDS as it does from herpesvirus infection. Genetic susceptibility or resistance may explain why most persons infected by the AIDS virus develop the full-blown disease while others (very few) simply remain unaffected despite a positive blood test. In fact, in a Scottish study of hemophiliacs inadvertently given blood contaminated with the AIDS virus, those who became ill most rapidly had inherited the specific HLA blood group A1 B8 DR3 (see chapter 9).

Early reports suggest some minor birth defects of the face in infected children. These features include a small head, wide-set eyes, a flat bridge of the nose, and a prominent, boxlike forehead. Since the majority of infected children have been black or Hispanic, variations of normal racial features may explain the above features. However, delayed development and brain dysfunction, which results in spastic paralysis of limbs, floppiness, or almost total paralysis, are all a direct result of infection with the AIDS virus.

Given that no cure is anticipated in the next few years, no

effective treatment is yet available, and no vaccine is in the offing, individuals should be extremely circumspect about their sex partners and should avoid promiscuity, use condoms, avoid drug addicts and intravenous hard drugs, and consider taking blood tests for AIDS. Surely no couple wants to bring a child into the world doomed to die of AIDS.

OTHER INFECTIONS

Any infection can, of course, occur during pregnancy, and if it is associated with high fever it may harm or even kill the fetus. Infections other than those mentioned above are not usually associated with malformed organs but may lead to brain, eye, or other organ damage. *Active* genital-herpesvirus infections, for example, may be contracted by the infant on its way through the birth canal; cesarean section is usually done to prevent direct contact with this virus. Mental retardation, sometimes coupled with microcephaly, hydrocephalus, or blindness, may result, as may additional physical-development problems. Congenital syphilis in the newborn may cause an array of devastating problems in multiple organ systems. Survivors may be left with severe disorders of the nervous system, including mental retardation, nerve paralysis, deafness, blindness, facial abnormalities, and other defects. Effective treatment is available but early diagnosis is important. Even then, some of these effects can be permanent, despite treatment.

X Rays during Pregnancy

It has been known for some time that exposure of the fetus to X rays can lead to serious abnormalities such as microcephaly, with associated mental retardation, bone defects in the skull, spine and eye abnormalities, cleft palate, and severe limb deformities. Any possible doubt about the harmful effects of radiation was removed after studies were made on the children of Japanese women who were pregnant in Hiroshima at the time of the atomic-bomb explosions. They showed a high incidence of mental retardation and microcephaly as the earliest manifestations of problems seen in those babies born alive.

The fetus can inadvertently be irradiated when the mother is exposed to X rays during medical procedures performed without realizing that she is pregnant. Such exposures do not cause chromosome (or other) defects, and genetic studies of the amniotic-fluid cells are thus not recommended. Usually, the fetus and, subsequently, the child will show no signs of abnormality. Only high X-ray dosage, which rarely occurs during diagnostic studies in early pregnancy, can result in microcephaly, with mental retardation and possibly multiple birth defects.

The relative risk to the fetus depends upon the dose of irradiation given to the mother during the first four months of pregnancy. If it was in excess of 10 rad (a measure of irradiation), then the risk of fetal damage is very high. A dose between 5 and 10 rad would be considered risky by some physicians, while most will hold that exposures of under 5 rad carry no clear risk of damage. Even multiple diagnostic X rays in the same woman — for example, of bowel, chest, and kidneys — usually do not amount to 5 rad.

Serious difficulty can arise from therapeutic radiation. High-dose X rays delivered in the treatment of cancer when a pregnancy is undiagnosed can clearly result in the fetal defects noted above. Thankfully, such situations are very rare.

X rays do have the propensity to change genes, and therein lies their greatest danger. Exposure of the fetus to X rays during diagnostic studies of the mother has been reported by some researchers to be associated with an increase in the risk of childhood cancer, especially leukemia, years later. Indeed, one study suggested that there is an increase of 40 percent in the risk of cancer, including leukemia, in children who have been exposed to X rays while in the womb. Not unexpectedly, some studies have found that the risk of childhood cancer increases with larger doses of X rays, as well as with earlier exposures during pregnancy. While it is true that other elements, such as individual susceptibility, may be important in the development of cancer following X rays, there is sufficient evidence to suggest that women should avoid unnecessary X-ray examinations especially during pregnancy.

Chemical Toxins

A dreadful consequence of twentieth-century industrialization has been pollution of our environment by over 55,000 toxic chemicals. The failure of people elected to public office to establish and enforce regulations is largely responsible for this modern scourge. Toxic wastes, which enter our water supply and food chain, contain endless cancer-producing chemicals, as well as many that cause birth defects. Celebrated legal cases against industrial polluters have been settled out of court, especially in areas where remarkable increases in the rate of leukemia and miscarriage have occurred. Unfortunately, proving that a particular chemical toxin caused a specific birth defect in a particular pregnancy is extremely difficult, aside from the matter of proving that a particular company is responsible for the mother's exposure to the toxin. While thousands of these chemicals can easily be shown to cause birth defects in laboratory animals, proof in humans is limited to a mere handful. While I will only briefly discuss the little that is known with certainty, every effort should be made to contain the continuing contamination of our bodies by industries out of control.

MERCURY POISONING

You would not have thought that mothers are likely to ingest mercury during pregnancy, if at all. Tragic experiences, however, have occurred in both the United States and Japan, where pregnant women have ingested mercury contained in contaminated food and subsequently delivered severely malformed and retarded children. The Japanese episode occurred after a factory began discharging its waste into the sea adjacent to a fishing village. The high content of mercury in the factory's waste material soon contaminated the fish, which were caught by the local fishermen and eaten by them and their wives. Many families were affected, but it took about seven years before a connection was established between the ingestion by pregnant women of fish with a high content of organic mercury and the birth of children with severe brain damage and other defects. This disaster occurred in Minamata Bay and the condition was subsequently called Minamata disease. Bread made from mercury-treated wheat seed similarly caused a major disaster in Iraq in 1971/72.

Today it is known that certain fish from inland lakes in the United States, Canada, and Sweden contain high concentrations of mercury dumped as the effluent from chemical plants and paper mills. We also now know that industrially caused acid rain leaches metals from soil into water more effectively than does normal rainwater, thereby sharply elevating the concentrations of, for example, mercury, which has shown up in fish from affected lakes in the Adirondacks and elsewhere in the United States. Since the fetal brain is thought to be at least four to ten times more sensitive to mercury toxicity than that of adults, deficits such as learning disorders, in addition to the less subtle defects described above, are also suspected in children exposed while still in the womb. Japanese studies strongly suggest that mercury may also cause major birth defects in multiple organs.

LEAD POISONING

Lead is an ancient and subtle poison. Skeletons from the oldest civilizations have revealed excessive concentrations of this metal. Its toxic effect on the developing embryo and child has even been held responsible for the decline of the Roman Empire! Apparently, Roman nobility stored wine and preserves in leaden casks and used lead compounds as sweetening agents.

We have learned poorly from history. In 1988, thousands of children in the United States were treated for lead poisoning. While important steps have been taken to combat this problem — which mainly stems from lead paint chipping off walls and woodwork and being eaten by children — the problem is far from being eradicated. Lead poisoning during pregnancy, however, is decidedly rare — and fortunately so. It has been known since the latter part of the nineteenth century that stillbirth, as well as congenital defects, occurred in the offspring of mothers exposed to high lead concentrations. More recently, though, it has been noted that the blood-lead concentrations in newborn babies of mothers who live in high-lead-exposure areas (for instance, near expressways) may be elevated. Since it is known that lead affects rapidly growing tissues, this observation is of some concern. Furthermore, it is now clear that low-level lead exposure is consistently associated with reduced performance on various IQ and performance tests.

POISONING BY PCBS (COLA-COLORED BABIES)

A so-called cola color, sometimes coupled with a black nose; low birth weight; undeveloped nails; teeth obvious at birth; a skin disorder — all together constitute the main features seen in babies exposed in the womb to *polychlorinated biphenyls*, or PCBs. PCBs are clear, oily substances used extensively in industry (for example, as an insulating material). Two remarkable episodes of poisoning occurred in Japan in 1968 and in Taiwan in 1979. In both instances, accidental contamination of cooking oil was involved; more than 1,000 persons in Japan and more than 2,000 in Taiwan were affected. Even though current use and disposal of PCBs are strictly regulated by the Toxic Substance Control Act of 1976, exposure to much smaller doses than those allowed for adults in the United States may still lead to subtle brain dysfunction in children who came in contact with the toxin while still in the womb.

OTHER POISONS

The astounding pollution of our water poisons the entire food chain. Not only does drinking water become a problem, but so does the eating of meat and fish, as well as all dairy products. The tens of thousands of chemical compounds released into our environment have the potential for causing cancer, birth defects, or both. Cases of flagrant toxic-waste dumping have occurred in the Love Canal area near Niagara Falls, in Woburn, Massachusetts, and in New Jersey. People in these areas have claimed high rates of miscarriage, cancer (such as leukemia), and birth defects. In these cases, great difficulties have been encountered in trying to prove that a specific toxin caused a particular cancer or birth defect in any one individual. Given the complexities involved in understanding the cause for either cancer or birth defects, such proof has not been achievable in the courtroom. Rather, clear and flagrant violation of laws governing toxic-waste disposal have led many polluters to settle out of court; consequently, the question of causation has mostly not been adjudicated. Since about 90 percent of all cancers are thought to originate in the environment (the exact causes are not recognized), and since there are no clues to the cause of about two-thirds of all major birth defects, there is serious concern about the role of toxins, which complicate our lives and possibly cause our deaths.

Among the agents that may harm the fetus are the defoliant known as Agent Orange, such herbicides as 2,4,5-T, various pesticides, cadmium, many disinfectants, and possibly some food additives. Fumes from gasoline, glue, and toluene, as well as many other occupational and home chemicals, including hair dyes and hair sprays, anesthetic gases, substances used in the printing trade and in smelting factories, and fat and organic solvents, are also thought to be able to cause birth defects.

While infections, X rays, and toxins during pregnancy may cause birth defects and mental retardation, drugs cause, measure for measure, many more anxieties. The drug question is complex and deserves separate and detailed consideration.

15

Drugs Spell Danger

You may have thought that no pregnant woman need be reminded of the dangers inherent in taking drugs. The consequences may be very serious, yet, despite all that has been said and written on the subject these past few years, women still take a remarkable four to fourteen prescription or nonprescription drugs during pregnancy. It is difficult enough to discover what one drug may do, let alone the potential effects of two or more on the fetus. Only a few drugs are definitely known to cause birth defects when taken during pregnancy. The reason for worrying about the almost endless list of available drugs is that the cause of birth defects is still unknown in about two-thirds of cases!

Proving that a particular medication does or does not cause birth defects is extremely difficult. A remarkable number of factors have to be considered before any safe conclusion linking cause and effect can be established. The stage during pregnancy when a drug is taken may be critical. Dreadful experience with the drug thalidomide, prescribed for nausea in pregnancy, mostly in West Germany and England, resulted in thousands of children being born without arms and legs or with catastrophic deformities of the limbs as well as other serious birth defects. Mothers who took thalidomide on the thirtieth day after conception had children with the most deformed arms and legs, while those who took the drug on the thirty-fifth day had babies who only had lower-limb defects. Inexplicably, some exposed twins were born with one having no limbs while the other had comparatively milder defects. Moreover, many women who took thalidomide had perfectly healthy babies! The timing turned out to be more important than dosage in producing these defects. No matter how much thalidomide was taken, the drug seemed to have attacked specific organs at specific times.

In contrast, the *quantity* of alcohol imbibed in early pregnancy may be more important in causing more-severe defects than the exact day of exposure. The effects of a particular dosage of a drug may vary from person to person. For example, when people are given anticoagulants (blood-thinning drugs) to break down or prevent clots in their blood vessels, some may experience excessive bleeding while others on the same dose exhibit no change in their bleeding times. As mentioned previously, inherited and other factors determine the way a drug is degraded in the body. In some individuals, certain drugs are broken down quickly and efficiently; but in others, the mechanism is much slower. This results in higher and accumulating drug concentrations, which may lead to toxic effects, on the one hand, and — who knows? — birth defects on the other.

Before a drug is tested on humans, it is first tried on laboratory animals. Thalidomide, for example, was tested for safety on pregnant rats and mice and appeared to have no ill effects. Later it was discovered that it did cause serious defects in the offspring of monkeys and rabbits. Clearly, it is unsafe to extrapolate the results of animal studies directly to humans. Important lessons can, however, be learned. We know that at least thirty different ways exist to cause cleft lip (harelip), cleft palate, or both in laboratory animals. It is now realized that these same defects in humans, which occur about once in 1,000 births, may also be caused by a whole host of other factors, including chromosome defects, single-gene diseases, and medications (such as for epilepsy).

Drug Interactions

Many pregnant women take more than one medication, thereby raising the question of whether interaction between drugs may occur and result in an even greater potentiation of their individual effects on the developing fetus. In pregnant rats, for example, the frequency of birth defects can be increased if aspirin is given together with benzoic acid, a widely used food preservative. Drugs may also interfere with the body's chemistry. For example, an anticonvulsant drug may bind any available zinc, thus rendering deficient the cells of the developing embryo at a critical time.

Similarly, such antiepileptic medication may compete with the body for an essential nutrient such as folic acid and again prejudice the developing embryo, resulting in defects such as spina bifida.

Maternal Illness

A mother's body constitutes the immediate "environment" for the fetus. Consequently, any illness, drugs, toxin exposures, and so on, may potentially be harmful. The high concentration of sugar and the associated disturbed body chemistry in the diabetic mother seem largely responsible for her increased risk of having children with serious birth defects. Fever, from any cause, may also be harmful in early pregnancy. A Japanese study showed that a statistically significant number of women who had fever during the first eight weeks of pregnancy later gave birth to a child with defects such as spina bifida. Is it the fever itself, or a virus or a drug acting in the presence of fever, or another cause of the fever that is harmful? Animal experiments in which the core body temperature was raised have shown that fever itself may harm the developing embryo/fetus and result in a higher frequency of birth defects.

Moreover, we know that some birth defects — certain heart defects and spina bifida, for example — may vary in their frequency according to the season of the year. Higher frequencies may, for example, occur in the spring or in the fall — implying, perhaps, the involvement of a virus in some cases.

Actually inheriting a susceptibility to infection may also be important. A high probability exists that a person who develops diabetes in childhood actually was genetically susceptible to a virus infection that damaged the pancreas, eventually resulting in diabetes. Hence, in juvenile insulin-dependent diabetes, what may be inherited is the susceptibility to infection, rather than the diabetes itself!

The body's immune system, which attacks invading infectious organisms, is probably also responsible for discovering whether the embryo is normal or not. As discussed in chapter 2, about 50 percent of miscarried embryos in the first three months of pregnancy have chromosome defects. Whatever the magical surveillance system is that allows the body to rid itself of a defective embryo, that system becomes progressively less efficient with advancing maternal age. Hence, it is likely that the increased frequency of chromosome defects among the children of older mothers is a reflection of the body's failing surveillance system. There may also be other explanations.

Drugs or Heredity?

Repeatedly over the years, I have seen couples who have had long-standing problems becoming pregnant. Tragically, I have seen some of these couples try for seven to thirteen years, finally achieving a single pregnancy, only to have a child with a serious defect. Often, in such cases, fertility drugs have been used — for example, to stimulate ovulation. Invariably, the temptation is to blame the drug for the defect. There is reason, however, to believe that, at least in some cases, fundamental defects exist in the egg — flaws that are causally related to the defects and that might even be hereditary.

Sometimes a rare or heretofore undescribed complex of birth defects occurs. Almost always, no such defects have occurred in that family. A currently impossible matter to resolve is whether the parents have each contributed a corresponding harmful gene to cause such a defect. (See the discussion of autosomal recessive inheritance in chapter 7.) Since it would be more probable than not that the mother took one or more medications during pregnancy, suspicion often automatically heads in that direction. But careful inquiry in some of these families reveals that the parents are in fact distant cousins, their great-great-grandparents having had the same name and having come from the same small town in another country.

Associations of genetic diseases within the same family may also be the root cause of certain defects. Nevertheless, a mother who delivers a child with Down syndrome may feel compelled to blame a powerful medication she was taking in early pregnancy. Only recently have we recognized that some genetic mechanism is probably more likely to be involved. For instance, her family history of Alzheimer disease may be the most significant factor: there is a higher frequency of Down syndrome within families in which at least one member has Alzheimer disease (see chapters 8 and 32).

Because so many factors exist that confound the analysis of whether a drug caused a defect, we rely on the science of *epidemiology* to determine whether any cause-and-effect relationships exist between defects and drugs or toxins. Epidemiological studies involve careful investigation of large groups, or populations, with detailed attention being paid to matching characteristics. These methods, when correctly used, have served especially well in debunking alleged associations between drugs and birth defects. Consequently, of the thousands of drugs available, only the relative few listed in table 16 (pp. 204–205) are definitely related to birth defects. Elaboration on some of these drugs is important.

Drugs for Epilepsy

Pregnant women who are being medicated for epilepsy run a risk of bearing defective children that is two to three times greater than that of nonepileptics. The risks — in the 7- to 10-percent range — include such defects as cleft lip and cleft palate; microcephaly, possibly with mental retardation; genital and kidney abnormalities; poor growth; and underdeveloped nails. Typical facial features have included a short nose with a broad, depressed bridge, tissue folds covering the inner corners of the eyes, widely spaced eyes, droopy eyelids, crossed eyes, a wide mouth, and a short, wide neck. Certain drugs of the chemical hydantoin group — for example, Dilantin (phenytoin) — have been implicated. Other anticonvulsant medications, such as trimethadione, appear to have additional typical effects, including V-shaped eyebrows, low-set, backward-sloping ears, and severe defects of the windpipe (trachea), voice box (larynx), and gullet (esophagus). The anticonvulsant valproic acid (for example, Depakene) has been causally linked to spina bifida. Reports suggest that about 1 in 100 epileptic women who take this medication during pregnancy will have a child with this severe defect — a risk ten times higher than nonepileptics. Various other defects — some serious — may result from these or other anticonvulsant medications.

Such problems are important since about 1 in every 200 people in Western countries suffers from some form of epilepsy. A remarkable 2 to 5 percent of the population have a seizure at some time. In 1981, 1.3 percent of all children born with birth defects had epileptic mothers. Moreover, such women were also found to be three times more likely to deliver low-birth-weight babies, who have lower levels of survival.

It is entirely possible that, at least in some of these instances, the fundamental cause of the epilepsy itself may be the critical factor leading to birth defects. It is known, for example, that epileptics have a higher rate of cleft palates and cleft lips than the general population, which suggests that there may be some inherited tendency, *apart* from the effects of drugs, to produce these defects. Certain inherited biochemical mechanisms may also account for some of the defects in the children of epileptic women. Enzymes necessary for the body to deal with such drugs may be deficient in some offspring, which would result in more overt defects. Enzyme differences may be the explanation in one very illustrative case: An epileptic woman, having had intercourse with two men in quick succession, conceived twins, who were later shown to be the offspring of two different men. Only *one* of the twins showed the defects associated with anticonvulsant medication — the so-called fetal hydantoin syndrome.

Other inherited biochemical mechanisms that may be slow to dismantle a drug in the body could possibly lie at the root of defects seen in only some offspring of epileptic women. There is reason to believe that we will soon be able to identify which women are especially susceptible to specific medications.

During pregnancy, the frequency and severity of actual seizures may damage the fetus by causing a lack of oxygen or by markedly raising the body temperature. Certain anticonvulsant drugs also have an effect on the body's immune system. This effect is probably related to an epileptic's known increased risk of leukemia or lymphoma, both of which are cancers of the blood-forming cells. Certain other tumors, such as neuroblastoma, have also been reported in infants of mothers taking the hydantoin-type medications.

If you are a female epileptic, it would seem prudent to consider seriously with your doctor which medication is best continued in the first three months of pregnancy. For seizure control, there may be no choice; it would, indeed, be extremely unwise to endanger your life by not taking a critically important medicine. Remember, with the latest prenatal tests and ultrasound studies (see chapter 23), the vast majority of serious structural defects can be detected early enough in pregnancy to allow the option of elective abortion.

Anticoagulants

Anticoagulants — drugs to stop blood from clotting — may be taken by women in their childbearing years for thrombosis (clot-

Table 16. Medications Conclusively Shown to Cause Birth Defects

Medication	Uses	Common Birth Defects and Consequences
Anticonvulsants (e.g., Dilantin [phenytoin], Tegretol, trimethadione, phenobarbital, valproic acid [Depakene])	Treatment of epilepsy	Multiple and variable serious birth defects; mental retardation; for valproic acid, spina bifida
Anticoagulants (e.g., dicumarol, or warfarin)	Prevention of blood clots	Underdevelopment of cartilage and bone (especially in nose); blindness; mental retardation; possible fetal or newborn bleeding
Anticancer drugs (e.g., busulfan, chlorambucil, cyclophosphamide, 6-mercaptopurine, aminopterin, methotrexate)	Treatment of cancer	Fetal death; multiple and variable serious birth defects
Antithyroid preparations (e.g., propylthiouracil, methimazole, radioactive iodine)	Treatment of hyper-thyroidism	Enlargement or destruction of fetal thyroid gland
Sex hormones (e.g., progesterone, estrogen, Progestoral, diethylstilbestrol [DES], methyltestosterone, Norlutin)	Oral contraceptives; formerly, prevention of miscarriage	Masculinization of a female fetus; for DES, vaginal cancers in adolescence, miscarriage, tubal pregnancy, and structural abnormalities of the genital tract
Antibiotics Tetracycline	Treatment of infections	Pigmentation of teeth and underdevelopment of enamel
Streptomycin	Treatment of infections (e.g., tuberculosis)	Deafness
Miscellaneous drugs Antimalarial medicines (e.g., quinine, chloroquine)	Treatment of malaria and certain heart rhythm disturbances	Miscarriage; marked reduction in the platelet count in the newborn; possible deafness or blindness
Thalidomide	Formerly used overseas for treatment of nausea and vomiting during pregnancy	Absent to profoundly de-formed limbs; multiple and variable serious birth defects

Table 16 (continued)

Medication	Uses	Common Birth Defects and Consequences
Antibiotics (cont.)		
Accutane (isotretinoin or other derivatives of viatamin A), etretinate	Treatment of serious acne or psoriasis	Hydrocephalus or micro-cephaly; deformed or absent ears; heart disease; abnormally small eyes; cleft palate; mental retardation; other defects and miscarriages
Vitamin A in excess	Harmful diet fads	Miscarriage; defects of head, face, brain, spine, and urinary tract
Penicillamine	Ridding body of excess copper	Hyperelastic skin

NOTE: Check with your physician about any medication with a different name that may be similar to those listed here.

ting) in their varicose veins. Less frequently, anticoagulants are needed by women who have undergone heart-valve replacement. Coumarin-type anticoagulants, such as Warfarin, taken during the first three months of pregnancy are associated with a characteristic sequence of birth defects that includes poor development of the nose and other cartilage, mental retardation, blindness, and possibly other serious defects. The risk of serious defects resulting from anticoagulants taken in the first trimester approximates 1 in 6. Further, about 1 in 6 such pregnancies can be expected to end in miscarriage or stillbirth. Despite this powerful medication, two-thirds of exposed pregnancies end with an apparently normal liveborn infant.

When the anticoagulant heparin is used in the first trimester, about 1 in 8 pregnancies has been noted to end in stillbirth and about 1 in 5 in premature delivery (with a third of these premature babies not surviving). Only two-thirds of pregnancies involving heparin exposure yield a normal outcome.

Clearly, any woman who finds herself pregnant while taking

anticoagulants should carefully consider the above risks and decide whether or not to continue the pregnancy.

Anticancer Drugs

In the late 1940s, a drug called methotrexate was used in women with tuberculosis or cancer to induce abortion. It was discontinued after major defects of the aborted fetuses were noted. Subsequently, however, at least eight mothers are on record as having taken this drug to induce abortion. Some of these mothers failed to abort and gave birth to offspring with serious malformations, including severe bony defects of the skull (including some missing bones), markedly malformed ears located low down in the neck, severe growth retardation, absent digits, and odd-looking faces. This drug continues to be used for both the treatment of leukemia and also for psoriasis. There are a number of other powerful drugs used to treat cancer, some of which have been associated with birth defects. Since it is uncertain whether any of these chemical agents cause a specific pattern of defects — that is, a recognizable syndrome — some of the disabilities may have occurred purely by chance. Nevertheless, there is an abiding need for women to be certain that they are not pregnant when taking these and other powerful medications.

Drugs for Acne and Psoriasis

The most recent discovery of a drug that causes grave defects in the developing fetus dates back only to 1983, following the marketing of Accutane (isotretinoin). While this drug proved to be a boon for individuals with very severe acne that was resistant to all other treatments, its danger to the embryo and fetus was recognized from experiments in animals. Advice against use in pregnancy was therefore explicit. Nevertheless, because of undetected pregnancies or the lending of this prescription "miracle cure" to friends (who became exposed while pregnant), over fifty infants have been born with severe birth defects linked to Accutane. Brain damage, heart defects, tiny and deformed ears, facial paralysis, cleft lip or cleft palate, and other defects have been reported. It seems that the critical period for exposure is between two and five weeks following conception. A similar drug, etretinate, which is used for the treatment of psoriasis, has also resulted in similar severe birth defects following use in early pregnancy. Particularly important among those defects were three cases of spina bifida. A substantial increase in miscarriages has been noted for both drugs.

Once again, use of these medications during pregnancy is absolutely to be avoided. An unplanned pregnancy that occurs while the mother is taking these drugs requires careful consideration of the options, which include abortion, if exposure took place between the second and the fifth week of pregnancy.

Both isotretinoin and etretinate are directly derived from vitamin A. Not surprisingly, megadoses of vitamin A taken by health faddists have resulted in at least eighteen births of babies with a whole range of severe, if not catastrophic, birth defects. It cannot be sufficiently emphasized that normal vitamin dosage should not be exceeded during pregnancy.

Sex Hormones

Sex hormones such as progesterone or estrogen might have been taken by a pregnant woman in one of three situations. First, she may have been taking them as oral contraceptives; if it failed, she would have become pregnant without realizing it, thus exposing the embryo. Second, some women formerly took sex hormones as a test (it is no longer in use) to determine if they were in fact pregnant, thereby simultaneously exposing the fetus to these drugs. Third, a woman who has had recurrent miscarriages might have been given sex hormones as a preventive measure. Today, sex hormones are only occasionally used for this purpose; once again, however, the fetus may inadvertently be damaged.

The main risk from taking these hormones in early pregnancy is that they may interfere with the development of the baby's genitals. For example, a female fetus exposed to these hormones in the first trimester of pregnancy may be born with masculinized, or "ambiguous," genitals. A clitoris may resemble a penis and the labia may have the shape of the male sac, or scrotum. While there have been various claims about progesterone/estrogen hormones causing heart, limb, and even esophageal defects, clear evidence of any such causal associations is still lacking.

In contrast, however, there are real risks from the sex hormone diethylstilbestrol (DES). This female sex hormone has not only been linked to rare genital-tract cancers in adolescent and young women, but also to structural abnormalities of the genital tract. Moreover, this hormone has also been associated with an increased rate of miscarriage as well as ectopic (tubal) pregnancy, and even with interference of normal fertility. Fortunately, most women exposed to DES prenatally have escaped serious problems. A female whose mother took DES during pregnancy fortunately has a low future risk of developing genital-tract cancer — probably in the range between 1 in 1,000 to 1 in 10,000.

Fertility drugs such as Clomid are commonly used to stimulate ovulation. Suspicions that the use of such drugs is associated with birth defects have now been negated.

Prudence would dictate that women not take sex hormones in early pregnancy, for any reason. Fortunately, even if taken inadvertently, the risks for any serious problem are extremely small.

A Drug That Rids the Body of Copper

A certain hereditary disorder, called Wilson disease, is characterized by high levels of copper in body tissues, especially in the liver and certain areas of the brain. Problems can be prevented by ridding the body of copper by using the drug penicillamine. This drug also has other actions, which include putting a damper on the immune system and lessening the effects of rheumatoid arthritis, a connective-tissue disorder (scleroderma), and some other disorders.

At least five infants have been born with hyperelastic skin after their mothers took penicillamine during pregnancy. Three other infants suffered brain damage, which may also be drug-related. While insufficient information is available to determine how risky penicillamine really is during pregnancy, its use should be avoided where possible. Fortunately, the vast majority of pregnancies in which penicillamine has been taken have ended with perfectly healthy children. The life and health of a mother with Wilson disease could be threatened by discontinuing penicillamine, and this would have to be weighed most carefully in decisions about pregnancy and whether or not the medication should be taken.

Antibiotics

Fortunately, as a class of heavily used drugs, antibiotics only rarely pose risks for the developing human fetus. Some tetracycline drugs taken in pregnancy may cause permanent staining of the child's first teeth, but usually not the second teeth, which emerge many years later. As little as one gram each day taken for three days during the third trimester can produce yellow staining of the deciduous (first) teeth. Tetracycline-stained teeth may appear gray-brown, yellow, or brown. This antibiotic taken after sixteen weeks of pregnancy may result in stained teeth.

When given at *any time* during pregnancy, the antibiotic strep-tomycin, which is used especially in the treatment of tuberculosis, may cause fetal nerve damage that results in deafness. This is not particularly surprising, since this drug may have a similar effect on the nerves that conduct hearing if given to an infant. Most exposures occur inadvertently when a woman fails to recognize that she has become pregnant. Pregnancy testing should always be done before taking a medication or undergoing a test that may harm the embryo or fetus.

Drugs and the Thyroid Gland

Disorders of the thyroid gland making it overactive (hyper-thyroidism) or underactive (hypothyroidism) affect about 0.82 percent of all women of childbearing age. Neither of these thyroid disorders causes birth defects, although there is a small risk, probably less than 1 percent, that such mothers will have a child with a chromosomal trisomy, such as Down syndrome. The real risks associated with these conditions come from the tests used to diagnose them and the drugs used to treat them during preg-nancy.

Radioactive iodine may be used to treat an overactive thyroid gland, and if pregnancy has begun without its being noticed, the fetal thyroid gland can be destroyed. Other medications could result in gross enlargement of the thyroid gland, so much as to even obstruct vaginal delivery in extreme cases. In such cases, the enlarged thyroid may even compress the windpipe and kill the infant. A woman who might be pregnant should be sure to have a sensitive *blood* test for pregnancy (yielding a positive result within two weeks of conception) before undertaking radioactive-iodine treatment. A baby born with its thyroid gland destroyed by radioactive iodine would exist as a cretin unless treated with thyroid hormone taken daily for life. Fortunately, thyroid-hormone supplements for treating hypothyroidism are safe.

Excessive intake of iodine during pregnancy, as salts or in cough or asthma medicines, may also result in enlargement of the fetal thyroid gland.

Drugs That Are Suspected But Not Proved as a Cause of Birth Defects

Many common drugs, including aspirin, caffeine, and Valium, have at one time or another been suspected as a cause of birth defects. For the vast majority of these drugs, the evidence presented has been weak. Comments about a few that are frequent causes for worry might be helpful.

TRANQUILIZERS

Various tranquilizers have long been suspected, but never proved, as causes of birth defects. Indeed, obtaining such proof is extremely difficult, as mentioned earlier. One of these drugs, Lithium, is suspected as a cause of birth defects, particularly of the heart and blood vessels, and especially affecting the right side of the heart. Lithium taken near term may produce severe toxic signs in the newborn — manifestations that are usually reversible. These may include cardiac irregularities, enlargement of the heart, thyroid, or liver, shock, intestinal bleeding, and other problems. Clearly, Lithium should be avoided during pregnancy where possible, but especially during the first three months and the last few weeks of pregnancy.

Only rarely is there an urgent need to take tranquilizers in pregnancy, and they are best avoided even when pregnancy may only be imminent.

BENDECTIN

Bendectin, a drug for the treatment of the nausea and vomiting of early pregnancy, has been in use for over a quarter of a century and has been taken by over 33 million women. Notwithstanding its enormous usage, poorly supported product-liability claims that Bendectin caused serious birth defects led the manufacturer to withdraw it from the marketplace in 1983. The overwhelming cost of insurance and the associated legal costs forced this decision.

The claims had been brought by mothers who had borne children with limb and heart defects as well as many other abnormalities. The vast majority of major studies that have been done to investigate any possible association of Bendectin with birth defects have revealed no association. Sadly, many mothers who took Bendectin in early pregnancy and who had a child with severe birth defects were not aware that they had an initial risk of 2 to 3 percent of having a child with a structural birth defect, and that in about two-thirds of such cases no clues to the cause have been discovered.

VAGINAL SPERMICIDES

A host of studies have tried to determine whether or not vaginal spermicides cause birth defects, but they have yielded conflicting conclusions. No definite causal relationship between the use (usually inadvertent) of these preparations in early pregnancy and birth defects has been proved. The latest studies negate any association.

MARIJUANA, COCAINE, AND OTHER STREET DRUGS

Trying to determine the cause of a birth defect in a drug abuser is especially difficult. Besides likely multiple drug use and possible chemical interactions between various substances, a range of health problems invariably complicate the picture. Even when only one drug has allegedly been taken, the presence of other contaminants cannot be ruled out. While such drugs as marijuana, heroin, LSD, and methadone have not been shown to cause birth defects, they are certainly known to endanger the fetus in other ways. Expectant mothers who smoke marijuana are mostly likely to have low-birth-weight babies who in turn have a multitude of complications that threaten their lives.

The addictive drugs, such as heroin or methadone, not only increase the risk of low birth weight but also greatly increase the risk of stillbirth. Surviving children are often born with an addiction to these drugs and experience a drastic withdrawal shortly after birth, although symptoms may sometimes last for as long as six to eight weeks. It is very distressing to watch withdrawal in a newborn. The infant is jittery and twitches, sometimes violently enough to cause skin abrasions. Limbs may become too rigid to flex, and diarrhea and vomiting may punctuate the baby's anguished, high-pitched cries. Fever and convulsions may depress breathing, which may in fact stop intermittently. Obviously, anyone who takes these drugs during pregnancy seriously harms herself and, worse still, her baby.

A clearer understanding of the effects of cocaine on the fetus during pregnancy and on the baby immediately after birth is only now beginning to emerge. One commonly encountered difficulty in ascribing any particular consequence to cocaine use is, once again, the fact that drug-abusing mothers frequently use other stimulants.

Mothers who use cocaine during their pregnancy can expect a whole range of problems and complications. To begin with, the rates of miscarriage and stillbirth are increased. Increased frequencies of kidney, urinary-tract, skull, and nervous-system birth defects have been reported. Many observers have documented an increased frequency of premature babies — in addition to infants who, although born on time, fail to grow properly while in the womb, have lower birth weights, are shorter, and have smaller heads. These cocaine-abused newborns regularly show abnormal sleep patterns, poor feeding, tremors, periods of not breathing, stiffness, and inadequate function of the eyes. Electroencephalograms (electrical tracing of brain waves) may remain abnormal months after birth (and perhaps longer), as may the eye problems. Indeed, a few such babies have had strokes in the earliest days of their lives! Bleeding behind the placenta and possible fetal distress during labor are two other recently recognized problems.

The outlook for long-term development of cocaine-abused babies is still to be documented. However, we already know that pregnant mothers who use the drug are "drug pushers": their own babies are the helpless recipients!

Certain drugs are clearly hazardous to the developing fetus. Proving that a drug actually causes a birth defect is exceedingly difficult. Since so many birth defects occur without a recognizable cause, it would be wise to avoid all unnecessary drugs while trying to become pregnant, as well as during pregnancy.

16

Maternal Illnesses and Habits Harmful to the Fetus

The mother constitutes the real immediate environment of the developing fetus. Therefore, her health and habits are critical to its survival and normal development. Problems arising from within the womb itself may also be harmful. In the previous chapters, we have seen that the origins of birth defects are either hereditary or acquired. In addition to environmental insults such as infection, toxins, X rays, and drugs, causes of acquired birth defects include the effects of maternal illnesses or disorders as well as specific habits. The overall effects of the factors discussed in this chapter rank as the most important, yet insufficiently recognized, conditions that interrupt normal fetal development.

Maternal Illnesses

DIABETES MELLITUS

Women with insulin-dependent diabetes have a two- to threefold increased risk — a risk of 7 to 10 percent — of having a child with a major birth defect, mental retardation, or a genetic disorder. Knowledge linking diabetes with birth defects extends back a century! We now know that the poorer the control of blood sugar immediately *before* conception and during the first three months of pregnancy, the more frequent the incidence of birth defects, and the more severe they are. Malformations of the spine (spina bifida), brain (anencephaly), and heart may occur from three to ten times more often than in nondiabetic women. Indeed, one defect of the lower spine and body may even result in the fusion of the lower limbs into a single limb described as a "mermaid deformity." This awesome defect, called sirenomelia, or defects of the lower spine and legs known as caudal dysplasia are so typical as to raise questions about whether the mother has undiagnosed diabetes if they turn up unexpectedly in a newborn.

Diabetes is common, affecting over 5 percent of the U.S. population — about 12 million people. The various types and causes are discussed in chapter 29. Suffice it to say here that insulin-dependent diabetes of the mother poses serious risks of birth defects. Mothers whose onset of diabetes was later than childhood and who did not require insulin to treat the disease do not appear to have an increased risk of having children with birth defects. The children of diabetic fathers also do not appear to have an increased frequency of birth defects. Mothers who become diabetic during pregnancy *probably* have no increased risk either. Oddly enough, women whose body chemistry suggests that they *may become* diabetic (that is, if tests show impaired glucose tolerance) are thought by some researchers to have a slightly increased risk of producing malformed infants. The available data, however, do not allow for any safe conclusion yet. Certainly, those women with the earliest onset, longest duration, and multiple complications of insulin-dependent diabetes clearly have the highest risks for their own health, for problems in pregnancy, and for serious birth defects.

There is much more, unrelated to birth defects, that insulin-dependent pregnant diabetic women should know. I have written more about this subject in *How to Have the Healthiest Baby You Can.* An increased stillbirth rate (4 percent) and large, heavy babies are two additional extremely important matters to be reckoned with in pregnancy with diabetes. Moreover, the infant of the diabetic mother may also be endangered. Although frequently oversized, these babies may be immature and face the other risks of prematurity: breathing problems (the respiratory-distress syndrome) and low blood sugar (hypoglycemia).

For these and other reasons, all women with insulin-dependent diabetes should seek and receive very careful medical attention, starting well before they *plan* to conceive. They should all receive genetic counseling, be advised of their risks of birth defects and other problems, and be offered the serum alpha-fetoprotein screening test (described in chapter 20), a detailed (level-II) ultrasound study, and the option of amniocentesis so that further prenatal-diagnostic studies can be performed (see chapter 21). Pregnancy should not be undertaken lightly by any woman with insulin-dependent diabetes. We do know, however, that with excellent control of diabetes beginning before conception, the frequency and severity of birth defects can be markedly diminished.

EPILEPSY

Epilepsy is discussed in chapter 15. I should emphasize here only that no expectant mother should stop her anticonvulsant medications without consulting her doctor and that when a pregnancy is *planned*, there are often opportunities to take medications that are less hazardous to the fetus.

SYSTEMIC LUPUS ERYTHEMATOSUS (SLE)

SLE is a chronic disorder that affects between half to three-quarters of a million people in the United States alone. This disease is remarkable in that it may affect virtually any or all tissues and organs, including skin, bones, joints, inner and outer linings of internal organs, kidneys, heart, intestinal tract, and the brain. A remarkable and unique disturbance in the body's immune system leads to this disorder through complex mechanisms in which the body's own defense system turns and attacks itself. An astonishing array of symptoms and signs may occur at various times, and may even disappear — sometimes for good.

Curiously, SLE is much more common among females, although some rare families have been described in which males preponderate. There is also a higher frequency, and indeed severity, of lupus among blacks, Chinese and other Asians, and American Indians.

Whatever triggers the body's immune system to turn upon itself remains a mystery. It is clear, however, that whatever those factors are — ultraviolet radiation, drugs and chemicals, and infection or abnormalities in the immune system are under intense investigation — genetic factors always have some role. Indeed, this disorder today is considered to be one in which an interaction occurs between multiple genes and one or more environmental factors. The occurrence of lupus in both siblings is remarkably more frequent in identical twins (57 percent) than in nonidentical twins (5 percent). Recent studies also show that the frequency among the brothers and sisters of an affected person ranges between 9 and 12 percent. In general, one parent with SLE is likely to have a 3- to 5-percent risk of having a child who might ultimately develop this disorder. Major progress using the new molecular techniques discussed earlier (chapter 8) will undoubtedly help solve the mystery of the causation of SLE and thereby provide a new and effective treatment.

While women with SLE appear not to have a problem in becoming pregnant, their risk of miscarriage, as well as of stillbirth, is considerably higher than others without this disorder. SLE does not usually cause birth defects, except for one rather characteristic condition called congenital heart block, which results from the blocked transmission of nerve impulses to the heart muscle. While affected children are otherwise normal, they may suffer fainting spells or dizziness, or, because they are subject to sudden death, may require the implantation of a permanent pacemaker.

This disorder can be diagnosed prenatally through electronic fetal monitoring. Such monitoring is important in order to facilitate early diagnosis and treatment.

FEVER

There is no doubt that induced high body temperatures in various animals during pregnancy results in serious birth defects. No such certainty exists for humans, mainly because their fevers are almost invariably due to concomitant infections. Some studies have clearly documented an increased frequency of serious birth defects, including eye malformations and anencephaly or spina bifida, following high fever in the early weeks of pregnancy. The problem is to distinguish between the effects of (1) a virus (or other organism) that causes fever, (2) the medications to treat it, and (3) the fever itself.

At one point, sauna bathing was suspected of causing certain birth defects. That idea petered out when it was realized that the sauna was a national pastime in Finland, where the rates of spina bifida are among the lowest in the world.

While academicians continue to wrestle with the question of whether fever or an infectious or other agent is the real cause of birth defects that follow high body temperatures in pregnancy, we know enough to be concerned and advise a detailed ultrasound study around the sixteenth week of pregnancy whenever high fever (above 102 degrees Fahrenheit) has occurred in the mother during the first eight weeks after conception. Normal test results will, of course, not exclude all possibilities of damage to the fetus, but could provide some reassurance.

IODINE DEFICIENCY

Large numbers of people in the world still live in iodine-deficient areas. Following the introduction of iodides into common table salt, this is no longer a problem in the United States. However, in certain areas of South America, Europe, and Africa, iodine deficiency remains a serious problem.

The child of a mother with iodine deficiency usually has physical and mental retardation, may be a deaf-mute, and may have an enlarged thyroid gland, among other features.

The problem of iodine deficiency is best and simply remedied before conception by providing a mother with ample amounts of iodine in her diet.

MISCELLANEOUS GENETIC CONDITIONS

Any genetic condition compatible with childbearing may complicate pregnancy, besides risking transmission of the same disease. Only a few common conditions are discussed here. In a general sense, if such a disorder markedly impairs the health of an expectant mother, miscarriage occurs much more commonly than in mothers without such problems. Mothers with cystic fibrosis, for example, clearly risk their own health and even their lives by becoming pregnant, since their respiratory problems invariably are aggravated as pregnancy progresses. Lung infections, heart failure, and diabetes may threaten their lives in pregnancy. There is also a higher risk of stillbirth. Purely because of the risk to the mother's health, pregnancy is mostly regarded as inadvisable if a woman has cystic fibrosis, although there are some exceptions.

On average, mothers with sickle-cell anemia lose close to a third of their offspring during the final weeks of pregnancy, during labor, or soon after birth. Close to 45 percent of their babies have low birth weights.

For over thirty years, we have known about the devastating effect of maternal phenylketonuria, or PKU (described earlier). Routine screening allows detection in the newborn and institution of a special, phenylalanine-free diet that prevents the development of mental retardation. Consequently, girls with phenylketonuria now have grown up and begun childbearing. By adulthood, most of these individuals have discontinued their unpalatable special diet and therefore have high concentrations of dangerous body chemicals that will damage the developing fetal brain. Hence, undertaking pregnancy without going back on the special diet will result in almost 100 percent of the offspring having mental retardation, microcephaly, heart defects, and, frequently, low birth weight. Affected mothers also have an increased likelihood of miscarriage. Women with phenylketonuria may plan pregnancies while on their special diet, and as long as they stick to it faithfully, their chances of having a healthy child may turn out to be similar to other women. More experience is needed before reliable predictions can be made, however.

Myotonic muscular dystrophy is a dominantly inherited disorder with multiple effects. Sustained contraction of muscles is the typical finding — resulting, for example, in an affected person having difficulty letting go of a handshake. Other features include frontal balding, cataracts, diabetes, atrophy of the testes or ovaries, and difficulty in swallowing. All muscles, including the limbs, the heart, and the internal organs, may be affected. Women with myotonic muscular dystrophy are frequently infertile and rarely become pregnant. If they do succeed, they experience three main problems. First, and most important, pregnancy makes their disease worse. Second, they suffer an increased frequency of miscarriage and premature labor. Third, all stages of their labor have been noted to be abnormal — prolonged, with inadequate progress because of weak muscles and more frequent hemorrhage. Besides the 50-percent risk of having an affected child, deaths of newborns, who may or may not have visible birth defects, are also more frequent. For reasons that are still uncertain, babies born with myotonic muscular dystrophy have greater problems when their own mothers (rather than their fathers) have had the same disorder.

Occasionally women with Turner syndrome (see chapter 3) may become pregnant. It is likely that most are mosaics who do in fact have cells with a normal chromosome complement as well as abnormal ones. Almost one-third of the pregnancies reported in such women have resulted in miscarriage. In addition, they also have an increased risk of bearing a child with some type of chromosome defect (the exact risk is unknown). Therefore, if a woman with this disorder becomes pregnant, prenatal chromosome studies via amniocentesis are usually recommended.

Women with Marfan syndrome (discussed earlier) have a higher rate of early miscarriage than other women do; they also deliver low-birth-weight infants more frequently. The most serious concern for an affected woman who becomes pregnant is the risk that her aorta could rupture. Careful study by a cardiologist and consultation with both a geneticist and obstetrician should enable most such women to bear children without threat to their own lives.

Hazardous Habits

Since a mother constitutes the immediate environment for her fetus, any hazardous exposure she risks could potentially interfere with the baby's normal development. While maternal illness might be unavoidable during pregnancy, other threats to the fetus can be controlled. Women who plan to conceive need to be educated about the importance of not becoming pregnant while pursuing certain destructive habits. Two of the most important are drinking alcohol and smoking tobacco. (The problems with illicit drugs are discussed in chapter 15.)

ALCOHOL IN PREGNANCY

Alcohol ingestion during pregnancy is today regarded as perhaps the third-most-common cause of mental retardation (after Down syndrome and the fragile-X syndrome). The mechanism by which alcohol wreaks havoc on the fetal body remains uncertain. The alcoholic woman may lack various essential dietary factors (such as folic acid), may be zinc-deficient, may suffer from infections more frequently, or may poison the fetus by passing on substances within the alcohol. Whatever the case, major indulgence in alcohol or even a single binge in early pregnancy may be sufficient to cause irreparable harm to the fetus. No one in fact knows the minimum safe dose of alcohol in pregnancy. The amount is likely to vary, since each person's body processes alcohol at a different rate; some individuals are more resistant to the effects of alcohol than others. An occasional glass of wine is not thought to be harmful, but, obviously, total abstinence is safer. Moreover, while the first three months are especially critical — this is when all vital organs are forming — the fetus continues to remain at risk for most of the pregnancy.

Pregnant women who drink hard liquor almost daily have a high risk of delivering a child with a combination of birth defects known collectively as the *fetal alcohol syndrome*. The major features of this remarkably common disorder — it affects between 1 in 600 and 1 in 1,000 babies — include mental retardation, growth failure, microcephaly, distinctive facial features, and other anomalies of the heart, genitals, bones, and joints. The faces of children with the fetal alcohol syndrome may not necessarily be diagnostic of the disorder. However, affected individuals typically have smaller eyes, folds covering the inner corners of the eyes, a small jaw, a flat midface, and possibly some minor abnormalities of the ears. Cleft lip or cleft palate and even spina bifida might occur. Later in childhood, hyperactivity, attention-deficit disorders, and learning difficulties may become obvious.

Women who have been drinking significantly during the first trimester of pregnancy should be apprised of the fact that they have about a 44-percent risk of having a child with mental retardation, possibly coupled with some of the other abnormalities mentioned above. The option to terminate such a pregnancy should be explained to such women.

SMOKING DURING PREGNANCY

Maternal smoking during pregnancy exposes the fetus to dangerous concentrations of the products of tobacco smoke. Recent estimates show that about 1 in 4 women smokers continue smoking during pregnancy, despite well-known serious consequences such as lower-birth-weight babies, higher rates of miscarriage, and increased frequency of death during labor or within twenty-four hours of birth, more frequent crib deaths, more childhood respiratory illnesses, and more frequent learning and behavioral problems in children whose mothers smoked while pregnant. Some researchers have raised the question of whether the exposure of children to their smoking mothers *after* birth contributes to some of these problems, or whether there is something about the child-rearing behavior of such mothers that leads to some of these problems.

(Parenthetically, a male who smokes tends to have a reduced sperm count, reduced sperm motility, and an increased frequency of abnormal sperm. These effects may be enough to contribute to a couple's infertility.)

There is certainly enough known about the dangers of smoking to make quitting the habit advisable — especially if a pregnancy is being planned.

Deformed Versus Malformed

Failure of a body part to form normally occurs when mechanisms of development go awry in the early weeks of pregnancy. Such malformations may be genetic in origin or may be induced by the many environmental factors discussed in this and the preceding chapters. Once normally formed, a body part may undergo change during the course of pregnancy from pressures or other factors that deform, disrupt, or damage. Such deformations are *not* genetic and would recur in a subsequent pregnancy only if some persistent abnormality of the womb exists. For example, pressure may be exerted upon one leg of the fetus because of an abnormally shaped womb, causing the baby to be born with a *deformed* — not *malformed* — leg. Clubfoot could be such an example. Pressure on the head or face could result in an asymmetric skull or distorted face.

Once in every few thousand pregnancies, a strand of connective tissue might stretch across the womb like a rubber band, ensnaring one of the baby's fingers or limbs. Occasionally in such cases a child may be born with one or more digits amputated; soon after birth, the placenta is delivered, followed by the digits. This mechanism is thought to result from a tear of the amniotic sac that envelops the fetus during pregnancy. The amniotic band may sometimes even stretch across the face of the developing fetus, disrupting the tissue and even damaging the skull or brain. Detection of this amniotic-band syndrome, as it is called, is only occasionally possible by detailed ultrasound study; its presence, of course, is usually not suspected.

Interference with the blood supply — for example, to an arm of the fetus — may also result in a deformation. A clot could form, or a shower of clots could be released from the placenta, thus blocking the main or smaller blood vessels leading to the developing arm. Deprived of their blood supply, the fingers or the entire arm may die, disappear, or shrivel up. These rare events — which, again, are not genetic — arise for reasons unknown, and do not usually happen again.

17

Genetic Counseling

Genetic counseling is probably as old as medicine itself. You will recall (from chapter 7) the accurate knowledge about hemophilia found in the Talmud. What is new about genetic counseling is the continually increasing availability of options that never existed before. There now exists medical technology to prevent birth defects through prenatal diagnosis, to detect carriers of genetic disease though they have no obvious symptoms, to diagnose certain genetic disorders at birth, to provide specific treatments in some cases, and even to predict inherited disorders in early pregnancy, many decades before any signs appear. These advances present you with vital options that you should fully understand. Remember, *by not knowing, you do not remove the chances, you remove the choices.*

What It's All About

Genetic counseling is communication — mostly between a doctor, prospective parents, and other affected or concerned persons — about the likelihood of hereditary disorders. The counselor's aim is to provide those seeking information with the fullest understanding of the disease in question and its implications, as well as the options available. The counseling process aims to help families through their problems, their decision making, their possible anguish, and their adjustments. The goal is definitely not to make decisions for them.

The primary hope is that counseling will provide enough understanding to produce rational decisions that will serve to prevent or at least decrease the recurrence of serious genetic disease or mental retardation in a family. While the vast majority of counselors would agree with these goals and hopes in general, a variety of opinions exists about how genetic information should be obtained and conveyed.

Who Needs Genetic Counseling?

The reasons why people seek genetic counseling can be grouped in general categories. They want to know

1. if they themselves have a genetic disease;
2. if they are carriers;
3. if they run the risk of having a child affected with a particular genetic disease;
4. what the implications are if a genetic disease has already been diagnosed in partners who are planning parenthood, and what the prognosis and treatment options are;
5. what help they can get in making a decision about the options of prenatal diagnosis, selective abortion, artificial insemination by donor, or adoption; and
6. what kind of help is available for an already affected child and where it can be found.

It would make sense for couples to seek counseling before or at the time of marriage, and certainly before conception — whichever comes first! In this way, a child with birth defects would not have to be born to make parents realize how they could have perhaps avoided such a birth, or at least how they could have benefited from prenatal counseling. Not infrequently, I have consulted with young couples who are contemplating marriage. A few of them have adopted a eugenic view: they broke off their relationships after discovering their 25- to 50-percent risk of having a seriously defective child with hereditary disease. That is mate selection on a true genetic basis — a practice that I suspect is still most unusual today.

In spite of all the new knowledge, at least 90 percent of the people in the United States, West Germany, and other Western countries who really need counseling do not receive it. This may be because they themselves do not realize that a family disorder is hereditary, and their physician has failed to recognize it or neglected to refer them for proper advice. If you are worried about a particular family illness, that alone is sufficient reason to seek counseling. You will usually obtain answers and, more often than not, you will leave reassured.

Who Provides Genetic Counseling?

For the most common genetic disorders, experienced physicians may often be able to supply the needed information. For the rarer problems, consultation with a medical geneticist would be advisable, even if only for a second opinion. Very worried people may find their anxiety unrelieved after consultation with their own doctor. A physician who is sensitive to their unallayed fears will at least suggest a second opinion from a genetic specialist. Sadly enough, this seldom happens, and people remain anxious and possibly ill informed until crisis or trauma takes them in search of the necessary expert diagnosis.

Medical genetic counselors are usually pediatricians or internists who have concentrated on hereditary disorders. Since 1982, the American Board of Medical Genetics has been granting specialty certification by examination for medical geneticists. The Canadian College of Medical Geneticists has a similar certification process. While some genetic counseling is provided by geneticists with doctorates in various sciences, they are not expected to know about the details concerning the diagnosis, treatment, and prognosis of all diseases. Likewise, those with master's degrees in genetic counseling have also not been trained in medicine. Both these groups have important functions in a medical genetics team in which there is a physician. A person seeking counseling should be cognizant of the important roles played by these nonphysicians. Medical geneticists most often operate as a team within a medical school/hospital.

It is expected practice for genetic-counseling physicians and others to provide the patient with a written summary of the counseling provided. This is a useful and important technique, especially since some of the information may be both pertinent and critical to the children in the family. For example, if one or two of them are actual carriers of a particular genetic disease, then that information is critical to them when they marry and begin childbearing. This might be some twenty-five years later, when the parents who originally went for counseling and the doctor have all died. The written summary is a valuable document and should therefore be kept with other important papers, such as the will or insurance policies.

The Counselor

Just as those who seek genetic counseling have a wide variety of backgrounds, personalities, beliefs, and so on, genetic counselors represent a wide variety of these features. They therefore have many different approaches to the counseling process. The genetic counselor may be influenced by special eugenic views, religious beliefs, age, qualification, training, and personal prejudices, as well as the state of his or her physical and mental health.

Also, geneticists have differing perceptions of their obligations as counselors. The vast majority feel that their role is simply to provide as much information and understanding to the patient as is possible, making every effort to explore the disease in question fully — by reviewing such considerations as the prognosis, possible treatment, and other available options, including artificial insemination by donor, adoption, prenatal diagnosis, and carrier detection. These counselors will also discuss the implications of a genetic disease on the family and raise for consideration all issues of concern, including the religious views of the individuals, the economic implications and emotional burden of raising an affected child, the possible effects on family life and on the other children, and the suffering or problems of an affected child. Moreover, they might be able to reassure the person who fears that there will be a social stigma attached to the disorder in question. Discussion of contraception and the effect of the whole problem on the parents' sex life may be raised, as well as the option of sterilization. Some counselors might even bring up the possible economic burden of the disabled child on society.

I recommend and practice the total-communication approach. I believe that everyone has a right to know and a freedom of choice. All available information should be given out freely, the whole subject explored, and all matters of consequence discussed. The flow of information should not depend solely upon the questions asked by those who are seeking advice. How could they, after all, anticipate the eventualities?

The Authoritarian Approach

A major difference of opinion exists among a few genetic counselors with regard to method. The directive, or coercive, approach is to dictate exactly what to do or not to do, such as: do not have an abortion; have a vasectomy; arrange for artificial insemination; have a tubal ligation; adopt children; practice total contraception; and so on.

The danger of the directive approach is the insinuation by the counselor of his or her own religious, racial, or eugenic beliefs (or other dictates of conscience) into the counseling process. I have repeatedly seen physicians of certain religious persuasions counsel their patients to avoid prenatal testing and selective abortion. Some of these physicians argue, of course, that their advice is in the best interests of the family. Or, they may hold that their authoritarian approach is best because so many parents are simply unable to comprehend all the factors and therefore cannot assess their risks and make the right decisions on their own.

Such an approach is, I think, to be condemned, not only on the grounds of counselor prejudice, but because it constitutes a moral affront to individual privacy. It is obviously a very personal matter whether you decide to have children or not, or to abort or not, and the doctor's role, in my opinion and that of the overwhelming majority of geneticists, is to provide you with all the information you could possibly need to make a balanced, rational decision on your own.

The noncoercive approach is not always easy, especially when the desire of a counselor to decrease the frequency of hereditary disease runs contrary to the desires of the parents. If the disease is a serious one, it is very tempting for the counselor to make every effort, albeit undictatorially, to persuade a couple not have any more children. While this may be a valuable effort for the good of society at large and also might spare the family great unhappiness, it opens up opportunities for the counselor to insinuate his or her own religious, racial, or eugenic views into what should be impartial counseling. A poignant case, which resulted in a lawsuit, concerned a doctor who acted according to his own religious beliefs to the grave detriment of his patient (chapter 6 gives the details).

Not infrequently, patients ask the doctor to be directive and to make a specific decision for them. "What would you do if you were in our position, Doctor?" I believe that every effort should be made by physicians not to succumb to the flattery of that question, but instead to spend more time with their patients, trying to help them discover their own priorities, which govern the decisions at hand. It is neither practical nor right for a physician to extrapolate personal beliefs, life-style, and similar choices onto patients. A doctor's primary responsibility is to understand, diagnose, and provide all the information available from years of research and from personal experience.

What Should You Look for in a Genetic Counselor?

What you expect may be very different from what others are seeking. A major factor is the degree of your personal involvement. If you are inquiring about your own risks of actually having Huntington disease because it runs in your family, your anxieties might be very different from those of someone who, prior to finding a mate, was simply trying to determine if he or she was a carrier of some genetic disorder. Should you have a child with muscular dystrophy or cystic fibrosis, your desire for advice and treatment might assume the proportions of desperation. Or, in contrast, you may have adjusted to the idea of a hereditary disease in your family and, having just married, be calmly considering your options.

The decision to seek genetic counseling may be one of the most important decisions of your life. This means that extreme care should be taken in choosing a genetic counselor. You are looking, ideally, for knowledge and humanity. You want not only expertise, but someone who is sensitive, aware of your anxieties and fears, and willing to give you an empathetic hearing. The best counselors know how to deal effectively with a wide range of expectations, educational levels, personalities, religious backgrounds, and emotional orientations.

Expectations of the Counselor

While you may define your expectations, so do counselors. A very important one is that both members of a couple, when appropriate, come for counseling together. The complex issues of guilt, culpability, family prejudices, serious differences of opinion between the spouses, pervasive ignorance, and fear — to mention only a few — make this very important, if not mandatory. No letter can replace face-to-face discussion. Moreover, with emotional chaos so often present, the problem of misinterpreted information reaching the absent partner and the possible under-appreciation of the true risk situation should make it abundantly clear that a couple must come for counseling together. Then both your expectations and those of your counselor are more likely to be fulfilled.

Risks and Odds

The counselor must keep in mind a whole range of factors. Simply telling the parents that they have a risk of recurrence for anencephaly of 1.5 to 2 percent in every subsequent pregnancy does not mean that the parents have understood or grasped the degree of probability. Some parents might regard a 2-percent risk as almost no risk at all, while others might consider it so grave that they decide to have no further children. To complicate matters even further, it is recognized that the appreciation of risk and the interpretation of odds vary among basic personality types. A pessimistic parent may reach conclusions very different from those of an optimistic parent with the same odds in mind. A basic difference in the attitude toward risk between risk-takers and more cautious persons is also well known. Moreover, attitudes toward risk do change with changing moods.

CHANCE HAS NO MEMORY!
An explanation indicating that a recessive hereditary disease has a recurrence risk of 25 percent might be totally misinterpreted. It means that for *each and every* pregnancy there is a 25-percent risk of the particular disease occurring again — and that this risk does *not* change with the number of children already born, or whether they are affected or unaffected. Some parents may make the mistake of thinking that since their first child already has the disease, they will be safe in their next three pregnancies! Genetic counselors have all, unfortunately, seen the consequences of such misunderstanding by parents.

Anxiety Block and Other Obstacles

Genetic counseling in the presence of overwhelming patient anxiety never proves to be very useful, since most people have difficulty assimilating all the information communicated by a doctor in moments of stress. Recognition of this anxiety block by both doctor and patient is very important and is best handled by scheduling a return visit some weeks later to reexplore all the issues, questions, and answers discussed in the earlier session.

Besides anxiety, denial of a diagnosis by one or both parents may serve as an even greater roadblock. Mothers or fathers may refute a fatal prognosis or possibly insist that the disease in question is not even hereditary. I have even seen one man, unable to accept the information that he was a proved carrier of a genetic disease, deny that he was the father of the child!

Imperfect reception of genetic counseling is common, possibly occurring in as many as 40 to 50 percent of families. One important study confirmed what medical geneticists have known or suspected for years: religion is the principal obstacle to wide application of new advances in genetics. For various religious reasons, many couples, regardless of the risk of producing a child with a serious or fatal genetic disease, simply push ahead, having more and more children.

The counselor is also sometimes faced with difficult intrafamily problems that complicate the counseling process. You will recognize that, in your own family or in others, communication between members is often less than ideal. You may know families, as I do, in which the parents elected not to tell their children about their first deceased child, or that there is still an older brother or sister or uncle with severe mental retardation tucked away in some institution. I recall one twenty-year-old patient complaining bitterly that when she was on the point of marriage she discovered through a neighbor that she had a brother, defective from birth, still alive, and living in an institution. Her parents had hidden this information from her for the two decades of her life.

These things happen even in the very best of families. In 1987, a London newspaper disclosed that two first cousins of Queen Elizabeth, who were listed in a leading directory of the British aristocracy, had actually spent decades as patients in a hospital for the mentally retarded. One still survives. These two first cousins were the children of the queen mother's brother John, and both were severely retarded. The queen mother is said to have expressed surprise that the two were still alive decades after their admission to the institution!

Suppression of the truth (for example, failure to disclose nonpaternity) is not only morally questionable but may even have legal implications. I have seen women who chose not to inform their sisters that they, too, ran the risk of having a child with a serious sex-linked disease such as muscular dystrophy or hemophilia.

Where to Go for Genetic Counseling

I have already suggested that, ideally, genetic counseling should be obtained from a team of experts located in a major hospital/medical-school complex. There are, however, many other excellent centers that provide genetic counseling. Moreover, a remarkable number of organizations devoted to the support of education and research in specific genetic diseases have been formed over the past decade. These valuable resources are listed in appendix B. Don't forget that your own pediatrician or internist may be able to provide you with all the information you need.

18

Artificial Insemination by Donor (AID)

One of the important options offered in genetic counseling is artificial insemination by donor, or AID. Not a new idea, the basic technique was mentioned in the Talmud as far back as the second century B.C.E. Professor S. J. Behrman of the University of Michigan School of Medicine discovered the following account of its use in animal breeding dating back to 1322: An Arabian is said to have placed a wad of wool into the vagina of his mare for one night. He extracted the wool in the morning, and the following night, apparently under cover of darkness, he placed it over the nostrils of a rival's prize stallion. The stallion immediately ejaculated into a cloth held by the hopeful Arab, which he then placed into his mare's vagina — and she soon produced a foal!

The first human use of AID is thought to have occurred in England between 1776 and 1799. AID was first accomplished in the United States in 1890, and since then about 500,000 children have been born using this method. A 1988 U.S. Office of Technology Assessment report estimated that 172,000 women have AID each year, resulting in some 65,000 babies.

Who Needs AID?

INFERTILITY IN THE MALE

Couples who have tried for a few years and failed to establish a pregnancy may be candidates for AID, though it would first have to be shown that the male and not the female partner has the deficiency for it to be a sensible option. Approximately 12 percent of all married couples are infertile (some estimates suggest that this amounts to about 3 million couples in the United States), with the male partner being biologically responsible about 40 percent of the time. I will not discuss here the gynecological reasons for infertility (a *problem* conceiving), nor the medical problems involved in male infertility or sterility (the *inability* to fertilize), which I cover fully in my book *How to Have the Healthiest Baby You Can.* Suffice it to say that if the semen contains an insufficient number of sperm or the quality of the sperm is poor (for example, if they swim poorly), AID can be considered.

Ejaculation of semen in men may under some conditions occur in a "backward" fashion; that is, the semen may be released into the bladder and not out through the penis. While this so-called *retrograde ejaculation* occurs only infrequently, it may be one of the rare causes of infertility, especially in diabetes mellitus. It is known that usually ejaculation is controlled through the brain but that it may on occasion be a completely reflexive action (for example, ejaculation may often accompany judicial hanging). Techniques are available to manage cases of infertility caused by retrograde ejaculation. In 1971, Dr. Richard B. Bourne and his colleagues first described the successful and most unusual use of artificial insemination in a case of retrograde ejaculation.

> The patient was a thirty-three-year-old man who had been diabetic since the age of eight years, and blind for three years. He had married four years previously, and he and his wife complained of their inability to establish a pregnancy. So far as the man could remember, he had been unable to ejaculate since the age of twenty-five years, although his orgasm had been normal.
>
> Retrograde ejaculation was confirmed by noting the absence of semen in a condom after intercourse and then seeing sperm in the first urine specimen passed after coitus.
>
> On the day his wife was due to ovulate, a catheter was inserted through the man's penis into the bladder in order to wash it with a special sugar solution. The catheter was removed and the patient later ejaculated; the semen was obtained by having him urinate as soon as it was possible after ejaculation. The urine sample containing the semen from the bladder was then spun in a centrifuge, and the semen (which collected at the bottom of the tube) was drawn into a syringe and used to inseminate the wife.

This procedure was repeated about eight times over a six-month period without success. Finally, it was realized that the urine was too acidic — sperm do poorly in an acid medium. The man's urine was thereupon alkalinized by giving him a dose of baking soda on the evening before and the morning of insemination. This technique succeeded, and the man's wife duly became pregnant and delivered a normal baby girl.

One unusual use of artificial insemination arose during the Vietnam War, when some wives in California were inseminated by sperm from their husbands overseas.

AID FOR GENETIC DISORDERS

When the male has a serious disorder such as Huntington disease, he and his partner run a 50-percent risk in every pregnancy of having an affected child. AID can be a godsend, since the risk can be eliminated by inseminating the wife with sperm from a donor who is not a carrier of the disease. Even if the male has not yet shown any signs of the disease, but knows that he may later in life, AID is sometimes chosen in order to safeguard the offspring.

A male with hemophilia would pass the gene to all his daughters, who would then be carriers and transmit the disease to half their sons — another example of an option for AID.

AID is also important as an option when both parents are carriers of some autosomal recessive disease such as cystic fibrosis or sickle-cell anemia. It should be realized, however, that simply by chance, the anonymous donor of the sperm might coincidentally be a carrier of a particular disease for which no laboratory test yet exists. (As I noted earlier, about 1 in every 25 whites is a carrier of cystic fibrosis.) If prenatal diagnosis is possible for one of these recessive disorders, and the partners are not opposed on principle to elective abortion of a defective fetus, then AID is not necessary.

BLOOD-GROUP INCOMPATIBILITY

Parents with incompatible blood types, especially those with Rh-negative problems, are specific candidates for AID. I have noted, from time to time, women who spontaneously abort severely affected fetuses in one pregnancy after another, and who then suddenly and inexplicably have a perfectly normal baby. It takes no geneticist to guess that some of these women have solved their problem through insemination that was not artificial — as lab tests have confirmed from time to time!

Selection of the Sperm Donor

Most frequently, sperm donors are medical students or young physicians in training, who are known to be of normal intellectual capacity, never to have had a birth defect, and to be in excellent health. In 1986, the American Fertility Society published rigorous guidelines governing the use of AID. They provide an extensive list of items to be checked when screening a potential donor's medical history and performing the physical examination, as well as a variety of tests that should be administered to exclude a whole range of infectious venereal diseases. These include microorganisms that cause AIDS (acquired immune deficiency syndrome), herpes, hepatitis B, gonorrhea, syphilis, and streptococcal, chlamydial, and trichomonas infections.

Current guidelines call for repeat testing of a donor at least every six months. I suspect that every couple would prefer to know that their donor has been tested for venereal diseases just before sperm donation. It is, of course, a matter of cost.

Usually, a complete family history of the donor is taken in order to diminish the chance that he could transmit a genetic disease. Specific guidelines to screen out genetic disorders in potential sperm donors are included in the recommendations published by the American Fertility Society. Here is a modified summary of them (note that these guidelines could apply equally to women donating ova):

1. The donor should *not* have (or have had) any genetic disorder or birth defect. (Even trivial birth defects, such as three or more coffee-stain-like birthmarks — café-au-lait spots — can have ominous significance; for example, neurofibromatosis may be involved.)
2. The donor should *not* have a family history of any genetic disorder or birth defect transmissible by a single gene or one that results from the interaction of genes and environmental factors (for example, diabetes, epilepsy, psychosis, and so on).
3. The donor should *not* be a carrier of a genetic disorder known to be prevalent in his ethnic group and for which reliable carrier tests can and should be done.
4. The donor should *not* carry a chromosome abnormality. (About 0.2 percent of people do. Chromosome tests are expensive and couples need to decide if they wish to pay to have the donor checked.)
5. The donor should be under forty years of age.
6. The donor should have Rh-negative blood if the prospective mother does.

7. The donor should *not* be used if he has a family history (involving parents, siblings, aunts, uncles, first cousins, or grandparents) in which there is or was major psychosis, epilepsy, diabetes, or *early* heart disease.
8. A permanent, confidential record of the donor's medical data should be kept. (Where necessary, important medical information can be provided *anonymously* to recipients. Moreover, if a genetic disease appears in the offspring, the donor should be contacted and advised of *his* risks for his own family.

If a potential donor belongs to one of the ethnic groups listed in table 7 (pp. 120–121) *and* there is a reliable carrier test available, that test should be performed. The most common examples are carrier tests for sickle-cell disease (in blacks), Tay-Sachs disease (in Ashkenazic Jews), and thalassemia (in Italians and others of Mediterranean extraction). Men over forty are best excluded as donors because of their slightly increased risks of transmitting dominant diseases caused by gene mutations and possibly even chromosome disorders. In one large study, chromosome analyses of 676 potential donors resulted in the exclusion of 2.6 percent for chromosome "abnormalities" (which clearly included some with normal variations). Careful screening of genetic family histories resulted in the further exclusion of another 3.4 percent.

The physical features of the donor, including aspects of his complexion, hair, eye color, stature, and so on, are noted in order to match him to the recipient parents. His blood type is determined (it should, as mentioned, be Rh-negative if the prospective mother's is) and his semen is examined to check the count and quality of the sperm. Usually, donors who are continually available to provide semen, even on weekends, are selected (although sperm samples can be frozen and stored for later use, as described below). Most commonly, they are paid $30 to $50 for each semen sample provided. Some view the payment of sperm donors as unethical, while others suggest that such inducement may lead to the concealment of pertinent family history about genetic disease. They point out that blood from paid blood donors, as opposed to voluntary ones, is of lesser quality and is infected more often.

The anonymity of sperm donors is assured by a code system. There is obviously a theoretical chance that children born into different families through artificial insemination with sperm from the same donor may marry each other many years later. This would really amount to marriage between a half-brother and a half-sister. The risk in such unions for having offspring with serious genetic disease would be very significant (see the discussion on incest in chapter 9). But if you are concerned about the potential consequences of AID offspring marrying and having children, contemplate for a moment the results of studies made in England and the United States that show that about 50 percent of children born in certain areas are illegitimate. The risks that these children might ultimately marry half-brothers or half-sisters would seem to be distinctly greater than the risks of AID offspring with shared genes intermarrying.

The Infertile Couple

Many couples who might benefit from AID do not approach their physician because one or both of the partners simply cannot handle the emotional aspects. Usually, the woman is extremely motivated to have a baby and is willing to try the method. Many men, however, are unwilling to go this route. There appear to be many reasons, the most common of which relate to the male's confusion between infertility and machismo. He may feel that his virility is at stake. Personal pride, aesthetic distaste, religious dictates, or moral aversion may be other reasons for not wishing AID.

Couples who approach a physician for AID are carefully interviewed together, and the insemination procedures and methods of donor selection are explained. Special care is taken to indicate to the male that having insufficient numbers of sperm or sperm of inadequate quality to effect fertilization has nothing to do with his sexual prowess or performance. If the partners fully understand all the issues involved, as well as the possible marital, psychological, and legal implications, then both are asked to sign a document to initiate the proceedings. This document absolves the sperm donor of responsibility for paternity, and also absolves the physician of any responsibility for subsequent birth defects or problems during pregnancy and delivery. The couple are, of course, assured of complete privacy and confidentiality. The sperm donors, as well as their wives, waive all rights to the resulting children. Moreover, they also waive the right to any legal action to discover the identity of the infertile couple.

Occasionally, it is necessary for couples who are ambivalent about the AID process to receive counseling from a clergyman, a psychologist, or a psychiatrist. Generally speaking, most physicians are unwilling to provide AID to infertile couples unless they are husband and wife, both partners explicitly approve of the procedure, and no marital instability is evident. However, 10 percent of physicians who provide AID have done so for unmarried women, including lesbians. This is obviously a very controversial practice, but in the realm of patient-doctor privacy. (It is common knowledge that single women who want to have a child usually succeed even if denied AID.) In any event, it is striking that even after the thousands of babies conceived through AID, the method has only rarely been an issue in marital breakups.

Success with AID

The fullest cooperation from both parents is required to ensure the maximum chance of establishing a pregnancy by AID. Very careful psychological evaluation of both prospective parents is necessary before initiating the procedure. The average success rate is about 72 percent, though it may take six to eight months to achieve a pregnancy. Nevertheless, persistent failures do occur in at least 15 to 20 percent of patients. As for birth defects in the offspring conceived through AID, few good studies exist. Because AID is confidential, the lack of good information is not surprising. Two studies — one French, the other Canadian — have concluded that there appears to be an increased frequency of chromosomal abnormalities after AID, but more research is needed before such a conclusion can be accepted. Meanwhile, it would be judicious for AID couples who achieve pregnancy to consider seriously such options as detailed ultrasound study, amniocentesis, and prenatal genetic studies. In a hard-won pregnancy, any procedure with risk should be weighed especially carefully. It should also be emphasized that another option for couples who are considering AID is adoption; the genetic counselor and the prospective parents should fully explore this possibility.

SPERM: FRESH OR FROZEN?

It is possible to use either fresh or frozen semen for AID. Obviously, having a group of individuals constantly "on call" to provide semen samples is not the best method, especially when trying to inseminate a woman who ovulates erratically. Techniques have thus evolved to freeze and store semen samples in liquid nitrogen at temperatures of minus 196 degrees Centigrade. In this way, samples are readily available (after thawing) and sperm from the same donor can be used repeatedly in the same patient daily or more often if necessary. There has been no clear evidence thus far to indicate which method — fresh or frozen — is the more successful.

Frozen-semen storage has some valuable applications. For instance, it can be used effectively in a man whose sperm count is low but not zero; his semen samples can be pooled in order to raise the total sperm count, then used to inseminate his partner. The frozen-storage system has been useful in other situations as well. A man who is planning to undergo vasectomy may wish to store a number of semen samples as a form of "family insurance" — to ensure that he can still father offspring at some later date (in case he remarries, for instance). In rare situations, a man may decide to provide a series of semen samples for storage before undergoing certain types of X-ray therapy that could render him sterile.

Recently, a study from the Centers for Disease Control concluded that it is unwise to use fresh semen, since such samples cannot be adequately examined on the day of donation. As noted above, a variety of harmful viruses can be found in semen. The report urged screening of both the donor *and* the semen to avoid venereal disease.

AID and the Law

Clearly, all the major participants in artificial insemination by donor are subject to legal purview: the donor, the recipient female, and her lawful husband or other partner, if any, the physician involved, and, finally, any child born as a result.

Two decades ago, the California Supreme Court ruled in a unanimous opinion that a child conceived through AID with the knowledge and consent of the woman's husband was truly the legitimate offspring of that marriage. Similar rulings have followed in other states. No American court, to my knowledge, has found AID to be an adulterous process. And a court in Scotland, as far back as 1958, concluded that even without the husband's consent AID was not adultery and not even grounds for divorce.

The recognition that children born as the result of AID are legitimate has been crucial from a legal standpoint, since otherwise such children would not be able to inherit and would have no right to claim the necessities of life from their nonbiological fathers. This is an important point in divorce cases in which one or more children have been conceived by AID. Claims for child support from the husband, who is actually not the biological father, have been repeatedly upheld in court.

As mentioned earlier, physicians who provide AID usually do so only for married couples, from whom written consent must be obtained. This signed agreement usually specifies that the husband will be regarded legally as the natural father, and further, that the couple will never seek to identify the sperm donor.

Should the child born through AID be told how he was conceived? You may be aghast even at the possibility! Well, ponder the question as you consider this case: A certain boy's father was dying slowly from Huntington disease, the dementing hereditary brain disorder. The boy became extremely distraught because of the 50-percent chance that he had for developing the same disease. To prevent the son from having a nervous breakdown, the boy's mother told him that he had been conceived through AID; his father, aware that he had the disease, had chosen to protect his future children by arranging for artificial insemination. Not only was the boy incredibly relieved, but he also expressed deep warmth and gratitude toward his dying father.

Most of the legal, ethical, and moral issues surrounding AID have been explored (among many other related subjects) in three volumes entitled *Genetics and the Law*, which I edited with my legal colleague G. J. Annas (Plenum Press, 1976, 1980, 1985). Only half the states have laws governing AID. Various perplexing questions remain. Should an unmarried woman be inseminated? Should a married woman be allowed to obtain AID without the consent of her husband? Should the use of frozen sperm be subject to regulations that limit experimentation on humans? While AID remains an important option for infertile couples or those who consider using the technique because of genetic-disease problems, a certain degree of religious, moral, and social conflict is bound to remain.

19

Newer Ways of Making Babies

"I've tried everything — had every test — followed every bit of advice — I'm frustrated, angry, in despair — isn't there anything you could do to help us have a baby?" I've heard this sad refrain of childless women repeated over and over for almost thirty years. No wonder there was so much excitement when the world's first "test-tube" baby, Louise Brown, was born in England in 1978. Millions of couples the world over sighed collectively at the thought that now, at last, there was a way to have their own baby. Indeed, since 1978, well over two thousand babies have been born through the technique better called *in vitro fertilization*, or IVF. IVF involves fertilization of an ovum by sperm placed together with it in a sterile plastic dish that is kept in a warm, moist incubator, followed by transfer of the fertilized ovum into the mother's prepared womb.

The actual procedure and laboratory techniques, which are based on surprisingly long experience in other species, are straightforward. Reasons for turning to IVF are usually not controversial, unlike some of the ethical, moral, and legal issues spawned by this new technology. A brief review of IVF today, with special emphasis on genetic considerations, provides insight into one of the more remarkable advances in medicine.

The State of the Art

As long ago as 1890, rabbits were used experimentally for transferring ova. In that year, Walter Heape recovered two embryos from an Angora doe rabbit and transferred them to a Belgian hare that had mated a few hours earlier. Six young were born, two of which were clearly distinguishable as Angoras. Since that time, successful transfers have been made in at least eight other mammal species, including pigs, mice, hamsters, rats, sheep, goats, cows, and horses. Initially, surgery was used to place the fertilized ovum into the womb of the recipient female animal. A high rate of successful transfer of fertilized ova — up to 90 percent — was achieved with this technique. Later, the less traumatic technique of proceeding through the female genital tract was studied. Successful transfers of fertilized ova through the genital tract and the cervix of cattle have been achieved in at least 50 percent of attempts. Similar success has been achieved with this approach in mice.

After studying the births of many hundreds of offspring in different animal species, researchers reported no induced congenital defects attributable to manipulation of ova and sperm — an extremely important finding. Nor is there indication of any increase in the frequency of birth defects in man as a consequence of IVF.

In 1978, two English physicians, P. C. Steptoe and R. G. Edwards, successfully accomplished the first human IVF and embryo transfer (ET), which resulted in the birth of Louise Brown. The first successful IVF in the United States took place in 1981 at Eastern Virginia Medical School in Norfolk.

Success did not come quickly, and was in fact dependent upon technical developments in various fields of medicine; their combined use culminated in IVF. Just over 100 years ago, chick embryos were successfully grown in tissue culture for the first time. An endless array of work followed, including Nobel-prize-winning efforts that eventually led to the sophisticated tissue-culture methods in use today. Throughout those 100 years, knowledge developed about the causes and treatment of infertility.

One of the technical facets of this work involved the development of *laparoscopy* — the viewing of organs within the abdominal cavity through a tiny telescopelike instrument. This was first accomplished in Russia in 1901. A long string of refinements followed that today allows retrieval of ripe ova under direct vision or by ultrasound for IVF purposes. Meanwhile, knowledge of hormone chemistry and function evolved, providing us with today's knowledge of how to stimulate the ovary to produce ova. Simultaneously, the necessary exquisite radioactive methods to measure the minutest changes in blood hormone levels enabled the precise monitoring of ovarian stimulation and harvesting of ova. In 1971, the same English physicians published their observations on the first human embryo observed in a culture dish after fertilization.

Who Could Benefit?

I have explored the many causes of infertility and approaches to its investigation and treatment in *How to Have the Healthiest Baby You Can*. About 10 to 15 percent of all couples in the Western world have problems becoming pregnant — some 5 to 6 million people in the United States in any given year. Infertility may be caused by a problem in either partner: women's disorders account for about 60 percent of the cases, and men's, 40 percent. The most obvious candidates for IVF/ET are women who have had their Fallopian tubes removed — for example, following a tubal (ectopic) pregnancy — or whose tubes have been irreparably damaged or blocked, possibly from chronic pelvic, venereal, or other infection. Other conditions, such as endometriosis, may also block an ovum's passage down the Fallopian tube. Moreover, males with a very low sperm count — a condition called *oligospermia* — may benefit from IVF, which allows the few available sperm to be concentrated around the ovum in a tiny dish.

Indications for IVF/ET have, of course, been dominated by the common causes of infertility. However, some very definite genetic reasons do exist and should be considered by all couples facing these dilemmas. Prime candidates are women who have a dominantly inherited disorder and thus a 50-percent risk of transmitting it to each of their offspring. There are many serious dominant diseases, such as those that affect muscle, brain, nerve, and connective tissue (see chapter 7). In these cases, a woman would depend upon a *donated* ovum fertilized by her husband's sperm, then implanted in her own womb; this would enable her to carry and deliver "their" baby.

IVF/ET can be a particularly appropriate option when there is a high risk of a genetic disorder that is not diagnosable prenatally; in such a case, selective abortion is not a possibility. Similarly, couples who do not wish to abort a pregnancy under any circumstances may choose IVF/ET as a way of selecting an ovum that is not likely to have the gene for the disease in question.

Women carriers of a sex-linked disease (such as Duchenne muscular dystrophy) who might be carrying an affected male fetus but do not wish to consider abortion could also select IVF/ET. Recently, two unexpected successes occurred in women with Turner syndrome, who are usually sterile. Two lucky women — one in Israel, one in the United States — were both able to bear a child successfully through IVF/ET.

There are also some women — and the number is very small — who are unable to have babies because they do not produce ova at all (a disorder called *primary ovarian failure*). Such a condition, incidentally, may be inherited as a chromosomal or sex-linked dominant disorder. These women — who might, for example, enter the menopause in their early twenties — could clearly benefit from IVF/ET, which in this case would require ova obtained from a donor.

Preparing for IVF/ET

After thorough evaluation of their infertility, qualifying couples who become candidates for IVF/ET need to recognize the enormous investment of time, money, and emotional energy involved in such an attempt to conceive. Expenses relate not only to the procedures, but to wages lost for missing work repeatedly and for the costs of traveling to and living near the IVF clinic.

Individual IVF clinics have different preparatory schedules. Some, for example, will require a woman to take her temperature daily for a few months before IVF is actually scheduled. This may help doctors to determine the exact times of ovulation. Eventually, the exact dates of ovulation will be confirmed by examination of temperature charts, a detailed menstrual history, hormone tests of blood and urine, vaginal-mucus tests that reflect hormonal changes, and ultrasound studies of the ovaries to determine the number of ripening ova present, their rate of development, and their exact location.

During this preparatory period, the male partner will be asked to supply at least two semen samples for analysis. These semen samples, as well as vaginal cultures from the woman, will be tested for infection, which if present will be treated before any other procedures are started. Given the enormous effort involved in achieving an IVF pregnancy, care should be taken to eliminate any preventable complications. For example, rubella immunizations should be completed at least four months before conception, and women would be advised to forgo cigarettes, alcohol, and any drugs not absolutely necessary.

PROCEDURAL CONSIDERATIONS

IVF clinics differ in their plans and schedules. Some will give the potential mother an injection of chorionic gonadotropin or other hormones to stimulate ovulation and to encourage more than one ovum to ripen. Detailed ultrasound study helps in the timing and execution of laparoscopy; a needle is inserted through the abdominal wall and through it the ova are gently sucked out of the ovary. Depending upon the clinic, some of the ova will be used immediately for IVF, while others will be frozen for later attempts, if necessary.

Once the ova have been successfully harvested, the male partner will be asked to provide semen for immediate fertilization. The semen, once obtained, is combined with the ova in a plastic dish containing a special nutritious broth and placed in a warm incubator. Following fertilization and division of a single fertilized ovum into two or four cells, it is transferred to the uterus. Transfer is achieved by sucking up the embryo into a very thin, sterile tube, which is then passed through the vagina and cervix into the uterus, where the embryo is deposited on the inner wall. Ideally, the embryo will attach itself there — the critical step called *implantation* — and begin to grow.

The extraction of ova for IVF is a minor surgical procedure that does have one potentially serious risk related to laparoscopy. There is about a 1 in 30,000 risk that the thin laparoscope might perforate the bowel, creating a life-threatening infection, or puncture a large blood vessel, producing serious bleeding. These are the main practical, although remote, risks. Fortunately, despite all this tampering with ova, sperm, and embryos, no increased frequency of birth defects has occurred in the over 2,000 babies now born following IVF/ET. Experience with pregnancies following use of frozen embryos is still very small, but so far no defects have been encountered.

HOPEFUL EXPECTATIONS

The chances of achieving a pregnancy, which of course is not synonymous with delivering a normal baby, depend upon various factors, not the least of which is the selection of patients and the technical skill of the IVF clinic. Results from 200 IVF clinics worldwide with a cumulative experience of over 11,000 pregnancies have provided some useful information. The chances of achieving a pregnancy following use of one embryo is about 9.5 percent; using two embryos at the same time, 15 percent; three embryos, 19 percent; and four or more, 25 percent. Worldwide experience suggests that using more than four embryos simultaneously places the mother at high risk of multiple pregnancy and an increased risk of miscarriage. Using three to four embryos, which is believed to give the best chance of success, has an associated multiple-pregnancy rate ranging between 14 and 24 percent.

A couple considering IVF/ET should be acutely aware of the different rates for establishing a pregnancy compared to the rates of actually delivering a baby. The rates just quoted are in contrast to the average rate of about 13 percent per menstrual cycle for actually delivering a baby. In their first 1,200 human IVF/ET procedures, pioneers Steptoe and Edwards in England reported 139 births — a rate of 11.6 percent. Generally, lower success rates have been achieved in mothers over thirty-five, those who have suffered recurrent miscarriages, those with high levels of estrogen hormones, and those who have never achieved a single pregnancy. Steptoe and Edwards were able to obtain one or more ova in 93 percent of their patients, and fertilization in 92 percent when the sperm sample was satisfactory. They were able to transfer normal growing embryos into the mother's womb some 95 percent of the time. Implantation proved more elusive, occurring in only 30 percent of the cases. Moreover, almost a third of these implanted embryos miscarried.

At this stage of IVF/ET development, doctors in this country recognized that the procedure results in more tubal pregnancies, miscarriages, and stillbirths than occur in natural pregnancies. Australian researchers, who have made major contributions to the IVF/ET field, report similar figures, as well as an incidence of premature deliveries that is three times higher than in the general population. Low-birth-weight rates are also higher, owing, of course, to premature births, as well as the increased number of multiple pregnancies. All major studies report no difference in the sex ratio of children born and no increased incidence of birth defects. There have been the same number of various types of birth defects as would ordinarily be seen from natural conception.

The average conception rate for each menstrual cycle for couples without infertility problems who have unprotected intercourse is about 45 percent. The natural successful pregnancy and delivery rate is, however, only around 25 percent. Clearly then, the birth rate following IVF has not yet reached the rate following natural conception. There are probably three main reasons: hormonal, immune-system, anatomical, or other abnormalities in the mother's reproductive system; technical procedural problems; and sperm abnormalities. The first reason is thought to predominate.

Embryo Transfer with Donated Ova

Sadly, there are a fair number of circumstances in which a woman is unable to produce ova. Her ovaries may have been removed after the rupture of cysts or during surgery for a previous tubal pregnancy or for other reasons, such as infection and abscess formation, in which case both a tube and ovary might have been surgically excised. Other women may have entered premature menopause — even in their twenties — while still others might have surgical scar tissue surrounding the ovaries or have genetic disorders that render ovarian function useless. The vast majority of these women have perfectly adequate wombs — certainly healthy enough to sustain a successful pregnancy. Three approaches in such circumstances are available and all have been successful.

A woman donor may be artificially inseminated with sperm from the prospective father. With exquisitely careful timing, the tiny embryo — only one cell — can be removed through a gentle wash of the donor's womb, performed by means of a thin tube inserted through the cervix. The fertilized ovum can then be placed in the womb of the mother, where, ideally, it will implant and develop successfully. Experience is small with this technique, whose attraction is that it avoids the potential dangers of harvesting ova with the laparoscope. It may, however, prove less successful than IVF because only a single ovum is recovered and used.

At present, a more likely route to success is donation of ova extracted through laparoscopy, fertilized in the laboratory with the father's sperm, and then transferred into the womb of the potential mother.

A third approach uses a frozen donated ovum fertilized by the husband's fresh or frozen sperm, with the embryo transferred to the mother's womb. Once again, experience with frozen ova is extremely limited.

Embryo Transfer to a Surrogate Mother

There are a fair number of women who, because they were born with a structurally defective womb or underwent hysterectomy, are incapable of bearing their own child. They and their husbands can, however, have their own genetic offspring together — by seeking out a surrogate mother. In these cases, after ova are removed from the woman and fertilized in vitro with the partner's sperm, the embryo is transferred to the uterus of a surrogate mother, who will actually carry the fetus and bear the child. The process is simple and has been repeatedly successful. Although reasonable enough in theory, in practice surrogate motherhood has presented a hornets' nest of legal and ethical dilemmas. Regardless of any contract struck between the genetic parents and the surrogate mother, experience has shown that it may be difficult if not impossible to insure that proper care will be taken throughout pregnancy, with special reference to diet, living habits, alcohol, drugs, and so on. Moreover, other contract abuses have already been experienced. One surrogate mother extorted money from the couple whose child she was carrying during pregnancy and again after delivery. That child was born with the birth defects associated with the fetal alcohol syndrome (see chapter 16).

Another case, with a different twist, involved a contract signed by a man with a woman who was to receive his sperm through artificial insemination and provide him with a child. Unfortunately, the child was born with microcephaly. The man disputed that the child was his and, sure enough, blood- and tissue-typing tests proved that he was right! Even though the contract had specified that the woman would forgo having intercourse with her husband for thirty days *after* the artificial insemination, the lawyers had neglected to restrict intercourse beforehand. The woman was already pregnant when the artificial insemination was performed!

In yet another, now famous case, a surrogate mother in New Jersey decided to keep the child and ignore the contract. She went so far as to run away with the child. Protracted court proceedings resulted in the child's being removed to the genetic father and his wife. Subsequently, a higher court ruled that the contract was invalid, and that visitation rights by the surrogate mother would be allowed. Meanwhile, the surrogate mother divorced, remarried, and had another child.

Legal, Moral, and Ethical Issues

A procedure that could result in a child having six "parents" clearly has the potential for raising endless issues. A couple who are both sterile could recruit a sperm donor and an egg donor and hire a surrogate mother who also has her own husband. The child born of such an arrangement would have biological parents, rearing parents, and surrogate parents. Since IVF/ET is here to stay and represents an extremely important option for couples with infertility, careful consideration of a few of the basic issues is important.

TINKERING WITH LIFE

IVF has the goal of assisting infertile couples to have their own child while respecting the sanctity of human life. IVF scientists have viewed the treatment of infertility as a clinical duty. Theologians, holding that life begins from the moment of conception, have responded with arguments concerning the moral status of the human embryo. The moral claim is that the individual's right to life may not be violated for any useful advantage or benefit to anyone else and that "experiments upon a human being" should be forbidden in the absence of consent (that is, consent from the embryo or fetus!). Such views reflect Roman Catholic dictates but not necessarily other Christian and Jewish views. Scientific arguments in opposition to the view that life begins at the moment of conception point out that a conception from sperm and ovum may result in a nonhuman product (such as a hydatidiform mole or a choriocarcinoma); that immunological studies show that embryonic cells are not recognizable as different to the mother; and that personhood cannot properly be attributed to the cleaving embryo until at least organ formation has begun. Papal pronouncements reject IVF because it affronts the morality of procreation, in which any interference is held unacceptable. Law in the United States as well as English common and statute law hold that the previable fetus has no legal personality, is not a person, and has no legal rights — a view that obviously includes the preimplantation embryo. Jewish moral tradition also demands total respect for fetal life and regards any violation as a moral evil. In this tradition, however, human life does not become inviolable and attain personhood until live birth has occurred. Morever, there are clearly biblical injunctions addressing the duty to preserve human life and health that can be interpreted as supportive of procedures aimed at treatment, including the remedying of infertility.

IVF PARENTS

The vast majority of couples who undergo IVF do so because of tubal or ovarian structural defects in the female. IVF is therefore regarded by the vast majority of people as a rather simple and natural treatment to overcome a structural or functional defect in one partner. Childless couples in these circumstances, including those of faiths who oppose IVF, do not regard themselves as morally bankrupt or evil. Cuddling their so desperately wanted child, such couples invariably become alienated from their rejecting religion.

The public-policy and ethical debate heats up in the face of requests for IVF by unmarried women. Existing laws protecting reproductive rights govern not only those within recognizable unions, but single women too. The wisdom of planned single parenthood is another matter entirely.

Beyond the issues that arise when single women apply for IVF services are those that surround requests presented by lesbian couples. Jurisdictions in various countries have adopted human-rights legislation that prohibits discrimination because of sexual orientation. While legislatures have the power to enact laws that specifically discriminate on grounds of marital status, they are likely to remain reluctant to do so.

SURROGATE MOTHERS

Unlike straightforward IVF, motherhood by contract has evoked a storm of criticism. A woman whose uterus has been removed or is defective may seek the help of a woman who is able and willing to carry and bear the genetic offspring of the infertile woman and her partner. Clear, simple, uncomplicated arrangements of this sort have occurred without difficulties. A typical example is a sister helping her own sister. A remarkable example is that of the South African woman who carried her daughter and son-in-law's fertilized ovum and gave birth to her own grandchild! In such cases, no contracts and no meddling legal intermediary complicate the family arrangement. The practice of hiring a surrogate mother, using a lawyer as broker, has raised a serious set of social problems, however. In this respect, commercialism has raised its ugly head. Babies conceived through surrogate contracts on a "fee-for-service" basis may be construed as "baby buying" — a practice prohibited in virtually all states. As mentioned, difficulties have already been encountered, and repeatedly so, when surrogates changed their mind and insisted on their right to keep the child. The emotional-bonding process of mother to fetus during pregnancy may be stronger than many realize. Indeed, the surrogate mother's refusal to give up the baby may reflect a deep maternal attachment, regardless of whose genes are involved.

In the absence of definitive legislation, considerable emotional distress has been engendered by courts applying established contract law. Courts faced with such contract dilemmas have simply upheld the written document and forced the removal of the child from the surrogate mother to the genetic parents. Some legal commentators maintain, however, that given her greater contribution and risk, as well as the need to provide certainty of identity and responsibility, surrogate mothers should be deemed the child's legal mother. There is obviously much controversy about this view and considerable opposition to it.

Certainly, many questions related to the selection of surrogate mothers remain in those few states that permit the practice. Should such individuals be married or single? Should they already have had a child? What medical or genetic conditions should be excluded before acceptance as a surrogate? Should certain health screening tests be done before finalizing a contractual agreement? Are psychological profiles of the surrogate mother necessary? Should any dietary or life-style directions (for instance, no alcohol during pregnancy) be dictated by a contract? How would such a contract be enforceable? Should specific prenatal diagnostic tests be included in the contract? What if a serious birth defect is detected? Does the surrogate mother have the right to abort or refuse termination? Do the contracting parents have a contractual responsibility to accept a child with defects? Should the surrogate mother ever be identified to the child? Should the surrogate mother be allowed to visit the child without specific limitations?

THE EMBRYO

Although no increase in birth defects has been associated with IVF, no reassurance can be provided to exclude every possibility. It would be expected that contracts specifically indemnify doctors from being held responsible for any such defects through "mishandling of embryos." Much more controversial is the focus on what should be done with surplus embryos. The consensus view is that embryos should not be sold, for that would make human lives simply another consumer product. Should unused embryos be promptly destroyed? Again, most argue that since spontaneous abortion occurs in some 78 percent of human conceptions, no great ethical dilemma should revolve around such disposal. Since embryos have been frozen and thawed later to yield normal children, questions have arisen concerning ownership. Can the IVF center sell such embryos to an infertile couple with the prior consent of the donating partners? Should all such embryos simply be donated to childless couples? Whose decisions hold sway? This is no idle speculation given the recent example of a wealthy American couple who died in a plane crash. They had stored two embryos in an IVF center in Australia. Questions that immediately arose included whether the embryo could inherit the estate, who the legal guardian would be, and whether the embryo could be adopted and implanted in another infertile couple (who would also be recipients of the estate).

Surplus embryos also present an extremely important opportunity for studying the fundamental, still secret processes that result in catastrophic birth defects. Can they be studied, with whose permission, and for how long? Australia and England are way ahead of the United States in considering and acting on these questions, mainly since President Ronald Reagan in 1984 vetoed an attempt to set up a national commission similar to those established authoritative bodies in England and Australia. Research on embryos — more accurately described as preembryos — up to fourteen days after fertilization has been permitted in England and Australia. The preembryo stage comes to an end at about fourteen days after fertilization, marking the beginning of development of a recognizable embryo rather than a cluster of cells.

Women undergoing sterilization by having their tubes tied have been asked whether they would donate their ova (rather than have them discarded) in order to help infertile couples or to avoid genetic defects. A high proportion of such women have indicated their willingness to donate their ova for research purposes.

The IVF Child and the Future

The physical and mental health of an IVF child is, of course, the prime objective. It is already clear that the frequency of birth defects as such is not increased in IVF-conceived children. There is also no special reason to think that such children will fare worse psychologically than, say, adopted children. Indeed, unwanted children seem far more likely to suffer than ones so fervently desired, conceived in love through these newer ways of making babies.

The entrance of monetary considerations through the surrogate process has drawn attention to the pressing need for legal guidelines. Such guidelines should be formulated to protect the interests of IVF children rather than those of donors, infertile couples, doctors, or IVF clinics. Complete, accurate, and proper record keeping is necessary and confidentiality is critical. Procedures to allow future access by the child to medical or genetic information is also important. Specific standards for donor selection, including required screening tests, should be developed forthwith. Legal recognition should be given to the genetic parents, and the sale of human embryos should be forbidden. Clear methods of regulation, licensing, and enforcement should be established to govern IVF-clinic operations.

IVF/ET is a valuable technology that has already helped a few thousand couples realize heartfelt dreams of having their own child. Given a normal womb, every woman should now have a chance of having her own child. Complex issues raised by these new advances should not cloud the main purpose: securing the health, safety, and very being of a child.

Prenatal Screening and Diagnosis

Ethics, Law, and Euthanasia

20

Detecting Unsuspected Fetal Defects and Pregnancy Complications through Blood Screening Tests in Early Pregnancy

Relatively few opportunities exist in pregnancy for avoiding the birth of a child with a serious birth defect. As it is, most such defects occur without known cause, are not suspected during the pregnancy, and have no available prenatal diagnostic test to effect their early detection. The two most common types of birth defects are critically important exceptions and are the main focus of discussion in this chapter. Neural-tube defects (NTDs), which include anencephaly and spina bifida (see chapter 13), and chromosome defects such as Down syndrome (see chapter 2) are in fact the two most common groups of serious birth defects that occur.

Anencephaly/spina bifida–type defects occur on average in between 1 in 700 and 1 in 1,000 babies born in the United States, although they are more common in certain areas (such as the hills of Tennessee). Moreover, 90 to 95 percent of all babies born with neural-tube defects and close to 80 percent born with Down syndrome are delivered without any advance warning of the fetal defect, to unsuspecting parents who are anticipating a happy event. A valuable blood screening test for a protein called *alpha-fetoprotein*, or AFP, can help detect such defects prenatally, however. This test is recommended for all pregnancies, and you should become well informed about it. In fact, California has by state law made it mandatory for every doctor caring for pregnant women to provide information about the test, to offer it in every pregnancy, and obtain the signature of any patient who declines. California authorities have wisely recognized that while certain death awaits all those born with anencephaly, severe defects associated with much pain and suffering will complicate the lives of the vast majority of children born with spina bifida.

What Is AFP?

AFP, a protein remarkably similar to albumin, is mainly made by the fetal liver; it is secreted into the fetal blood and then into the fetal urine, which passes directly into the amniotic fluid. AFP may enter the mother's blood from the amniotic fluid or directly from the fetal circulation. It is astonishing that the concentration of AFP is about 50,000 times higher in the fetal blood than in the mother's. The concentration in the fetal blood is about 200 times higher than it is in amniotic fluid. This is important, since even a few drops of fetal blood — for example, accidentally mixed in during amniocentesis — can raise the concentration of AFP in the amniotic fluid and lead to a false interpretation of test results.

Curiously, the fetal liver shuts down manufacturing of AFP by the end of the first three months of pregnancy. Nevertheless, given the high concentrations in the fetus, AFP continues to be valuable for diagnostic purposes in the amniotic fluid until about 24 weeks of pregnancy. AFP concentrations in maternal blood are best measured for screening purposes between 16 and 18 weeks of pregnancy. The AFP concentration rises weekly (between 6 and 19 percent per week) until about 28 to 32 weeks of pregnancy. Maternal-serum-AFP (MSAFP) screening is therefore highly dependent upon accurate assessment of the fetal age. If the pregnancy is assessed to be further along than it really is, or is underestimated, results may not be reliable (see below).

How Is Screening Done?

A maternal blood sample is obtained between 16 and 18 weeks of pregnancy, the serum is separated from the blood cells, and measurements are performed most often (but not always) using extremely sensitive tests. Only reliable laboratories with solid experience using this screening test in pregnancy should be used since very important adjustments need be made to account for the week of pregnancy, weight of the mother, her age, her race (blacks have about 10 percent higher AFP values), and whether she is a diabetic taking insulin. Consideration of all these items is extremely important if reliable results are to be obtained. Blood samples can be sent by overnight mail to any laboratory with proficiency in this subject. We receive hundreds of such samples each month in our own Center for Human Genetics at Boston University School of Medicine.

Who Needs the MSAFP Test?

All women in all pregnancies should be offered this screening test. Those with increased risks for having a child with either a neural-tube defect or a chromosome defect and who will be having an amniocentesis anyway should nevertheless still have this test as an aid to detecting complications of pregnancy (see below). Remember, MSAFP is a screening, *not* a diagnostic, test, and therefore should not be relied upon to exclude one of these defects.

How Is the MSAFP Test Used?

The birth defects I am discussing are detected most efficiently between 16 and 18 weeks of pregnancy. Blood samples obtained earlier than 16 weeks might *normally* have low AFP values. Hence, a sample drawn too early might require that a woman have a second blood sample drawn. Most centers do not advise a second blood sample in the face of a higher reported value of MSAFP. For both high and low reported values, an ultrasound study is recommended immediately to get an accurate assessment of fetal age. If the pregnancy dates are correct and the value is truly high, and no fetal defect is seen by ultrasound, an amniocentesis is recommended forthwith for further AFP studies and chromosome analysis of the amniotic fluid. This same sequence of tests is recommended for those with low MSAFP values.

High MSAFP values are those that exceed the 95th or 97th percentiles, while low values fall below the 3d percentile. Care should be taken to distinguish between routine office ultrasound studies and detailed ("level II") studies done by a specialist specifically to examine for fetal defects using machines of greater power and resolution. Faced with a high MSAFP level, the recommendation is clearly to have a level-II ultrasound study immediately. If no fetal defect is found on ultrasound study, an amniocentesis is performed then and there.

AFP is then measured in the amniotic fluid. High levels directly imply the presence of a leaking fetal defect, such as spina bifida. (Note that about 5 percent of such defects are skin-covered and hence do not leak. These biochemical tests would therefore not help in such cases.) In addition to neural-tube defects, leaks may occur following failure of the abdominal wall to close (a condition called *omphalocele*), or through the kidney (a fatal disorder called *congenital nephrosis*, most commonly found among those of Finnish or Scandinavian descent). There are, unfortunately, a host of other defects that might leak, but each is either unusual or rare.

In an effort to more accurately identify the origin of a fetal-AFP leak, the amniotic fluid is also measured for a nervous-system enzyme called *acetylcholinesterase*. In cases in which fetal blood has spilled into the amniotic fluid either from a fetal defect or because of the amniocentesis needle, the acetylcholinesterase helps sort out the confusion caused by the high concentration of AFP in fetal blood and resolves more than 90 percent of these potential dilemmas. The presence of this brain enzyme in amniotic fluid most often implies a leak from the nervous system, but given the immaturity of the barrier between the fetal brain and blood, the enzyme might also circulate in fetal blood and therefore emerge into the amniotic fluid when other defects besides those of the nervous system are present. While the amniotic-fluid-AFP test may sometimes yield a false-positive result (almost always due to a previous fetal bleed into the amniotic fluid), it is extremely rare in an amniotic-fluid sample not contaminated by fetal blood to have a false-positive acetylcholinesterase result. In my experience with over 100,000 pregnancies examined in our laboratories, the combined use of the AFP and acetylcholinesterase assay together with level-II ultrasound yields diagnoses in over 99 percent of these defects.

What Is the Value of MSAFP Screening?

There are three broad categories of benefit when MSAFP screening is used between 16 and 18 weeks of pregnancy. Let us briefly examine these three groups.

MSAFP SCREENING FOR NEURAL-TUBE AND OTHER BIRTH DEFECTS
Between 3 and 5 percent (depending upon which upper limit of normal is used) of pregnancies screened have a raised MSAFP value. Women with elevated MSAFP values have about a 6-percent likelihood that their fetus has a major birth defect, and all will be offered the level-II ultrasound study and amniocentesis as described above. In some of these women, a multiple pregnancy will be found to be causing the elevated MSAFP values; two or more fetal livers effectively produce more AFP. Detection of more

than one fetus at this early stage is useful, since more rest and medical attention are necessary in multiple pregnancy. Not infrequently, however, the ultrasound following a high MSAFP finding reveals that the fetus has died.

Some women have declined MSAFP screening because they think that the test presents only the option of abortion if a fetal defect is found. It is really important to realize that there are other options for women who are opposed to aborting a defective fetus. If such women really care, they would be advised to have MSAFP screening, because detection of these defects allows the best possible care for their child. A recent experience of mine illustrates this point very well.

J.K. had a routine MSAFP screening test at 16 weeks of pregnancy. An elevated level prompted an immediate level-II ultrasound study. J.K. and her husband were shocked to learn that the wall of the fetal abdomen had not closed and that some bowel hung out into the surrounding amniotic fluid. The ultrasound also revealed that this was gastroschisis, a defect rarely associated with other serious defects in the same fetus.

J.K. and her husband, who were opposed to abortion, were relieved to hear that the defect was surgically remediable after birth. Fortunately, prenatal AFP screening enabled a prenatal diagnosis and proper planning of the birth. Elective cesarean section was arranged to avoid passage of the baby's exposed bowel through the nonsterile birth canal, which would cause infection in the newborn. Immediate surgery to close the defect after birth was successful and uncomplicated. Without MSAFP screening, a beautiful child might have been lost to infection contracted in the birth canal.

In this case and those involving spina bifida, early prenatal detection will facilitate planning of elective cesarean-section delivery in a sophisticated medical center where specialists are ready and able to perform emergency surgery on the newborn immediately after delivery. In this way, the best results for the child can be achieved. Moreover, the babies of such women may be saved by attention rendered following recognition that theirs is a high-risk pregnancy.

Insulin-dependent diabetic women in pregnancy normally have lower MSAFP values. Normal levels of AFP for these women must first be established by a laboratory or, alternatively, formal adjustments must be made for the expected lower values. This is particularly important since the risk of serious birth defects (especially anencephaly and spina bifida) in such pregnancies is significantly raised and is generally in the 7- to 10-percent range. I believe that insulin-dependent diabetic women should not only have MSAFP screening and a level-II ultrasound, but that they should also be offered an amniocentesis study.

MSAFP SCREENING FOR IDENTIFYING HIGH-RISK PREGNANCIES

Fortunately, the vast majority of women with elevated MSAFP values are later found to have normal ultrasound, prenatal chromosome, and AFP results. However, in our Center for Human Genetics we find that this group has over a sevenfold increased risk of fetal death or stillbirth, over a fivefold risk of infant death in the newborn period, about a fourfold risk of low birth weight, and double or triple background risks for premature delivery, insufficient amniotic fluid, placental abruption (partial sheering of the placenta off the uterine wall with bleeding behind it), and toxemia of pregnancy (high blood pressure, water retention, and protein in the urine). Viewed in terms of probability, in our center women with an elevated MSAFP value have a 1-in-32 risk of stillbirth, fetal, or neonatal death, a 1-in-21 risk of premature labor (due to ruptured membranes), a 1-in-29 risk of toxemia, and a 1-in-10 risk of at least one of five major complications (too little or too much amniotic fluid; placental abruption; toxemia; and premature ruptured membranes).

Early identification of a pregnancy at increased risk will facilitate more careful observation, more frequent visits to the doctor, additional ultrasound studies, hormone and stress tests, electronic fetal monitoring during labor, and a heightened sensitivity to the need for cesarean section should any problem arise during labor.

MSAFP SCREENING FOR CHROMOSOME DEFECTS

Since about 80 percent of all babies born with Down syndrome are delivered by mothers *under* thirty-five years of age, for almost all of them there would have not been a good reason for amniocentesis. Therefore, any test that can recognize an increased risk of chromosome defects would be valuable. In 1983, the association between *low* MSAFP values and fetal Down syndrome was recognized. We now know that mothers with a low MSAFP have a 1- to 2-percent likelihood that their fetus has Down syndrome or another chromosome defect. An ultrasound study ordered immediately after discovery of a low MSAFP value may reveal that the pregnancy was not as far along as previously supposed and that the blood sample was therefore obtained too early — in other words, that the low value registered was normal. If, however, the ultrasound confirms that fetal age was correct or that the pregnancy was even further along than suspected, amniocentesis specifically for chromosome analysis would be recommended forthwith. MSAFP screening in this way enables the detection of about 20 percent of all pregnancies in which the fetus has Down syndrome. Obviously, this is not particularly impressive but it is much better than no test at all. New advances combining MSAFP and certain hormone measurements are expected to raise this detection rate to about 60 percent in the years ahead.

In assessing low values, adjustments also need be made for maternal weight, since the heavier the mother, the lower the serum AFP. Since blacks have about a 10-percent-higher MSAFP level, adjustments for race are also necessary.

Limitations

MSAFP screening for neural-tube and other major birth defects is now established as a routine part of pregnancy care. Close to 100 percent of cases with fetal anencephaly can be detected through screening, but only 80 to 90 percent of spina bifida cases can be detected. Screening for low MSAFP values to detect chromosome defects is still in the research arena but is slowly beginning to emerge as a useful test.

While it is true that the higher the amniotic-fluid AFP value is, the more severe a leak is likely to be, the severity of a defect such as spina bifida cannot be predicted by such measurements.

Anxiety about Screening

As noted above, the vast majority of women with a high MSAFP level do not have a defective fetus. These mothers may complain about the "unnecessary" anxiety. However, a significant proportion of them have a higher risk of pregnancy complications and will therefore benefit from more careful and frequent obstetric attention. Screening for low and high MSAFP enables detection of women with high MSAFP levels who did not know they had about a 1-in-17 risk of carrying a defective fetus. Avoiding anxiety by not being screened does not remove their chances; it removes their choices.

I have repeatedly witnessed the emotional chaos that engulfs a family following the birth of a child with spina bifida or another major, prenatally detectable birth defect. Any anxiety generated "unnecessarily" during pregnancy is totally dwarfed following such a birth. The ensuing emotional and economic family burden coupled with the suffering of the child weigh heavily in favor of MSAFP screening.

The Future

We have made considerable progress in our research on MSAFP screening between 9 and 12 weeks of pregnancy. I feel certain that we will soon make it possible to screen much earlier than the currently recommended 16 to 18 weeks of pregnancy. Better still, research in progress may yield clues to the cause of defects such as spina bifida, which is likely to assist in prevention rather than avoidance of such catastrophic defects. Recent developments in that direction provide reasons for hope.

21

Prenatal Diagnosis: Amniocentesis and Chorion Villus Sampling

"Doctor, is my baby normal?" The universal question of hope and uncertainty. A plea for a happy beginning. Virtually every woman worries during pregnancy, hoping and praying that the baby she will have will be normal and healthy. This anxiety is shared to a greater or lesser extent by the father of the expected child. We are all comforted by the knowledge that about 96 in every 100 babies are born without evidence of major birth defects. But the 3 to 4 percent of babies who are *not* born sound are the root cause of our anxieties during pregnancies.

Indeed, many informed couples today sensibly concern themselves about their risks of having a defective or deformed child before they start a family, having become aware that it is possible, in an increasing number of situations, to prevent tragedies.

This chapter introduces you to what is at present your main option in the avoidance of birth defects, mental retardation, or genetic disease. This option is prenatal diagnosis — a technique that allows for the accurate diagnosis of serious or fatal genetic disease in the fetus early in pregnancy. Once such a diagnosis is made in the fetus, you may choose to have an abortion and try again with another pregnancy.

Dozens of questions must be running through your mind. How are these prenatal tests done? When? Who needs to have such studies? Who does these tests? Where? Are they accurate? What if the fetus is defective and I'm opposed to abortion? I will try to answer most of the questions you are likely to be thinking about in a sequence of six chapters on prenatal diagnosis, concluding with a discussion of some of the overriding moral, ethical, legal, and social issues in chapters 25 and 26.

Background Developments

The diagnosis of genetic disease in the fetus has only a very short history. During the early 1930s it became possible to introduce a needle into the womb to sample the fluid in which the fetus grows. Some three decades later, in 1961, the first diagnosis of a genetic condition was made with the fetus still in the womb. The disorder was that of Rh blood-group incompatibility between fetus and mother, which could possibly result in the death of the fetus or newborn baby from hemolytic anemia. Research made rapid advances in the understanding of this disorder, which was first treated in the early 1960s by exchanging the blood of the fetus still in the mother's womb with blood of the correct blood group. Subsequently, it became possible to prevent this disorder by immunizing susceptible mothers who were Rh negative. Studies on the fetus with blood-incompatibility problems are, however, mostly confined to the last three months of pregnancy. In contrast, prenatal studies to detect — and, increasingly, to treat — other serious genetic diseases are done between 14 to 16 weeks of pregnancy.

The ability to determine the sex of the fetus by simply staining a cell was first successfully applied in 1955 to cells from the amniotic fluid derived from the fetus. This work had been based upon the serendipitous though critical observation made by Dr. M. L. Barr and Dr. E. G. Bertram in Canada, in 1949. They had been studying the nerve cells of cats and noticed that after staining with certain dyes, cells from the female had a darkly stained blob on the outer edge of the nucleus. This blob, subsequently called the *Barr body*, or *chromatin mass*, was not present in male cells, thereby allowing us for the first time to differentiate male from female cells by means of a very simple technique. In 1960, success was achieved in demonstrating the fetal sex in pregnant women who ran the risk of having children with sex-linked disorders such as hemophilia or muscular dystrophy.

In the mid-1960s, technological developments in actually growing human cells in laboratories made it possible to analyze the chromosomes in cells from the amniotic fluid. Experience in diagnosing diseases using chromosomal analysis slowly began at that time. In 1968, Dr. Carlo Valenti, in New York, first prenatally diagnosed a fetus with Down syndrome. He used a small needle passed through the abdominal wall to obtain amniotic fluid from the womb — a procedure called *amniocentesis*. Efforts at that time had begun in Denmark to obtain amniotic fluid or placental tissue through a small instrument passed through the vagina and cervix into the womb. Problems with infections resulting in fetal loss led to that procedure being discontinued. The Chinese and the Russians persisted and were successful at obtaining tiny pieces of the developing placenta for diagnostic genetic studies. This procedure has become known as *chorion villus sampling* (CVS), because tiny fingerlike projections (each called a *villus*) of one enveloping fetal membrane (the *chorion*) are plucked off for study.

Amniocentesis for prenatal diagnosis of genetic disorders gradually became a standard of expected care. Chorion villus sampling, which enables an even earlier fetal diagnosis in pregnancy, is still being studied to determine its safety and accuracy. Let us examine amniocentesis in detail, as well as the current status of chorion villus sampling.

Amniocentesis

COUNSELING BEFORE AMNIOCENTESIS

Ideally, each woman having an amniocentesis for prenatal genetic studies should be counseled before the procedure. This may simply be done in a careful discussion with her obstetrician, with her partner present. Such an arrangement is probably acceptable if the amniocentesis is for a routine hazard such as advanced age of the mother. The obstetrician should, of course, be expected to know the risks of the procedure and be able to discuss all the relevant aspects. When the disorder in question is rare or a complicated new test such as DNA analysis is to be used, a consultation with a medical geneticist is advisable.

It is important to realize the limited scope of prenatal diagnosis. Even though there are over 4,000 hereditary disorders that are each due to a particular single gene, fewer than 200 can be detected prenatally. Fortunately, all recognizable chromosome disorders can be diagnosed, as can various other birth defects, such as spina bifida; moreover, prenatal fetal sex determination in cases of sex-linked disorders at least provides some options. But couples who undertake prenatal diagnosis can at best come away with reassurance that the most common groups of genetic disorders (chromosome and neural-tube defects), together with any other specific disease being tested (for example, thalassemia), have been excluded.

INFORMED CONSENT
Amniocentesis is a minor surgical procedure, and as such requires the prior understanding and consent of women who agree to undergo the procedure. As would be the case for elective abortion, the physician is required to inform the patient carefully about the very small risks of the test (which are discussed below). In addition, both the doctor and the laboratory doing the studies should ensure that the couple involved know that the amniotic cells may not be cultivated successfully in about 1 percent of cases, that a second amniocentesis may therefore be needed, that the answers provided are not 100-percent guaranteed, and that all possibility of genetic disorders or birth defects cannot be excluded by the test. Moreover, the patient should be advised that the result provided might pertain to only one fetus if a multiple pregnancy is missed. Written consent is usually required.

WHAT IS AMNIOTIC FLUID?
A few days after fertilization of the egg by the sperm, the new embryo attaches itself to the wall of the womb. The embryo becomes enveloped in a thin-walled sac. This sac — very much like the yolk of an egg — slowly fills with fluid secreted mainly by the embryonic and, subsequently, fetal tissues. Ultimately, fetal urine adds to the volume of fluid within the sac inside the womb. This liquid is the *amniotic fluid*, in which the fetus is suspended and grows.

At the site where the embryo attaches to the womb, a circular, flat structure develops; called the *placenta*, it provides direct communication between the fetus and the mother. The placenta, while serving as a safety barrier to some of the toxic agents eaten by the mother, also lets through nutrients, infectious agents, and certain drugs. At the same time, some substances from the mother's circulation also contribute to the amniotic fluid that engulfs the fetus. The amniotic fluid contains a variety of chemical substances, including proteins, carbohydrates, and fats, as well as cells that originate from the fetus.

HOW ARE THE STUDIES DONE?
The amniotic fluid contains cells that peel off the skin of the fetus or are passed from inside the mouth or lining of the lung, through the intestinal tract, or in the fetal urine. When these cells are

examined after the amniotic fluid is obtained for study, some are found to be dead. Many, however, are living cells, and it is these that can be grown in laboratories. The cells are placed in sterile dishes or flasks containing a rich broth and kept in a warm incubator. The broth is changed a few times a week, and the growth of these cells is observed closely under the microscope. One cell divides into two, then from two into four, from four into eight, and so on, until there are thousands to millions. This usually takes two to three weeks, at which time the cells are broken open in order that their chromosomes can be examined. A cell has to be in a particular phase of division called metaphase (see chapter 2) when the chromosomes are spread out, for the examination and chromosome determination to be successful. Between fifteen and thirty cell "spreads" are usually examined in each case. With these techniques, then, it is possible to analyze the chromosomes of the fetus, determine its sex, and check the activities of a whole array of different enzymes.

A result is generally obtained from the first amniotic-fluid sample in about 99 percent of cases. Nevertheless, a number of things may go wrong along the way. The obstetrician or the laboratory may contaminate the fluid, although total loss of a sample because of contamination is very unusual.

Even without contamination, the amniotic-fluid cells may simply not grow. This happens in about 1 percent of cases. A repeat amniocentesis because of a failed cell culture then becomes necessary.

Some physicians do not ensure safe delivery of the sample to the laboratory, but I believe that the patient should be assured that due care will be taken in its transport. Generally, the best results are achieved when the sample reaches the laboratory within hours of its being obtained. Where there is no choice, the sample may be left in a refrigerator (at 4 degrees Centigrade) overnight. Our lab at Boston University has, on occasion, successfully grown amniotic cells sent from countries 10,000 miles away — samples that have spent five days in transit at room temperature. Bloody amniotic fluid samples are obtained in 5 to 10 percent of cases; this will not usually hinder the study, except perhaps to slow down cell growth a little, unless the sample is extremely bloody.

HOW AMNIOTIC FLUID IS OBTAINED

Amniocentesis is usually performed by an obstetrician, who first injects a local anesthetic into the skin somewhere between the pubic hairline and the navel, then inserts the amniocentesis needle into the womb through the abdomen (see figure 30). Some women prefer to have a needle inserted only once, and do not even have the local anesthetic. The amniocentesis is performed in the doctor's office or on an outpatient basis at the hospital. The recommended time for the procedure is when the mother is 14 to 16 weeks pregnant. The reason for this timing is that the enlarging womb emerges from the pelvis into the abdomen of the pregnant mother at 12 weeks of pregnancy. Before that time, it cannot be felt through the abdominal wall. Between 14 to 16 weeks, the womb can easily be felt between the pubic bone and the navel. Therefore, it is relatively easy to insert a needle into the womb at this stage of pregnancy.

THE USE OF SOUND WAVES (ULTRASOUND, OR SONAR)

Ultrasound, or sound wave, test studies immediately precede and sometimes accompany the amniocentesis. This technique, which is believed to be safe, involves the passage of high-frequency sound waves through the womb and makes it possible to locate the fetus and the placenta accurately. In this way, the needle can be inserted into the womb, avoiding the fetus and the placenta,

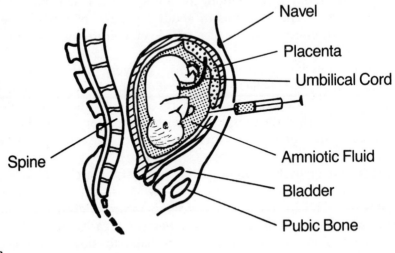

Navel

Placenta

Umbilical Cord

Spine

Amniotic Fluid

Bladder

Pubic Bone

Figure 30

An amniocentesis needle properly entering the amniotic cavity, without penetrating the placenta

thereby decreasing the chance of fetal damage or bleeding. Unfortunately, in 40 to 50 percent of cases, the placenta lies along the front wall of the womb, and the needle may have to pass through it.

The ultrasound study performed as part of a routine amniocentesis enables assessment of fetal age, but is not aimed at determining structural details. Hence, while it is expected that certain severe defects, if they exist, will be detected, study with more sophisticated machines that provide better visualization are necessary if most structural anomalies are to be detected. This type of examination is usually performed by an ultrasound expert, who could be either an obstetrician or radiologist. Repeated ultrasound studies, every few weeks, may be valuable in tracing the pattern of head growth. Such measurements also allow for further assessment of fetal organs and body growth. Some hereditary disorders are associated with poor fetal growth, which doctors call *intrauterine growth retardation.*

Twins, triplets, and so on will normally be detected by ultrasound, at least by 16 weeks of pregnancy, and frequently before 10 weeks. Women who have required drugs to stimulate ovulation and become pregnant need careful monitoring by ultrasound, since they are at risk for multiple pregnancy. Obviously, recognition of twins, triplets, and so on is critical when prenatal diagnosis in the face of a high risk is to be undertaken, so that amniotic-fluid samples are obtained from each fetus. If high-resolution ultrasound is not available and genetic risks are high (5 percent or over), I would advise couples to travel to the nearest major center for such studies. A lifetime of chaos can result from failing to realize the significance of this advice and not bothering to make the effort.

TIME AND TIMING

At present, to grow cells and analyze the chromosomes of the fetus, 2 to 3 weeks on the average are required before the results become available. For some of the rarer hereditary biochemical diseases, up to 6 weeks may even be necessary. It is therefore important that the amniocentesis be done at the best time, which is between 14 and 16 weeks of pregnancy, to provide early reassurance or an opportunity for the parents to elect abortion. Many sad experiences have occurred when this test has been done too late or not at all.

Amniocentesis during the third trimester is frequently done to assess fetal lung maturity before elective cesarean section or for monitoring Rh or other blood-group incompatibility. Amniotic-fluid study may also be important if a fetal defect is detected by ultrasound. Chromosome study might reveal, for instance, that the fetus has a grave defect such as trisomy 18. If, as so often happens in these situations, fetal distress owing to a lack of oxygen develops during labor, an immediate decision is necessary for or against cesarean section. The child in this situation is destined to be profoundly retarded because of the genetic defect, so, despite the risks to the fetus, couples should be given the option to *avoid* cesarean section and its attendant complications for the mother.

While there have been some instances of a physician forgetting to suggest amniocentesis altogether or offering it too late, much more often the pregnant woman comes for her first obstetric visit too far along in pregnancy. In many instances, women may not seek their first appointment with the doctor until they are almost five months along. If couples are concerned about having children with birth defects, they should see their doctor before conception for necessary immunizations, tests, treatment, and genetic counseling, and again immediately after the first missed menstrual period.

HOW SAFE IS AMNIOCENTESIS?

Obstetricians have been performing amniocentesis for decades. Almost forty-five years ago, substances were being injected into the amniotic fluid for fetal X-ray studies. Many amniocenteses have been performed on women in the last three months of pregnancy in the course of managing Rh disease, with complications to the mother or fetus occurring only rarely.

It would be reasonable to believe that inserting a needle into the womb does carry some hazards for both mother and fetus. Theoretically, at least, it may be possible to pierce a blood vessel and cause some bleeding, to introduce infection, to precipitate premature labor and therefore cause a miscarriage, or to damage the fetus.

In October 1975, the United States federally funded National Institute of Child Health and Human Development Collaborative Amniocentesis Registry Project reported its findings after a four-year study. Nine collaborating centers, including my own, had studied 1,040 women who had undergone amniocenteses and another matched group of 992 pregnant women who had not undergone the procedure. Care was taken to match women for many important differences, such as age, the number of previous pregnancies, race, income group, husbands' occupation, level of education, and so on.

Those women who had amniocenteses had no major complications themselves. Some of them did have minor "complications," which included transient cramps, vaginal spotting (bleeding), or leakage of a little amniotic fluid.

There was no significant difference in rates of miscarriage or fetal death in women who had had an amniocentesis compared to those who had not. In fact, the data showed that the women who had amniocenteses had a lower rate of miscarriage than those who did not have the prenatal-diagnosis procedure. (One must take into account the likelihood that women who have amniocenteses are better informed, take better care of themselves, choose their doctors more carefully, and have doctors who in turn are probably more up-to-date and practicing better medicine.)

Even though the Collaborative Amniocentesis Registry Project observed no increased risk to a pregnancy because of amniocentesis, current experience, which involves many more cases, suggests that there is a slight risk of miscarriage, albeit very small: less than 0.5 percent, or 1 in 300 to 400 at most.

No significant damage to the fetus was observed in babies whose mothers had amniocenteses, although a skin scratch or puncture mark by the needle has been noticed in a handful of cases. The Collaborative Amniocentesis Registry Project analyzed the babies born in both its control groups again at one year of age, and no differences in the frequency of birth defects or of motor or mental retardation were found. The conclusions of this careful study indicate that amniocentesis is a safe though not completely risk-free procedure. Similar studies in Canada have come to the same conclusion. By the way, it also became clear that about 3.2 percent of women who do not go through the testing nevertheless miscarry or have a fetal death during the last six months of pregnancy.

A major British collaborative study concluded that miscarriage risks were about 1 percent. However, various problems with this study essentially invalidate the conclusions. Most recently, a large Danish study of women under thirty-five came to a conclusion similar to the British study's. The reason for this higher risk remains unclear.

A BALANCE OF RISKS

How would you decide whether the slight risk of amniocentesis causing fetal loss or damage is outweighed by the risk of having a defective child? This decision-making could be described as balancing chances, or more properly, balancing benefits. If, for example, the mother is aged forty and pregnant, the approximately 2- to 3-percent chance of a fetal chromosomal disorder is between six to ten times the chance of miscarriage through amniocentesis. In this situation, parents are likely to opt for the test. If the mother is thirty-five, the risk of having a chromosomally abnormal child (for example, a Down-syndrome baby) approximates 0.8 percent. She therefore is about two to six times more likely to have a chromosomally defective child than to miscarry as a result of prenatal testing by amniocentesis. However, the decision of a couple who have been trying for a pregnancy for five to ten years may obviously differ from that of a couple who already have had six children. The worry about losing the long-awaited offspring would likely dominate the decision of the first couple. Ironically, infertile couples who have tried hard to achieve a pregnancy may have slightly higher risks for fetal defects and therefore need to consider carefully prenatal tests such as amniocentesis. I have repeatedly witnessed heartbreaking cases in which a couple took eight, ten, and even thirteen years to conceive, only to find that the fetus — or, worse still, the child at birth — had a catastrophic defect!

Who, then, most needs amniocentesis and prenatal studies? The Rh-negative-sensitized mother with an Rh-positive husband is naturally going to need help. There is, however, a large group of mothers whose plight is more subtle and less well known. I devote the next two chapters to them.

Chorion Villus Sampling (CVS)

Early developments in CVS were noted above. Steady development continued throughout the late 1970s, and a major collaborative study in the United States by the National Institutes of Health (NIH) is nearing completion. The major goal of this study is to determine the safety and accuracy of CVS.

THE PROCEDURE

CVS is a painless procedure with only occasional slight discomfort. It involves passing a thin plastic tube (catheter) through the vagina and cervix while the progress and location of the tip are guided by ultrasound images. Once the correct site of the developing placenta has been reached, a tiny piece of tissue is sucked into the tube and thereby torn off the chorion (membrane).

The procedure is immediately preceded by ultrasound study to be sure of the exact fetal age and that all is otherwise well. Experience has shown that 9 to 11 weeks is the safest period for CVS. Because there is concern about infection resulting from passage of the catheter through the birth canal, major trials are under way to discover whether CVS could be safer when performed like amniocentesis (transabdominally).

RISKS

Just as for amniocentesis, the risks can be fetal or maternal. Serious risks to a mother are posed by infection, though it is still uncertain what degree of risk exists. It is thought to be small, but a worrisome few infections have occurred. While no expectant mother has died — and over 30,000 procedures have been done worldwide — there have been 1 or 2 near-catastrophes. One mother, some days following CVS, collapsed, went into shock, and almost died. She had developed a raging infection in the womb, and in order to save her life, the entire womb (which contained the fetus) was surgically removed. Kidney failure followed and dialysis was necessary for a while until she eventually recovered — no longer able to have a child! Remember, however, that such a serious complication has happened once in about 30,000 cases. The infection rate in experienced centers is about 1 in 500.

Vaginal bleeding seems to occur in 10 to 15 percent of women following CVS. Some bleeding into the biopsy site occurs in about 4 percent. It is still unknown how often Rh-negative mothers can be sensitized by fetal bleeding after CVS, and whether protective injections of immunoglobulin is safe for the fetus (see chapter 9).

The major risk of CVS is fetal death. In the most experienced centers worldwide, the risk of fetal loss is thought to be about 2.3 percent following one catheter insertion, 2.8 percent after two insertions, and 10.7 percent after three. Enormous difficulty is, however, encountered in trying to determine an accurate figure. There is a considerable natural loss of pregnancies that may occur unrelated to the procedure. All manner of confounding factors that cause or contribute to miscarriage also need to be considered. For example, advancing maternal age, the number of previous miscarriages, the amount of alcohol consumed, the number of cigarettes smoked, the presence of infection, and on and on, all increase the likelihood of fetal loss. This is why such a careful major study is being done. The hope is to dissect out the risk of the CVS procedure from other factors that could cause fetal death.

The latest NIH-study data show no significant difference in the rate of miscarriage between 2,278 women who had CVS and 671 who underwent amniocentesis. The number of women in the amniocentesis group is too small to make any final conclusion about which procedure is safer for *both* mother and baby. Current knowledge *suggests* that amniocentesis is somewhat safer, but late-pregnancy elective terminations following amniocentesis still need to be considered in determining benefits and risks. Moreover, the current NIH data reflect studies by very experienced centers. It is likely that risks of miscarriage or maternal complications following CVS would be more significant, at least early on, in an inexperienced center. Success with cultivation of cells obtained from CVS reached 97.7 percent compared to 99.1 percent following amniocentesis.

Meanwhile, early data show no increased risk of birth defects in children born subsequently. The risks of prematurity or of premature labor also do not appear elevated.

WHO NEEDS CVS?

The indications for CVS are the same as for amniocentesis as discussed in this and the following chapters. The one key exception is detection of neural-tube defects such as spina bifida; in this case, amniotic fluid is required for diagnosis.

The tissue obtained through the catheter (or via the transabdominal route) can be used for diagnosis of chromosome and biochemical genetic disorders. The tissue cells can be cultivated for about two weeks and then harvested for the various analyses. Some centers also take a small piece from the tissue sampled and directly analyze the chromosomes. This often works, and results may emerge within five hours or at least by the next day!

WHO SHOULD NOT HAVE CVS?

Specific contraindications are now being recognized, and most centers have a list of situations in which they will not do CVS. This list usually includes vaginal infection, an inaccessible canal through the cervix, vaginal bleeding, previous failure of CVS, and a pregnancy already over 12 weeks.

HOW ACCURATE IS THE DIAGNOSIS?

Accuracy questions also dominate the collaborative study in the United States and others in Canada and Europe. Thus far in the NIH study, no errors have been made in chromosome analyses. The key concern relates to the fidelity of the cells sampled. Difficulty has arisen repeatedly when chromosome results from immediate study, from cultivated cells, and from the fetus if aborted or amniotic-fluid cells have not been consistent. These chromosome "discrepancies" have varied in all directions. For example, in one case the direct immediate analysis showed normal chromosomes, the cultivated cells revealed Down syndrome (trisomy 21) in *all* cells, and fetal tissues showed eighty-eight normal cells and only one with trisomy 21. This problem of mosaicism and other discrepancies is not uncommon in CVS experience and is currently thought to arise in 1 in 100 women studied.

Prenatal diagnosis so much earlier than amniocentesis at 16 weeks allows safer abortion, however. The latest push is to examine the safety of much earlier amniocentesis (at 12 to 14 weeks). Whatever the sampling procedure, technology now allows us to detect a rapidly increasing number of serious genetic diseases. Let me now discuss these possibilities, which *all* prospective parents should consider.

22

Prenatal Diagnosis: Chromosome Disorders

There are many reasons why you (or your partner) should consider prenatal genetic studies. You may have a family history of a certain hereditary disease; you may have previously had an affected child, you may be a carrier of a chromosome disorder; you may be thirty-four or older; you may have an abnormally low or abnormally high maternal-serum alpha-fetoprotein (MSAFP) screening result; and so on. The reasons for prenatal studies can be grouped in four general categories: couples at risk for having a child with (1) a chromosomal abnormality, (2) a sex-linked disease, (3) one of the hereditary biochemical disorders of metabolism, or (4) a hereditary disfiguring birth defect. I will only consider the chromosomal-disorders category here; the rest are discussed in the next chapter.

Certainly, the most common reason anywhere in the world for prenatal studies is to determine the chromosomal constitution of the fetus. And small wonder, since 1 (0.65 percent) in every 156 children born in Western countries has some chromosomal abnormality. About 1 in 250 has a *serious* chromosome defect — over 20,000 in the United States alone each year. On the other hand, it is remarkable to what extent nature takes care of its own mistakes through the process of miscarriage (see chapter 2). About 97 percent of Down-syndrome embryos and fetuses are spontaneously aborted.

Between 40 and 60 percent of all fetuses studied after miscarriage show serious chromosomal abnormalities. Those miscarried in the fourth month have been found to have them in 25 percent of the cases; after four months, in only 3.5 percent of the cases. Careful analyses of stillborn babies revealed that 6 to 11 percent of them had chromosomal disorders. Our bodies, it would seem, have a mechanism to counteract such errors, though they slip up from time to time. But now, with the help of advanced medical technology, what nature has overlooked can be diagnosed early in pregnancy and parents given the choice whether to stop the pregnancy or not.

Unfortunately, since neither cure nor meaningful treatment is possible for chromosome disorders, abortion or continuation of pregnancy are the only options. Those against abortion might still opt for amniocentesis, since once forewarned that the fetus is seriously defective they might not wish to jeopardize their own health (through cesarean section) if fetal distress occurs in labor. Since this complication is common when the fetus is defective, and since operative delivery is almost invariably done to ensure that the baby sustains no brain damage, some women would rather choose vaginal delivery given certain severe/profound retardation in the child. Regardless of the choices made, all should respect the painful decisions forced by these sudden dilemmas.

The decision to abort is *solely* a parental one, which should remain as uninfluenced as is humanly possible by the physician or anyone else.

Indications for Prenatal Chromosome Studies

The reasons why you would need to consider prenatal chromosome studies are (1) advanced maternal age; (2) a low or high MSAFP level (see chapter 20); (3) you or your mate, although healthy, carries a chromosome defect; (4) you or your mate previously had a child with a chromosome defect; or (5) for some other miscellaneous indication. A closer look at these reasons is important.

ADVANCED MATERNAL AGE
Some 75 to 80 percent of all babies with Down syndrome are delivered by mothers *under* thirty-five years of age, who account for almost 95 percent of all births. This means, of course, that mothers thirty-five and over have one-fifth to one-quarter of all Down syndrome babies, but that most are delivered by younger mothers.

For reasons that are not yet exactly understood, older mothers give birth to children with chromosomal abnormalities much more often than younger mothers do. Since the body is so remarkably efficient at ridding itself naturally (by miscarriage) of about 97 percent of Down syndrome embryos and fetuses, the likely reason for the increasing rate among older women is a progressive breakdown in the body's surveillance (probably a flaw in the immune mechanism), which allows more chromosomally defective embryos to survive. The actual age-related risks of fetal Down syndrome in particular and chromosome defects in general are listed in table 17. As you can see, at 16 weeks a thirty-seven-year-old pregnant woman has a 1-in-149 risk that she is carrying a fetus with Down syndrome, and a 1-in-82 risk that she is carrying a fetus with some type of chromosome defect. At birth these risks are somewhat less, owing to the deaths of chromosomally defective fetuses between 16 weeks and term.

Table 17. Chances of Chromosome Defects in the Fetus at the Time of Amniocentesis (±16 Weeks)

Mother's Age (in years)	Down Syndrome (Trisomy 21)		All Chromosome Defects	
33	0.24%	1:417	0.54%	1:185
34	0.31	1:323	0.65	1:154
35	0.40	1:250	0.80	1:125
36	0.52	1:192	0.99	1:101
37	0.67	1:149	1.22	1:82
38	0.87	1:115	1.52	1:66
39	1.12	1:89	1.90	1:53
40	1.45	1:69	2.38	1:42
41	1.87	1:54	2.99	1:33
42	2.41	1:42	3.76	1:27
43	3.11	1:32	4.75	1:21
44	4.01	1:25	6.00	1:17
45	5.18	1:19	7.60	1:13
46	6.68	1:15	9.65	1:10
47	8.62	1:12	12.26	1:8
48	11.12	1:9	15.59	1:6
49	14.35	1:7	19.86	1:5

SOURCE: Based on a study by Professor Ernest Hook of Albany Medical College (New York).

Increasingly, geneticists now begin to recommend amniocentesis at thirty-four (rather than thirty-five) years of age, given the fact that the risk of carrying a chromosomally defective fetus at that age exceeds the risk of fetal loss related to the prenatal diagnostic procedure. Risk considerations notwithstanding, and regardless of the woman's age, fully informed couples should be able to decide whether they wish to have amniocentesis performed. Insurers, however, may not be willing to cover these "nonrecommended" procedures because of costs.

Patients and doctors alike have viewed advanced maternal age as a risk associated mainly with Down syndrome. But though this is the most common disorder, four other chromosomal disorders are also correlated with increased maternal age: trisomy 13, trisomy 18, triple-X syndrome, and Klinefelter syndrome (see chapters 2 and 3).

Danish studies show that women under thirty-five whose partners are fifty-five or older have about a 1-percent risk of having a child with a chromosome disorder. In Denmark, such couples are offered amniocentesis for prenatal diagnosis. Although confirmatory data are not available, prenatal studies should seriously be considered if recommended when the male is fifty-five or older and his partner is under thirty-five.

LOW OR HIGH MATERNAL-SERUM AFP

There is an increased risk of Down syndrome and other trisomies when the mother has a very low level of serum alpha-fetoprotein (a subject discussed fully in chapter 20). This indication for amniocentesis is likely to become more frequent than that of maternal age. High MSAFP is much less often associated with a chromosome defect, but couples are advised to have the chromosomes checked in such cases as well.

A PARENT WHO CARRIES A CHROMOSOME ABNORMALITY

Most of us really do not know if we are a carrier of a chromosome abnormality. At least 1 in 500 people does carry such an abnormality, which when transmitted could result in birth defects and mental retardation. Population testing for chromosome abnormalities is currently not feasible, but you might consider having a blood test because of a family history of chromosomal abnormality, a previous affected child, repeated miscarriage, infertility, or if an actual defect is suspected.

If a child has Down syndrome (the result of an extra number-21 chromosome), blood tests of the parents are usually not obtained. If the child has a chromosomal translocation or mosaicism (see chapter 2), the parents' blood is tested to determine whether the condition was inherited, and from which side of the family the problem may have originated. Once a familial disorder is detected, brothers, sisters, and other relatives on that side of the family should be tested. Prenatal genetic studies are best offered in subsequent pregnancies when one parent has been shown to have a translocation. About 2 to 3 percent of all infants with Down syndrome arise through a translocation (mostly between chromosomes 14 and 21). One of the parents is a carrier about half the time, and the other cases arise spontaneously.

Parents who transmit a translocation have higher risks of having a child with a serious chromosome defect, so amniocentesis is recommended for all such individuals. For certain translocations, risks are higher if females are the carriers than if males are, although it is not known why this is so. Various factors influence these risks besides gender: which chromosome is involved, the site and size of the break, and whether the break is brand-new or inherited. Since hereditary forms of chromosomal abnormalities can be diagnosed in the fetus, most of these parents can, through prenatal diagnosis, avoid the birth of an affected child.

The importance of acting on knowledge about family history is exemplified by the following case of mine:

Joanne was eighteen and unmarried. When she found herself pregnant by her fiancé, she came for genetic counseling. Her pregnancy had already advanced to 16 weeks. The reason for seeking counseling was that her sister had Down syndrome. She was therefore worried that her child might inherit it. It was not known at that time which kind of Down syndrome her sister had: the very uncommon hereditary form or the common nonhereditary type.

Because there is a very significant risk of carrying a fetus with Down syndrome for those individuals who carry the hereditary form, amniocentesis and prenatal genetic studies were elected immediately, rather than waiting for blood chromosome tests to determine if Joanne was a carrier.

Some two weeks later, the results showed that the fetus had the hereditary type. This eighteen-year-old prospective mother and her fiancé then elected to have an abortion, which was done forthwith, and the affected fetus was found as predicted. Blood studies for chromosomal analysis, performed before the abortion, had also shown that Joanne was a translocation carrier of the disease, and later we confirmed that her sister and mother were carriers of hereditary Down syndrome too.

Less than one year later, now married, Joanne again presented herself for prenatal studies — this time three months pregnant. An amniocentesis at 16 weeks was performed and we were able to show on this occasion that the fetus had normal chromosomes and was a male.

On the very day that this normal result was provided, however, Joanne indicated that she had been exposed to what sounded like a classic case of German measles (rubella) when she was three months pregnant. Immediate blood studies disclosed that she had indeed contracted the infection. Another amniocentesis was performed between 21 and 22 weeks of pregnancy and the German-measles virus was found in the amniotic-fluid cells. The parents elected — for the second time — to abort a pregnancy. Studies on the aborted fetus showed severe fetal infection by German-measles virus, confirming the prenatal diagnosis. [As mentioned earlier, children born after rubella infection in the womb may have major birth defects, including mental retardation, cataracts, heart defects, stunted growth, and deafness. Amniocentesis is not usually recommended for exposed mothers who might be susceptible to rubella.]

Some months later, Joanne arrived again, pregnant for a third time. Amniocentesis this time revealed one amniotic-fluid cell with a ring-shaped chromosome — which is decidedly abnormal — in about sixty cells analyzed. My colleagues and I indicated that we did not know the meaning of this, but that we did not believe it likely that the fetus was affected, although we could not be sure. The young couple, having already been through so much, decided to continue the pregnancy. They had a daughter, as predicted by the test results, and she looked like a sound baby. Chromosome studies on both blood and skin revealed a normal pattern!

A PREVIOUS CHILD WITH DOWN SYNDROME

Parents who have had a child with the common trisomy-21 form of Down syndrome have a 1- to 2-percent risk of having another similarly affected child or one with some other disorder in which the number of chromosomes is affected. Even with this small risk, I have personally seen at least half a dozen families who originally had a child with trisomy-21 Down syndrome and subsequently had another similarly affected baby. Amniocentesis is therefore recommended if the parents have had a child with trisomy 21 — and, in fact, in every pregnancy that follows the birth of a child with too many or too few chromosomes.

MISCELLANEOUS REASONS FOR PRENATAL CHROMOSOME STUDIES
Many situations brought up frequently by young couples are not considered suitable for prenatal studies. If either the mother or father has experimented with hallucinatory drugs, or if the mother has been a heroin addict and is now being treated with methadone, amniocentesis is not recommended. Nor are prenatal studies suitable at this point for patients exposed to rubella, chickenpox, other infectious diseases, or even radiation during pregnancy. Our present inability to provide a reliable diagnosis makes the testing neither feasible nor sensible.

Women with an overactive thyroid are justified in seeking prenatal studies, however, because of evidence suggesting that they may have an eightfold risk of bearing offspring with some sort of chromosomal abnormality. Women with underactive thyroids similarly have an increased risk and are candidates for prenatal chromosome studies.

Also, in the case of a woman who is carrying twins, there is a 1-in-6 risk that one of the pair will have a chromosomal abnormality. Amniocentesis and prenatal studies are clearly indicated.

A young pregnant mother might request an amniocentesis on the basis of anxiety (perhaps her neighbor just gave birth to a child with Down syndrome). She may be allowed to have the study, but her health insurer probably will not cover the cost.

Accuracy

Data from the United States Collaborative Amniocentesis Registry Project found an accuracy rate for prenatal diagnosis of 99.4 percent. This was based on some 6 cases out of 1,020 in which an erroneous diagnosis was made.

In 3 of these, the fetal sex was determined incorrectly, mainly because diagnoses were offered before a sufficient number of cells were available for study. Admixed mother's cells caught on the needle may have entered the amniotic fluid, thereby confounding the analysis. It was not a serious matter, in that none of these three couples ran any risk of a sex-linked disease.

Sad to report, however, there were 2 cases in which normal chromosomes were reported, but infants with Down syndrome were later born. In one of these, it is entirely possible that the sample of amniotic fluid sent to the laboratory belonged to another patient, who was having an abortion on the same day the amniocentesis was performed.

The inaccurate diagnosis in the sixth case was of a biochemical disease, galactosemia, for which treatment is available. The parents luckily elected against abortion. Soon after birth, the baby was found not to have the disease.

The error rate for prenatal diagnosis now is probably between 0.2 and 0.6 percent. The vast majority of mistakes probably reflect maternal cells mixed in with the amniotic-fluid sample. Very occasionally, a mosaic of importance could be missed, and even more rarely a failure to analyze the chromosomes correctly may result in the missed diagnosis of a serious disorder. Fortunately, such cases are rare. In fact, considering the accuracy of laboratory tests in general, and given the complexity of prenatal genetic studies, it is striking to see accuracy rates as a general rule between 99.4 and 100 percent. This is a remarkable achievement.

Problems and Pitfalls

MOTHER'S CELLS IN THE MIX

It may come as a surprise to some that a few cells of the mother may literally adhere to a rough edge on the beveled part of the amniocentesis needle. Although these needles are high-precision instruments (even though they are discarded after a single use), the minutest rough edge or even slightly ill-fitting inner stylet probably accounts for maternal cells being mixed with the amniotic-fluid specimen. For every 1,000 amniocenteses done, about 3 will have *some* maternal-cell admixture. Maternal and fetal cells are not easily distinguished under the microscope when cell cultures are examined. Luckily, however, at least half of such admixtures will be noticed when chromosomes are analyzed and male fetal cells are observed. Occasionally, the mother's cells may outstrip the fetal cells in culture and the results provided, unbeknown to anyone, will give no indication about fetal health. In addition to this rather straightforward pitfall, other rare problems can intrude, as in the following example.

A thirty-six-year-old nurse had an amniocentesis because of her age. I reported that the fetal chromosomes were of a normal male.

The telephone call was not long in coming after a baby girl was delivered. Dismayed by the apparent error, I rapidly reexamined the slides and workbooks and decided that a mixup was well nigh impossible. The parents were quick to respond to my request for a blood sample from the baby to confirm what we all hoped would be normal female chromosomes. Remarkably, we detected that their little girl had the chromosomes of a male. (This rare condition is called *gonadal dysgenesis,* and this type was inherited as a sex-linked disorder.) The child underwent abdominal surgery to remove tissue that resembled testicles; this was necessary because of the increased risk of cancer in such cells. Only a tiny womb was present. The child is being raised as a girl, who, while normal, will be sterile and will require hormones to develop breasts.

Situations that increase the likelihood of maternal cells being admixed in the amniotic fluid include those in which multiple needle-sticks are made in an effort to get fluid and in which very bloody fluid is obtained. In addition, there are extraordinarily rare cases in which a person is born with normal male chromosomes in some cells and normal female chromosomes in others (chimeras, discussed earlier). Fortunately, chances are only about 1 in 1 million for such occurrences.

When determination of fetal sex is critical — for example, in the face of sex-linked diseases, which affect males only — our lab routinely measures the amniotic fluid for sex hormones and we carefully consider the ultrasound diagnosis of sex while waiting for the chromosome report. It is obviously preferable to have all three point to the same sex.

MOSAICS: REAL OR PSEUDO?

When different cells from the same fetus show different chromosome complements, mosaicism is said to be present. Between 1 and 3 in 1,000 individuals are born mosaics. One intermingled set of cells most often has normal chromosomes, while the second set may have any one of scores of different chromosome defects. The more normal cells present, the healthier the person will be. Hence, if an individual is born with only some cells showing trisomy 21, that person will still appear to have Down syndrome but will likely function closer to normal. Sex-chromosome mosaics are especially common.

The prenatal diagnosis of mosaicism is usually made when two sets of cells with different chromosomes are detected in at least two different cell-culture containers. Problems arise, however, when cells with an abnormal chromosome set are found in only one container. This is referred to as *pseudomosaicism*. Cells with the same chromosome abnormality are found in only one container about once in every 157 cases studied. The almost invariable outcome, regardless of which chromosomes are involved, is a normal baby. The origin of these abnormal cells is uncertain, but some may come from the placenta and have no meaning for the fetus whatsoever.

I must emphasize that it is never possible to rule out true mosaicism totally. The good news, however, is that babies born with true mosaicism that was missed during *properly performed* prenatal diagnosis are very likely to appear and develop normally.

DIAGNOSIS OF THE UNEXPECTED

Mothers in their mid-thirties properly concerned about having a child with an extra chromosome, which results in serious to profound mental retardation (as in Down syndrome), are not infrequently confronted with the unexpected detection of a chromosome defect with consequential but not catastrophic implications. Typical examples are the sex-chromosome defects described in chapter 3. I vividly recall the dramatic irony that one couple confronted. Our lab detected that the fetus was an XYY male with all the implications (including criminality), as outlined in chapter 5. The amniocentesis, as usual, had been obtained for advanced maternal age (the mother was thirty-eight). Both parents were FBI agents! They did elect to terminate that pregnancy, but not infrequently couples have decided to continue.

THE UNINTERPRETABLE OBSERVATION

About once in every 2,000 amniocentesis studies an unexplained extra chromosome fragment — technically called a *supernumerary marker* — is found in all or some of the fetal cells. The origin of such a fragment is invariably obscure. Some are, in fact, inherited directly from one parent, who may have such a fragment in only some cells. If such a parent is normal and healthy, it is extremely likely that the child, too, will ultimately be the same. If the fragment has arisen afresh, about 18 percent may have associated birth defects or mental retardation. Curiously, these marker chromosome fragments increase in their frequency as mothers age. Luckily, these findings are uncommon and even when they do occur are mostly associated with an uneventful outcome.

WHOSE CHROMOSOME IS IT, ANYWAY?

The subtle and sophisticated nature of the chromosome analysis, involving as it does such picayune examination, quite often raises a question about the normality or otherwise of at least one chromosome. Any such observation automatically sparks the immediate request for blood chromosome studies of each parent. From time to time, and more often than you might think, it becomes obvious that half the fetal chromosomes have been derived from someone other than the husband. The case of the sailor described on page 20 illustrates this point.

TWINS

Painful dilemmas are inevitable when one of a twin or triplet is detected prenatally as having a serious genetic disorder. Since the frequency of twins rises with maternal age and peaks in the upper thirties, the situation may arise in which a chromosomal or other genetic defect is detected at the very time these studies are recommended. In the past, couples who already had had a child with the serious disorder in question or who clearly recognized its ominous significance almost invariably terminated the pregnancy. In these dreadful circumstances, couples have often asked about the possibility of saving the normal fetus. In the last few years, technical skill has made it possible to save the healthy one by causing cardiac arrest of the other. Over thirty such cases are known. Usually, the dead fetus is slowly absorbed by the mother's body, leaving only a tiny sac of tissue. In the vast majority of cases, the healthy twin has survived to bring joy to the parents.

TECHNICAL FAILURES

Growing fetal cells successfully in plastic containers has been possible for barely two decades. In fact, the special broth, or culture medium, in which cells grow has become reliable in only the past few years. Not surprisingly, the relatively few amniotic-fluid cells that are living (the vast majority are not) may not grow satisfactorily, thus making chromosome analysis impossible. When amniotic fluid is bloody, cell growth might be slowed markedly.

Even in the best of laboratories, cells may occasionally fail to grow. Fortunately, growth-failure rates are mostly below 1 percent. When the cells fail to grow adequately, a second amniocentesis is recommended — usually within fourteen days of receipt of the first sample. Since the cells grow in a sterile broth, infection may occur in the laboratory at any time and ruin the culture. As a precaution, a number of different cultures are always set up at the beginning and the containers are kept in different incubators. What happens when there is an electrical failure that wipes out the current feeding these critical incubators? Presumably, they are directly wired to an emergency electrical generator to take care of just such eventualities.

During the past decade, a number of different centers have reported sudden and remarkable failures in cell growth. Thorough investigations revealed that chemical substances toxic to cells have been inadvertently incorporated into some batches of syringes or containers in which the amniotic fluid was obtained or kept. Thankfully, these episodes have been few and far between, but they have been incredibly distressing to all concerned when they have occurred.

You can imagine a woman's dismay and chagrin when a laboratory reports that the amniotic-fluid sample submitted for prenatal studies was actually a urine specimen! This happens about once or twice in every thousand cases in our experience, invariably when the preferred full bladder, which makes ultrasound visualization easier, is mistakenly needled and urine rather than amniotic fluid withdrawn. A second amniocentesis is the only option in such instances.

Considering the requirement for virtual perfection in doing prenatal diagnostic tests, couples should determine where their amniotic-fluid sample is to be analyzed. Some are dismayed that the living cells are unnecessarily being transported halfway across the country overnight to a commercial laboratory that offers their health plan a "good deal." "Cell-growth failures" in such cases should at least evoke concern.

While chromosome studies still constitute the most common reason for amniocentesis, other diagnoses increasingly are being sought. These will be considered next.

23

Prenatal Diagnosis:
Hereditary Disorders of Males,
Other Biochemical Diseases, and
Disfiguring Birth Defects

The chromosomal disorders discussed in chapter 22 are mostly associated with risks below 10 percent and often below 5 percent. In contrast, the risks of having children with the disorders I am about to discuss are much higher and range between 25 and 50 percent for individuals who carry certain defective genes. If you were faced with a 25- to 50-percent risk of having a child with a serious or fatal genetic disorder, you might, as very many prospective parents have done in the past, have chosen not to have any (or any more) children.

Since 1968, prenatal diagnosis of these more complex disorders has become critically important. It allows couples the opportunity to have unaffected children selectively, sparing the parents the agony of losing a child, and, more important, sparing an affected child the pain and suffering of early death, disease, or serious deformity.

Prenatal Studies for Sex-linked Disorders

I discuss sex-linked inheritance in chapter 7 and note that sex-linked diseases usually occur in males only, while in some *sex-limited* diseases, females only may be affected. There are over 250 recognized sex-linked diseases, affecting every organ system. Despite these many sex-linked diseases, at present prenatal diagnosis can specifically be made in fewer than forty. They are individually rare and some are named after physicians who described them. Examples include hemophilia A and B, Duchenne muscular dystrophy, fragile-X syndrome, Fabry disease, Hunter syndrome, Lesch-Nyhan syndrome, and Menkes steely-hair syndrome.

Fabry disease is a biochemical disorder caused by a missing enzyme. A complex fatty substance accumulates in the body because of the missing enzyme (which would ordinarily break this compound into pieces) and causes kidney and blood-vessel problems that lead to high blood pressure, kidney failure, and strokes. Most patients, after many years of symptoms, have died in their thirties and forties owing to a lack of specific treatment. The question is, if you were the parent, would you elect to terminate a pregnancy in which the fetus was diagnosed as having Fabry disease?

Hunter syndrome is also a biochemical disease that involves a missing enzyme; in this case, the complex substance that accumulates is called mucopolysaccharide. Accumulation of this material in all parts of the body leads to multisystem problems. For example, deposits of this substance in the brain will lead to mental retardation; in the heart, to cardiac failure; in the joints, to severe limitation in movement; in the liver, to marked enlargement; and so on. Death in almost all patients occurs invariably by the twentieth year — and usually much earlier — after prolonged suffering. Unlike Fabry disease, Hunter syndrome prevents any sort of a normal existence in the intervening years before death.

The *Lesch-Nyhan syndrome* — also a biochemical disorder caused by a missing enzyme — is an extremely unpleasant disorder characterized not only by profound mental retardation and features of brain damage (stiff limbs with peculiar movements), but also by self-mutilation. Indeed, some of the patients whom I have seen have so severely bitten themselves around the mouth and limbs that I and other physicians have been left in a state of shock upon realizing that these were self-inflicted injuries. Given good care and attention, these patients may live on many years in their profoundly retarded state. They often require restraining — for example, tying their hands — to prevent them from mutilating themselves.

Affected children with *Menkes steely-hair syndrome* have hair that feels similar to steel wool; in addition, they are retarded. The basic defect in this condition concerns the way the body handles copper.

Only a few of the sex-linked disorders can now be diagnosed in the fetus. (This limitation, I'm happy to say, is slowly beginning to change.) At present, the only recourse parents have in the case of sex-linked diseases that are not diagnosable prenatally is to determine the sex of the fetus. If a female fetus is found, the parents can be reassured (a critical exception is the fragile-X syndrome, discussed in chapter 6). However, if it is determined that there is a male fetus present, there is a 50-percent chance that it is affected. Since there is no way of being certain, the parents must decide simply on the basis of the high risk whether to take a chance or terminate that pregnancy. Consider for a moment the unusual plight of Sharon:

> Sharon was twenty-six, unmarried, and four months pregnant when she came for genetic counseling. She had had one son from a previous pregnancy, who had been affected by a disease of his immune system. He died in infancy from overwhelming infection. Thereafter, Sharon had been counseled that this disease was sex-linked, which meant that *she* was the carrier, and that every time she had a male child, he would have a 50-percent chance of having the disease. She simply refused to accept that she was "responsible" and consequently sought out a different father for *each* of her next two babies.
>
> During the second pregnancy, she again refused prenatal studies and ultimately had a baby boy who also died from infection during the first year of life.
>
> Her third partner convinced her to have prenatal studies in her third pregnancy. This she eventually did, and the results showed a female fetus. This time she gave birth to a healthy girl and married the father.

There are some unusual sex-linked diseases that are confined to females. Disorders of this kind — such as *incontinentia pigmenti*, a skin disorder associated with brain damage — can be managed by determining whether the fetus is a female. In this group, virtually all females will be affected, and the parents could selectively elect to have unaffected boys.

Hemophilia A and Duchenne muscular dystrophy are two of the more common sex-linked diseases that are familiar to most people. But there are so many others that great care must be taken by both the doctor and the family in obtaining an accurate family history. *Renpenning syndrome,* in which there is mental retardation without any other physical signs, is confined to males. I well recall one family — whom I saw before the days of prenatal diagnosis — in which the child being examined already had two brothers with mental retardation and three mentally retarded uncles with the same disorder. The only way to suspect sex-linked inheritance is for the physician carefully to analyze the family pedigree.

Increasingly, it is possible to perform tests to detect the female carrier of such diseases. For example, it is now possible to detect almost all carriers of hemophilia and Duchenne muscular dystrophy by a blood test using DNA techniques. For muscular dystrophy, a muscle enzyme (creatine phosphokinase) that leaks into the blood is also often measured to give a higher probability of recognizing a carrier. Unfortunately, the carrier-detection tests for both hemophilia and muscular dystrophy do not provide answers in 100 percent of cases — because of recombination, the gene shuffle discussed in chapter 8. A negative result causes uncertainty and leaves the question basically unanswered. Fortunately, carrier-detection tests are steadily becoming possible in more of the sex-linked and other disorders.

Prenatal Studies for Hereditary Biochemical Disorders

Many hundreds of different hereditary biochemical disorders of metabolism are known. About 1 in every 100 children born has one of these biochemical disorders. Many of them do not cause mental retardation, or impair the child's normal development or general health to any great extent, if at all. Many others, however, cause severe mental retardation, seizures, stunting of growth, and early death. Close to 150 of these biochemical disorders can now be diagnosed in the affected fetus early in pregnancy. The first diagnosis of a biochemical disorder in the fetus while in the womb was made in the late 1960s; the disorder was Tay-Sachs disease.

HOW ARE THE DIAGNOSES MADE?

Cells from the amniotic fluid are placed in small dishes containing a nutrient broth, and are kept in a special warm, moist incubator. They grow slowly, and after a period of two to three weeks (or, occasionally, as long as six weeks), there are enough cells to work on. Each of the cells having the genetic blueprint will show the specific biochemical defect — for example, deficient activity of an enzyme — thereby enabling a diagnosis to be made.

It is obviously wonderful news to hear that the fetus does *not* have the particular biochemical disease in question. But could you imagine the dismay if, instead, the child unexpectedly had Down syndrome. I therefore believe strongly that once the suspected biochemical disease has been excluded, the chromosomes from the same sample should also be studied to exclude other defects.

Who, then, needs an amniocentesis and prenatal genetic biochemical studies? There are presently three clear indications:

1. When both parents are found to be carriers of a particular hereditary biochemical disease in which accurate prenatal diagnosis is possible.

   (The best-known example is Tay-Sachs disease. Prospective parents can be tested before conceiving. Many couples in which both partners are carriers can thus be identified *before* they have affected children. They subsequently can have each pregnancy monitored by prenatal studies, and can opt to terminate a pregnancy when an affected fetus is diagnosed. In this way, these couples will be spared the torture of seeing a child develop normally for six to eight months and then begin to retrogress and deteriorate: have seizures, go blind, fall into a vegetative state, waste away, and die, usually between two and five years of age. Without prior screening, it may not be discovered that a person is a carrier unless his or her sibling has an affected child and the person then goes for testing.)

2. When the parents have previously had a child affected by a biochemical genetic disorder that can now be diagnosed prenatally.

3. When carrier detection is either inconclusive or impossible in parents at risk but the prenatal diagnosis can be accurately made.

TREATMENT IN THE WOMB

The ultimate goal of making an early prenatal diagnosis is to treat the fetus successfully while it is in the womb. Aborting the whole organism is obviously not the ideal approach, though it is the only way available at present for the vast majority of serious or fatal genetic diseases.

For a few disorders, such as Rh disease, treatment of the fetus directly or through the mother has now succeeded, however. The first prenatal diagnosis of a *biochemical* disorder that was treatable in the womb was the rare disorder *methylmalonic aciduria*. This disorder causes failure to thrive, vomiting, lethargy, biochemical disturbances, poor muscle tone, and eventually mental and motor retardation. Treatment of the fetus through the mother during pregnancy is carried out by giving her intramuscular injections of massive doses of vitamin B_{12}. This method secures the child's health at birth, when a special low-protein diet is started. In this way serious illness, mental retardation, and early death have been averted.

Considerably more common is *congenital adrenal hyperplasia* (CAH), inherited equally through a gene from both parents (autosomal recessive). About 1 in 5,000 to 13,000 whites and 1 in 7,550 Japanese are born with CAH — nowhere near the remarkable 1 in 282 among the Yupik Eskimos. Various forms of this disorder occur, each due to a deficient though different enzyme along a stepwise pathway that finally results in the production of "cortisone." Signs of the most common form are masculinization of the female genitals (a baby girl may be mistaken for a boy), excessive growth, early appearance of pubic hair, and enlargement of the penis or clitoris. Critically important in about two-thirds of affected children is the occurrence of a life-threatening crisis one to four weeks after birth, characterized by vomiting, diarrhea, and salt loss leading to collapse and even death if not diagnosed and treated with "cortisone." Where needed, surgical correction of the female genitals is possible, and normal growth, puberty, and fertility can be achieved through lifelong medical treatment with cortisonelike supplements. Today, both carrier detection and prenatal diagnosis are possible for most families, using DNA techniques (see chapter 8) combined with special blood-group linkage studies (see the HLA discussion in chapter 9).

The very first inherited biochemical disorder found to cause mental retardation was *phenylketonuria* (PKU). Since that description in 1934, we have learned that PKU occurs in about 1 in 14,000 newborns in the United States and as frequently as 1 in 4,500 in Northern Ireland. Transmitted by a recessive gene from each parent, all problems are the result of a deficient liver enzyme. An affected *untreated* child will develop irreversible mental retardation. Hence, in most Western countries, blood screening of newborns is done to make an immediate diagnosis and institute the special low-protein diet through which mental retardation can be averted.

Today, new DNA techniques have made both carrier detection and prenatal diagnosis of PKU possible for most families. Despite the availability of effective treatment after birth, prenatal diagnosis remains a serious option. The reason is that the special diet is unpalatable and very restrictive, difficult to enforce in early childhood, and needs to be continued for as long as possible. Usual practice was to discontinue the diet at four to seven years of age. Recent studies, however, show intellectual deterioration, loss of IQ points, learning difficulties, and behavior problems after the diet is discontinued.

A steadily increasing number of women with PKU are entering the childbearing years. If they become pregnant, the chemical products that accumulate in their blood damage the fetal brain and other developing organs. Their risk of having a retarded child or one with a heart defect or microcephaly approaches an incredible 100 percent! Only a mere handful of cases are known in which the diet was adhered to strictly well *before* conception (and through pregnancy) and a healthy child born.

Galactosemia is another treatable hereditary biochemical disease where prenatal diagnosis is possible. If the fetus is affected, special lactose-free dietary treatment of the mother started early enough will almost invariably avert early death or mental retardation, cataracts, and liver damage. There are a few other very rare disorders — such as tyrosinemia, homocystinuria, maple-syrup-urine disease, and propionicacidemia — where prenatal diagnosis and early treatment may be critical to save life or prevent mental retardation or other consequences. Progress in actual prenatal treatment for genetic disorders can be anticipated, provided that fetal research is not even further interdicted by state legislation. A few other disorders are now being conquered by early diagnosis and treatment in the womb. Continued support for medical research will undoubtedly provide more and more opportunities for early treatment or prevention, reducing the need for abortion — the major option today.

Prenatal Studies for Gross Physical Birth Defects

The defects in this category — and there are very many — are gross physical deformities in the child. In these cases, the chromosomes appear normal, and no biochemical or genetic deficiency can be demonstrated. Typical abnormalities in this group include missing or deformed hands or limbs, gross malformation of the head (anencephaly), a very big head (hydrocephalus), heart defects, and scores of other anomalies. Three approaches are useful in tackling the prenatal diagnosis in this group of diseases.

SOUND WAVES (ULTRASOUND, OR SONAR)

The use of sonar has moved from a research-and-development phase to accepted use as a new diagnostic tool. This technique involves the passage of sound waves through the uterus, thereby enabling accurate measurement of the fetal head size, length, thighbone, and hence the fetal age; location of the placenta; detection of twins or triplets; and diagnosis of some major disfiguring defects.

If twins are detected by sonar and the risks of genetic disease are high (over 10 percent), it is possible to outline one amniotic sac by instilling a radio-opaque substance into the amniotic fluid through the amniocentesis needle. Under X ray and sonar control, amniotic fluid can be obtained from the second sac. This can be a difficult technical procedure, and an experienced doctor is needed to perform it. In the presence of multiple fetuses, a high genetic risk, and an inability to obtain fluid from two or more sacs, the parents may have to decide whether to keep or terminate the pregnancy based on the statistical risks of genetic disease alone.

When twins are detected before the amniocentesis, parents should give extremely careful consideration to the next step: the actual amniocentesis. Amniotic fluid from one sac will be obtained and studied, and a result will be provided. What if the fluid around the other twin cannot successfully be obtained? The first result may be normal, but what if the other twin is abnormal? Or vice versa? It is extremely rare (but not unheard of) for both twins to be affected, but it is not rare for one to be affected. The dilemma is not eased, because the option to terminate the pregnancy may mean abortion of a normal fetus!

The prenatal diagnosis of malformation of the brain and head (anencephaly) has been made in early pregnancy by the use of ultrasound. The same technique has also successfully been used in the prenatal diagnosis of polycystic disease of the kidney in later pregnancy. Hydrocephalus and microcephaly — both very often associated with mental retardation — have been diagnosed in the fetus early in pregnancy. The prenatal diagnosis of both these disorders is possible in most but not all cases by serial ultrasound measurements at two- to three-week intervals until twenty-four weeks of pregnancy.

X RAYS

The bone structure of the fetus is at a very early stage of development around sixteen weeks of pregnancy. Therefore, there is a very limited use of X rays to visualize bone abnormalities. Between twenty to twenty-four weeks of pregnancy, it is possible to obtain, in some cases, useful diagnostic information. For example, there is one serious hereditary bone disease called *osteopetrosis* in which the bone density is greatly increased. The prenatal diagnosis of this disease has, in fact, been made at twenty-four weeks of pregnancy using X rays. Other serious hereditary bone diseases, such as various forms of dwarfism, may be diagnosable in the same way. I would recommend special heed be taken of your family history before seeking this kind of study.

FETOSCOPY

This technique involves looking directly at the fetus through a tiny telescopelike instrument whose caliber is the size of a large hypodermic needle. This *fetoscope*, as it is called, is introduced into the uterus via the abdominal wall under local anesthesia and is manipulated under direct ultrasound guidance. A long, thin needle is passed through the fetoscope and fetal blood is obtained by piercing a placental blood vessel under direct vision through the fibre-optic part of the device. Any disorder that requires fetal blood for diagnosis or confirmation of diagnosis, rather than amniotic fluid cells, can be diagnosed prenatally with this technique. Rare conditions such as the hereditary hemolytic anemias or other blood and immune-system disorders can be detected this way. The common hemolytic anemias — sickle-cell disease and the thalassemias — can now be diagnosed using the DNA of amniotic fluids cells. Urgent fetal-blood sampling is sometimes necessary to determine the significance of an uncertain prenatal diagnosis made through amniotic-fluid-cell study. Examples include questionable chromosomal mosaicism and the fragile-X syndrome.

Detracting from more frequent use of fetoscopy is its real risk to the fetus. In the best of hands, 3 to 6 percent of pregnancies will be lost because of the procedure and 10 percent will deliver prematurely. It is, therefore, usually reserved for situations where risks are in the 25- to 50-percent range.

The fetoscope also enables biopsy of the fetal skin and liver, for diagnosis of certain rare hereditary diseases. Moreover, using it as a telescope facilitates direct visual diagnosis of a few serious genetic diseases that involve typical facial or other features. Examples include conditions associated with abnormal or missing ears, hands, digits, or limbs. Fortunately, newer, higher-resolution ultrasound has largely replaced much of the earlier need for the risky fetoscopy. This valuable technique, however, enables treatment of the fetus by direct transfusion of blood or other therapeutic products.

Much experience is required to perfect this technique, and this is usually accomplished by fetoscopy being done with consent just prior to elective abortion. In certain states, such as Massachusetts, antiabortion legislation has prevented development of fetoscopy and even deprived couples who wish to *treat* the fetus of that right!

DIAGNOSING FETAL DISEASE THROUGH THE MOTHER
There are a few disorders of the fetus that can be diagnosed by testing the mother — without an amniocentesis! Rh incompatibility is the best known (see chapter 9). Screening mothers' blood alpha-fetoprotein levels facilitates detection of a host of serious birth defects (see chapter 20).

The prenatal diagnosis of hereditary disorders has represented the most significant advance ever in the avoidance of mental retardation and serious or fatal genetic disease. There are many hereditary disorders, however, that cannot yet be diagnosed prenatally, and it is important for all prospective parents to recognize such limitations. Nevertheless, many parents who face higher-than-normal genetic risks are and will be blessed by healthy children whom they would otherwise not have had at all — solely because of advances in prenatal diagnosis and the consequent options and opportunities.

24

Predicting or Choosing the Sex of Your Baby

Throughout recorded history, people have tried to predict the sex of a coming child. All manner of magic has been evoked: astrology, numerology, dreams, examination of the entrails of sacrificed animals, the pattern of flights of birds, and other supernatural techniques. The political importance of producing a male child has dominated society — from royalty down to those who till the soil. Not surprisingly, world literature is replete with descriptions of how to tell if a girl or a boy will be born. Only in recent times have accurate predictions become possible, however.

Modern Methods of Sex Determination

Fetal sex is determined most accurately — virtually 100 percent of the time — by chromosome analysis of amniotic fluid cells obtained by amniocentesis. Chorion villus sampling yields tissue from which the same determination can be made at nine to eleven weeks of pregnancy, with slightly greater risk of fetal loss, and perhaps slightly less accurately. Cells obtained by either of these two methods can be subjected to DNA analysis using sequences found only in the Y (male) chromosome and results provided within twenty-four hours. Sex determination using ultrasound around sixteen weeks of pregnancy is about 95 percent accurate.

The most pressing reasons nowadays to determine the fetal sex are related to the risk of having a child with a hereditary disease confined to males only (see chapter 7). There are also a few very rare disorders — transmitted in a different way from the sex-linked disorders — that cause the parents to elect to have only males.

The use of amniocentesis and prenatal genetic studies simply for family-planning purposes is a most inappropriate use of a very expensive technology. Moreover, there is a certain repugnance shared by many for a process that aims to abort a pregnancy simply on the basis of individual family gender preferences.

Selecting the Sex of Your Future Child

A more rational approach to family planning that allows selection of the sex of your children in order to avoid recurrence of a hereditary disease or defect has been heralded by new and exciting progress in medical genetics. It became possible around 1970, by using a simple staining technique, to show that about half of all sperm bear a Y chromosome (and will therefore make a male), while the other half bear an X chromosome (and will therefore make a female).

In 1973, more progress was made in separating out male-determining from female-determining sperm. The techniques used were based upon prior knowledge that the former have a superior swimming ability compared to the female-determining sperm. This apparently reliable technique allows for the separation of 85 percent of swimming (and, presumably, functional) male-determining sperm. The female-determining sperm, or the X-bearing sperm, are heavier and therefore slower, since they contain up to 4 percent more genetic material (DNA) than the Y male-bearing sperm. Success rates of 85 to 100 percent in achieving the desired sex have been claimed in very small studies. However, more carefully executed studies have *not* supported these claims. Furthermore, the high degree of success claimed by manufacturers of commercial kits to predetermine sex remains suspect.

Success with the technique would obviate the need for consideration of abortion by couples facing risks of offspring with sex-linked diseases for which prenatal detection is not yet possible or preferred. Such couples could then choose simply to have female offspring. Moreover, couples who desire to select the sex of their children for less drastic reasons could also do so without the need to resort to abortion because the "unwanted sex" was found in the fetus.

MEANWHILE, BACK AT THE RANCH!

Medical technological breakthroughs tend to take time. Until the technique just described is perfected, information currently available might help you in trying to select the sex of your children:

Ovulation usually occurs twelve to sixteen days before the beginning of the next menstrual period. The egg can be fertilized for some six to twenty-four hours after ovulation. The sperm are able to achieve fertilization up to five days after intercourse or artificial insemination.

A variety of factors may influence fertilization through an X-bearing (female) or Y-bearing (male) sperm. The acidity or alkalinity of the vaginal (and cervical) secretions may be important. The woman's body temperature usually rises at ovulation, and this, too, may affect the two different types of sperm.

The age of the sperm may be relevant. The less frequently coitus occurs, the "older" the sperm, and, it seems, the more likely that males will be conceived. During the Second World War, in both England and the United States, appreciably more males than females were born, perhaps bearing out the theory.

The best information, today, points to a great likelihood of having males if coitus occurs four or more days before the woman's temperature rises, or one or more days thereafter. The *opposite* has been noted following artificial insemination.

SOME IMPLICATIONS OF PRESELECTING THE SEX OF YOUR CHILDREN

The ratio of the two sexes at birth is not quite as equally distributed as might be expected. There is, in fact, a slight excess of males, which led a distinguished biologist some years ago to remark that we were heading for a world shortage of marriageable females. Most evidence currently indicates that if couples were offered the opportunity to select the sex of their children, the result would be an increased proportion of males in the population. Undoubtedly, different social classes and ethnic groups would have different preferences. It is extremely likely that in less developed countries male children would still seem to be a more positive acquisition to a family as a help in working the land, earning money for food, performing necessary rites at the graves of their parents and ancestors, and so on.

In a study involving 100 pregnant women, a female fetus was detected in 46. Some 29 of these women elected to abort simply for that reason. Only 1 woman out of the 53 found to be carrying a male fetus chose to abort. (In one of the cases, the fetal sex was not determined.) The study was done in China, at the Tietung

Hospital of Anshan Iron and Steel Company. Its aim was to assist in family planning.

Many parents appear to desire one child of each sex. The opportunity to choose the sex of one's children might ultimately reduce both family size and the population, since couples with children of the same sex are more likely to have additional children than are couples who have achieved both a boy and a girl. A similar tendency has also been noted: that women who have had only daughters have more children than those with only sons. Information available clearly documents that today, as throughout history, couples have consistent preferences for male offspring and tend to prefer a boy to be the firstborn.

What if sex preselection does become routinely available? Certain ominous predictions have been made in the light of these recognized preferences. Some have predicted that an excess of males in the population would lead to an increase in prostitution, homosexuality, marriage between males, and an increase in the number of males who never marry.

You might assume that most parents, given the opportunity of preselecting the sex of their child, would avail themselves of it. But current attitudes of married women in the United States indicate that they would not wish to be able to choose the sex of their children. These attitudes may well be modified, however, when the technology is proved to be simple, accurate, and socially acceptable.

25

Ethics, Morality, and
Prenatal Diagnosis

The significant clinical benefits of every major advance in medical research have inevitably raised social and legal issues. Kidney and heart transplantations are but two recent examples. The "new genetics" — best exemplified by prenatal diagnosis, carrier detection, and DNA technology — has, as expected, made waves that disturb both the family and society at large. The acceptance of these new genetic advances depends enormously on personal belief and morals, and equally on theology and the laws of the land. Questions and dilemmas have arisen. The impact has directly implicated parents, physicians, the fetus, and our social institutions.

The Parents

All parents, given the choice, would undoubtedly choose to have a child who is normal and healthy. No such choice, however, existed until the late 1960s, when it became possible to predict accurately whether or not a fetus had certain serious genetic defects and to do so early enough for the parents to elect abortion if a fatal or untreatable defect was found. When the United States Supreme Court decision in 1973 made abortion legal, the last obstacle was removed in making the crucial prenatal-diagnosis test a practical option for parents.

This hard-won right of women is threatened continually — as it was in 1987 by a bill passed by both houses of the Illinois legislature. This bill, amending the state's Right of Conscience Act, would have specifically allowed physicians to refuse involvement in prenatal diagnosis that might lead to an abortion. The language of the bill made it clear that a physician need not counsel a woman, or offer her prenatal diagnosis, or refer her to another physician — *even* if she had an increased risk of having a defective child! No individual, state legislature, or federal government should be allowed to force couples to have a retarded or otherwise genetically defective child. This was the effect of the Illinois legislature's "right-to-life" bill. Fortunately, Governor James R. Thompson vetoed the legislation. Such bills, if enacted into law, threaten the very freedom we all hold so dear in a democracy.

A PRIOR COMMITMENT

There are parents solidly opposed to abortion, who would rather not know whether a fetus is defective or not. Others, concerned about the morality of abortion, often have an amniocentesis and prenatal studies anyway, hoping that no defect will be found and that the question of abortion will not come up. No *prior* commitment to abortion is necessary, since all parents have a basic right to know not only about their own personal health, but also about the health status of their future child. Those parents who are ambivalent do not invariably have to address themselves to the question of abortion. When confronted with the bad news that the fetus indeed has a serious genetic defect, the overwhelming majority of parents nationwide (but not all of them) have opted to terminate the pregnancy.

ABORT A DEFECTIVE FETUS?

The fundamental philosophy of prenatal diagnosis is to reassure parents at risk that they may selectively have *unaffected* children. In only about 3 percent of all at-risk pregnancies studied is a fetus found so defective that mental retardation or serious or fatal genetic disease is certain; hence, in only about 3 percent of all such pregnancies is there a need to decide about abortion. However, because of prenatal diagnosis, thousands of couples who formerly were too petrified to take a chance on account of their risks now have their own healthy children. Indeed, the number of children born because of the availability of prenatal diagnosis and abortion is far greater than the number of pregnancies that are terminated because of serious fetal defects.

Those who would rather not abort and who are willing to accept high risks of having a defective child have every right to stick to their beliefs, though they may agonize for the rest of their lives over the fact that they have condemned a child to pain, deformity, and often, eventually, the horrors of institutionalization. (I cannot help but reiterate here the opinion handed down by the Rhode Island Supreme Court that "a child has a legal right to begin life with a sound mind and body.")

You might expect that those individuals who choose not to abort obviously defective fetuses would themselves take care of their suffering offspring, who may be blind, deaf, and profoundly retarded, who may leak feces and urine and be incapable of being toilet-trained, and who may require attention day and night and special care for a lifetime. In reality, many such families who are "morally" against abortion eventually place their affected children in institutional situations equal to a horrible living death. Thus, with a moral twist, they force everyone else (society) to assume responsibility for the physical and financial burdens entailed in their "moral stance."

I have repeatedly seen the misery, despair, and absolute desperation of parents who try to take care of extremely defective children in their homes. Yet those who institutionalize their children are almost invariably racked by guilt. Either way, I respect their right to choose. I definitely do not subscribe to the view, held by an increasing number of people, that parents who elect not to have prenatal diagnosis, and subsequently have children with defects that could have been prevented, should be financially responsible for the care of such children. (The argument is that sooner or later we all pay through taxes for the care of the defective.) I do believe, though, that prospective parents of a defective child should pause to reflect on the possible future that stretches ahead of them, the suffering of the child and the burden on their other children after their own deaths, before turning down prenatal diagnostic studies. But, as I have said, I do not believe it is the responsibility or the right of either the physician or society to "tell" the parents what to do.

WIFE VERSUS HUSBAND

Difficult situations have arisen in which a father, against abortion, has found himself in disagreement with his wife, who wishes to abort a defective fetus. You can imagine — if you have not already seen it — a woman who is a carrier of a sex-linked disease, such as muscular dystrophy, deciding that she will not continue the pregnancy if the fetus is a male, while her husband, believing he is the decision-maker, insists that the pregnancy not be terminated. The courts have repeatedly addressed this question. In one Illinois case, the court denied a husband's motion to prevent his wife from having a routine abortion. The court declared that the right of privacy "was broad enough to encompass a woman's decision whether or not to terminate her pregnancy." This stand is in accord with the established legal principle that every adult of sound mind has a right to determine what shall be done with his or her body.

A recent encounter I had remains unique in my twenty-two-year-long experience with prenatal diagnosis. A woman who had undergone prenatal genetic studies because of her age was informed that the fetus had Down syndrome and was advised to come in immediately for genetic counseling. Repeated calls by her own doctor urging her to see me were eventually successful. She first made and then canceled appointments repeatedly, before finally arriving in her twenty-fourth week — too late for abortion. She and her husband entered my office with him crying. She had told him the diagnosis for the first time on their way in, having earlier misled him into believing that the test results were normal. She, it turned out, was against abortion, a course she knew he would have urged.

MATERNAL LIABILITY

A pregnant woman is, of course, responsible for the fetus in her womb, but only recently has she been regarded as *liable*. United States courts in eleven states have ordered cesarean sections, overriding pregnant women's refusals (usually on religious grounds) to undergo recommended surgery. These cases mostly involved situations in which the placenta covered the exit of the womb — a condition called *placenta previa*, which threatens the lives of both mother and baby at delivery.

This new judicial philosophy expands the potential liability of mothers who negligently or purposefully damage or abuse a viable fetus — that is, one at least 28 weeks old. Pregnant women who are abusers of drugs or alcohol are currently the main focus of concern. In fact, late in 1986 a California woman was arrested and criminally prosecuted for willfully failing to provide adequate care for her fetus and thus causing the death of her newborn as a consequence of her drug abuse. An English court, subsequently upheld by the House of Lords, ordered the removal of a baby from its drug-addicted parents, after the child was born suffering the effects of drug addiction. Extension of child-abuse laws to include the fetus in the last three months of pregnancy appears to have begun. This trend is neither contradictory nor in opposition to a permissive attitude to abortion *before* fetal viability is reached.

A SPECTRUM OF SEVERITY

Abortion foes persistently claim that terminating a pregnancy because the fetus has a serious defect denigrates those living with handicaps and, further, makes society intolerant of imperfection. In my view, however, the prime constitutional principle to be heeded is that all couples have a fundamental right to procreate and found a family without interference by government or any religious group. No one should have the right to force a couple to have a seriously defective child by disallowing abortion. Enabling couples to avoid lifetime tragedies in no way detracts from our heavy responsibilities to care for the handicapped.

Prenatal-diagnosis cases do arise in which disorders of variable severity are detected and in which no predictive measure of the degree of disability is possible. One of the best examples is the XYY male fetus, which can have the unpredictable and variable features discussed in chapter 5. Another might be sickle-cell disease, which is routinely detectable sufficiently early in pregnancy to allow parents the option to abort. Between 5 and 10 percent of those with sickle-cell disease do not survive past five years of age, after suffering constantly throughout their brief lives. Many others, however, will survive to thirty and beyond with intermittent pain, infections, and crises. Some 5 to 10 percent of affected children will suffer strokes. Abortion choices are difficult given such unpredictable severity of a disease, yet are even more complex when treatment but not cure is available.

The best example in which effective treatment is available but necessary for a lifetime is hemophilia. Since excessive bleeding is the problem, hemophiliacs have traditionally required many blood transfusions. The AIDS epidemic has resulted in many hemophiliacs contracting AIDS through transfusions. This complication has made couples faced with a certain prenatal diagnosis of hemophilia even more seriously consider abortion in the last few years. The inability to guarantee that the blood supply is 100-percent safe from AIDS viruses suggests that these dilemmas will continue for some years.

Many other prenatally detectable disorders may be treatable to a greater or lesser extent (see chapter 33). Some, such as obstructed kidneys, require surgery in the womb with uncertain results; others, such as phenylketonuria, require long-term difficult diets. Decision making in such cases is complex and requires sophisticated medical and genetics expertise.

The excruciating problem encountered when one co-twin is detected with a grave defect is discussed in chapter 21 in the context of selective feticide.

EARLY DIAGNOSIS, LATER DISEASE

New gene-analysis techniques (described in chapter 8) now enable the prenatal diagnosis (as early as 9 to 11 weeks) of fatal or serious genetic diseases that will manifest themselves *decades* after birth. The first example is the fatal disorder Huntington disease, which mostly begins to show itself through signs of brain degeneration in an affected person who is between twenty-five and fifty years of age. Other examples of late-manifesting diseases in which early prenatal detection is or will soon be possible include myotonic muscular dystrophy, neurofibromatosis, adult-type polycystic kidney disease, and certain familial heart disorders that involve high blood cholesterol (or fats).

Couples have begun to face extraordinary dilemmas in trying to decide whether to request a prenatal diagnostic study, and whether to abort a pregnancy with a defective fetus or have the child and hope that progress over a few decades will yield an effective treatment. For diseases that are dominantly inherited, such as Huntington's, even more painful implications exist. In such cases, if one parent has the disease, there is a 50-percent risk that each child has the disease also, although it may not yet have shown. For such a person at risk who in turn starts a family, a prenatal test that finds an affected fetus means that this parent is also doomed.

Who, then, would want to know this information when no treatment is possible in the near future? Those who hear good news have a new lease on life, while the unlucky ones may be able to plan their lives and families with clear knowledge of the implications. Depression and suicide among individuals at 50-percent risk for Huntington's are known to occur much more frequently than in the population at large. Will such "no-hope" knowledge help or hinder people in their quest to survive and procreate? (Faced with no treatment option, those at risk for Huntington disease who are asymptomatic are advised to take extreme care in deciding whether or not to have a fetus tested. *Experienced* geneticists should be consulted before irrevocable personal decisions are made.)

One other issue — that of confidentiality and disclosure — confounds the new prenatal-diagnosis possibilities. The nature of the DNA tests discussed earlier (chapter 8) will for years to come require simultaneous study of blood samples of some close relatives. How results are kept confidential and to whom disclosure is made turn out to be far from obvious. For example, suppose that an individual with Huntington disease dies and his sister's blood is obtained for the study needed by the deceased man's pregnant daughter. Say the result shows that the sister also has the disease, which has not yet become obvious. Moreover, she does not wish to know the result — even though pressed by her own children, only *some* of whom want to know their risks. The family pressures can be enormous; and even if the sister's request is honored, her children might be able to discover the information from their cousin.

DETERMINING THE SEX OF THE FETUS

Many laboratories, including our own, have been approached repeatedly for prenatal sex determination on the basis that if the fetus is not a particular sex, then abortion would be sought by the parents. I have taken the position that prenatal sex determination for "family-planning" reasons is morally inappropriate, and have therefore always declined to provide any such service. This position is not really counter to my fundamental philosophy that parents have a right to know everything about their fetus, including its sex. In both China and India, with their crushing population problems, such uses of prenatal diagnosis are not frowned upon.

The Physician

A MATTER OF TRUST AND COMMUNICATION

On occasion, the parents of an affected child may expressly forbid their physician to contact or communicate with other family members who may be at risk for having offspring with the same disease. In muscular dystrophy, for instance, the mother's sisters may have an increased risk of being carriers.

Would you consider the physician justified in going above the heads of the family, searching out the mother's sisters, and informing them of their risks and available options such as prenatal diagnosis? Do you believe that the physician acting in this way on behalf of the sisters, who may be at risk, can or should be sued for breach of confidentiality and trust for divulging private information? The obvious precedent for abridging the physician's duty of confidentiality may be found in statutes that require doctors to report persons with certain contagious diseases, such as meningitis. In such cases, laws confer "absolute" immunity on the physician.

In an enlightened society, you would expect communication within families sufficient to ensure that children would be made aware of serious genetic disease in their own (possibly deceased) siblings or other close family members. Unfortunately, as discussed earlier, it is not rare for parents to hide from their own children facts concerning an institutionalized or deceased defective child or close relative. Do you believe that such children, who may themselves be carriers of a genetic disease have a legal right to be informed by their parents about disorders that affected their siblings or other family members? What legal recourse do these individuals have, and what is the physician's role in divulging their family history to them?

The law is not especially helpful in this regard, but in many states a physician's license can be revoked for willfully betraying a professional secret. If you were engaged to be married and were worried about possible genetic disease in your future spouse's family history, would you expect your spouse-to-be's physician to divulge that medical background to you? How would you feel if you were the one with the disease in your family and your physician revealed it? And what about the ethics and legality of such action by the physician?

The Fetus

QUESTIONS, QUESTIONS

A veritable cascade of questions has enveloped all discussions concerning the fetus and the rights of parents. Who can dispute that parents have a right to have children? Does this right continue if they have a high risk of producing defective offspring? If the mother, for example, has a disease called phenylketonuria, she will damage almost 100 percent of her offspring: virtually all of them will be mentally retarded or have serious birth defects. There are those who argue that parents in such situations have no right to continue to produce defective children.

But does society, acting for the good of all, have the right — or indeed the responsibility — to adjudicate or even legislate for parents at risk for producing genetically abnormal offspring? Should society restrict the number of offspring (as is now happening in China), or even offspring of a particular color? Does a group have any right to dictate on religious or metaphysical grounds that parents should not prevent the birth of a defective offspring? Can it be usefully argued that the fetus is not merely a part of a woman's body, but a separate being over whom she may be presuming to exercise divine judgment? Then, if the fetus is indeed considered to be a living being with equal rights, does not every fetus (or child) have the inalienable right to be born free of physical and mental defect? Does the fetus therefore have societal rights, and if so, at what point during pregnancy?

THE FETUS AND THE LAW

Two major problems complicate all considerations of the legal status of the fetus. First, the fetus has been determined by the United States Supreme Court not to be a "person" as defined by the Constitution. Second, there remains a problem concerning when life begins. Certain religions are clear in their belief that life begins at conception. In efforts to define more specifically when life begins, medicine chose the concept of "viability," defined as the ability to exist outside the womb. A fetus after 24 weeks along in pregnancy is considered viable if able to exist on its own outside the mother's body.

What is already abundantly clear is that the fetus does have legal rights. Many cases have been adjudicated, concluding each time that the fetus has property rights and may inherit. Some courts have vested the fetus with legal rights as of conception, provided that it is subsequently born alive — even if only one gasp of breath is taken. Indeed, legal actions have been brought against individuals responsible for a father's death by children who were still in the mother's womb at the time. Other courts have recognized such fetal rights only if injury to the parent occurred at the time the fetus was in the "viable period." (Other aspects of genetic defects, the fetus, and the law are discussed fully in the next chapter.)

WHEN DOES LIFE BEGIN?

While the courts have already adjudicated cases concerning the legal rights of the fetus, society, through its lawmakers, has begun to debate and legislate. Nevertheless, neither the courts nor society will ever be able satisfactorily to settle the deep moral and ethical dilemmas that have beset the recent advances in prenatal diagnosis. Again and again, it is asked: When does the fetus take on an individual existence? Is it at the moment of conception, or perhaps when the fertilized egg implants in the wall of the womb? Is it at the moment the heartbeat is established, or when brain-wave activity is demonstrable? Or, is it only when the fetus is viable — after 24 weeks — and is able to exist outside the womb? While it will always be impossible to reconcile the views of those who believe life begins at conception with the views of those who interpret life in terms of its quality, dignity, and humanness, it would seem reasonable to allow people to pursue their own beliefs without being forced to follow the religious or other dictates of those with whom they disagree.

Human nature is such that if something good can be done, such as prenatal diagnosis, then it usually will be done. However, what you or I may call good and right, someone else may call morally or ethically wrong. How, then, can guidelines be established that reflect our humanistic concerns? Professor Joseph Fletcher, an acknowledged ethicist, now retired from the University of Virginia, believes that "if human rights conflict with human needs, let needs prevail. . . . [R]ights are nothing but a formal recognition by society of certain human needs, and as needs change with changing conditions, so rights should change too."

Society

What may be good for you and your family may not necessarily be good for society as a whole. A balance of benefits is constantly being struck between individual needs and society's goals. The ability to avoid genetic disease by prenatal diagnosis has developed against the background of rapidly changing cultures all over the world. While in some countries the specters of famine, disease, natural disasters, and war still loom large on the horizon, many other nations are caught up in their public concern for population growth, women's rights, the consequences of illegal or legal abortions, the number of unwanted children, the question of euthanasia for the congenitally defective newborn and irreparably brain-damaged individuals, as well as the soaring costs of long-term medical care.

Historically, Western society has taken very definite steps to secure the best public health for the good of all. In the United States, many compulsory requirements have been issued by the individual states. There are numerous statutes in a variety of jurisdictions that require persons to be vaccinated or immunized against smallpox, measles, German measles, and poliomyelitis. A state requires testing for venereal disease and Rh (blood incompatibility) disease. A state may even demand treatment for venereal disease, tuberculosis, and for newborn eye infection (neonatal ophthalmia). A state may reserve the right to incarcerate those who are mentally incompetent. Most states prohibit all mating of persons as close as or closer than first cousins. Others currently have compulsory sterilization laws, and a significant number of jurisdictions retain legal authority to sterilize institutionalized mental defectives. A few states even permit the sterilization of certain convicted felons.

DETECTING CARRIERS OF HEREDITARY DISEASE

The 1970s saw a rash of state legislation governing the question of screening for carriers of hereditary disease. So hasty have some of these efforts been that in one state, Georgia, lawmakers inadvertently required the impossible — immunization for sickle-cell anemia! To provide the largest number of options, it would be most sensible to provide young people at the time of marriage with tests to determine if they are carriers of certain genetic disorders. In this way, the best options are available to prevent the occurrence of any tragedies. Currently, the only test widely demanded at marriage is the one for venereal disease, such as syphilis. States are rapidly moving to make AIDS testing mandatory also. Wide-scale screening programs for disorders such as sickle-cell anemia or Tay-Sachs disease have been systematically replaced by expectations that doctors (at risk of malpractice) would offer carrier tests to patients in specific ethnic groups at risk.

CONFIDENTIALITY AND INSURANCE

Detection of a chronic fatal dominant genetic disease such as Huntington's, which is associated with a 50-percent risk of transmission, would be of crucial interest to a life- and health-insurance company. Given such a family history, could a company demand specific gene tests before providing coverage? After all, insurers have always requested urine, blood, electrocardiogram, and other tests, and have lately begun AIDS testing, too. The infrequent occurrence of fatal hereditary diseases may make gene testing unnecessary, but clearly life-insurance companies prejudice the sick and could easily extend their actuarial caution to the asymptomatic possessor of a fatal gene.

COSTS AND BENEFITS IN PRENATAL DIAGNOSIS

You may feel that it is futile to consider the benefits of preventive programs when treatment is not yet available and abortion is the only real option. Some facts are, however, unsuppressible. The already overburdened taxpayer often feels that where prevention is possible, it should be pursued. The burgeoning health-care costs of institutionalization for the defective individual represent a clear, chronic financial drain. The projected cost of lifetime care in an institution for one individual with Down syndrome far exceeds $4 million. The estimate for lifetime home-care of a child with spina bifida is in the hundreds of thousands of dollars. The taxpayer is understandably concerned. Each year in the United States, close to 20,000 children are born with some kind of chromosomal abnormality, and about 6,000 are born with anencephaly or spina bifida, the two most common birth defects. Institutional-care costs are presently about $85,000 per year in Massachusetts for each patient. Society, in other words, makes a commitment each year in excess of billions of dollars for these two birth defects alone. The expenditure by society over twenty years at current costs (which, of course, will not apply) will have grown to over $40 billion!

Hence, willingly or not, we are forced to consider the economic aspects of prenatal diagnosis, remembering that these techniques are mainly applied to disorders that involve irreparable mental defect or fatal or serious genetic disease. Taking into consideration the cost of amniocentesis and prenatal studies, as well as the cost of elective abortion, estimates have been made of total "avoidance" costs. If you consider only women who are age thirty-five and over, in the United States alone, then the necessary costs, provided all such women had the procedures, would be less than $150 million. As I have said, mothers thirty-five and older, while constituting only 5 percent of the childbearing population, give birth to about 20 to 25 percent of all children with Down syndrome. Projections for the cost of lifetime care for defective offspring born to mothers aged thirty-five and over show that society would be committing over $2 billion in one year — many times the cost of avoidance through prenatal diagnosis and elective abortion.

26

Prenatal Diagnosis:
Lessons from the Law

Anguished and angry parents who have a baby with a serious birth defect, or who later discover that their child is retarded, may seek help from the courts to punish the doctor, nurses, or hospital and to obtain a monetary award for the costs of care and pain and suffering. Many such cases evolve because of poor communication between the physician and the couple concerned. When a warm, empathetic relationship exists between a physician and a couple, the likelihood of a malpractice suit is very small. In contrast, the noncommunicating physician who evades questions, who appears hostile and unconcerned when questioned, and who transmits complex explanations hastily and in technical jargon, clearly invites a legal suit. Increasing degrees of anguish, anger, frustration, and thoughts of revenge against the medical establishment are seen after fetal death, stillbirth, death soon after birth, delivery of a child with a serious birth defect, and discovery of a child with mental retardation following brain damage during birth.

Couples who decide to bring a malpractice action should, however, give the most careful thought before undertaking this route. Few realize the emotional exhaustion, the frustrations encountered in the process, and the subtleties or nuances that distinguish a rational medical decision from negligence. What some parents might regard as an overt act of reckless negligence by the physician may ultimately be regarded by a jury as non-negligent — a rational, though not optimal, act by the physician, who therefore is not liable. The couple bringing suit must also be able to prove that the negligence they claim directly "caused" the birth defect, brain damage, or other problem. Emotionally overwhelmed, some couples may fail to recognize the realities — that a tragic outcome to pregnancy is not rare and that culpability does not necessarily attach to anyone. We all have a risk of having a child with a birth defect or mental retardation, though couples are encouraged to take as few chances and make as many informed choices as possible.

Couples who find a successful attorney whose expertise is medical malpractice may discover that he or she turns down their case. In the United States, over 75 percent of the cases are rejected if the chance of winning is regarded as slim. Since the contingency fee is prevalent in this country, medical-malpractice lawyers are keen to pursue those cases in which large monetary awards are possible — especially those in which success is more likely than not — rather than cases in which the potential reward is relatively small. If a fetal death or stillbirth occurred, a modest award at best can be anticipated in contrast to the considerable sums the parents of a *surviving* retarded or defective baby might receive.

The contingency-fee system allows parents to bring suit without any initial cost to themselves. If their suit should be successful, usually about a third of the award will be taken by the lawyer, who will have spent on average much more than 400 hours on the case. In addition, expenses for court costs, expert witnesses, medical examinations, and so forth will have to be subtracted from the remaining portion of the award. The consequence is that a large portion of the award does not in fact end up benefiting the surviving child.

While revenge may have been sought, couples I have spoken to after *successful* legal marathons — usually of two to five years' duration — frequently have mixed feelings. Some have regretted that they took legal action, while others have been satisfied in receiving adequate financial awards to assist in the long-term medical care of their damaged or defective child. Either way, no award will remove their hurt, correct a birth defect, remedy mental retardation, or right a wrong.

Claims of Negligence

When a patient sues a doctor for malpractice, an injury allegedly stemming from the physician's treatment or lack thereof has occurred. Such malpractice claims charge negligence by the physician. For a patient to succeed with a cause of action for negligence, four interrelated elements must be proved:

1. *Duty:* The patient — called the plaintiff — who is suing the doctor must demonstrate that the physician in question owed him or her a clear duty or obligation. This duty, which is recognized in law as having arisen through the physician-patient relationship, requires that the doctor act in accordance with the normal standard of care established by the medical profession. In general, this means protecting patients from unreasonable risks.

2. *Breach of Duty:* The plaintiff must demonstrate that the doctor actually failed to act in accordance with the normal standard of care — whether the physician simply omitted to do something important, or actually did something that violated the normal standard of expected care.

3. *Cause:* The plaintiff must be able to show that the act of the physician could reasonably be interpreted to have "caused" the resulting "injury." (This would not mean that the physician would be held responsible, for example, if the German-measles virus caused brain damage in a fetus and the child subsequently was born with mental retardation and other birth defects. Rather, the physician might be liable for not having counseled the patient about the risks of such an outcome and for not having informed a patient about the option of selective abortion. Hence, the legal phrase above simply refers to *a* cause and not necessarily *the* cause of the mental retardation in such a case.)

4. *Damages:* The plaintiff must show that whatever the physician did or did not do resulted in actual loss or damage, which may be physical, financial, or emotional, and which may have affected the patient or a near relative such as a spouse, heirs, children, and so forth.

For any plaintiff to win a case, there must be proof by a "preponderance of the evidence" that all of the above four elements exist. In other words, the judge or the jury (depending upon which court the matter is heard in) must simply find that the proof of the fact at issue is more probable than not.

Enormous advances in medical genetics have brought prospective parents new expectations in childbearing. As recently as the mid-1960s, it was not possible to make a prenatal diagnosis of a genetic disorder in the fetus. Soon after that time, such tests became possible and steadily have become the normal standard of care. At the same time, parents-to-be have expected not only to be informed about tests they should consider, but also to be fully informed about any greater-than-normal risks they might face. Failures by physicians to provide proper care in these contexts have led to a rapidly escalating series of lawsuits. Physician negligence in birth-defects-related legal actions has largely been pursued in cases that charge "wrongful birth" or "wrongful life."

Wrongful Birth

Malpractice claims against physicians for "wrongful birth" are brought *by the parents* of defective newborn infants. Such causes of action are well recognized in the United States. The basic claim made in such cases is that if it had not been for the physician's negligence, the defective child would not have been born.

Wrongful-birth claims are often based on the contention that a physician failed to provide or recommend genetic counseling so that a couple could learn of their increased risks, regardless of any action they may or may not have taken thereafter. Alternatively, if the physician failed to offer, perform, or recommend carrier-detection tests or prenatal diagnosis, liability could accrue. Obviously, had the couple been given simple blood tests and thus learned that one of the partners carried a harmful gene for a specific disorder, they might have avoided pregnancy altogether, had prenatal diagnosis, or considered other options. For instance, a couple at increased risk because of advanced maternal age may not have been informed about their risks or about the availability of prenatal diagnosis through amniocentesis; they subsequently might deliver a baby with Down syndrome. In a wrongful-birth suit, these parents could claim that if the physician had not been negligent, they would not have tried to conceive, or would not have continued the pregnancy.

In early cases, judges gave various reasons for denying the claims of parents in wrongful-birth suits. First the courts claimed that "every child is a joy and a blessing." Then the point was made that it was not the parents who were harmed or injured, it was the child who suffered. In denying such claims, the courts pointed out that the emotional pain and suffering of the parents is not usually legally compensable. Eventually, however, parents' claims came to be successful and the courts' reasons grew clearer and more rational. The courts held that (1) for every wrong committed, there should be a legal way to redress that wrong; (2) physicians are responsible for disclosing important information to parents-to-be; and (3) accurate genetic counseling, carrier-detection tests, and prenatal diagnosis are to be regarded as legal standards of care.

Wrongful Life

A malpractice action against a physician brought *by the defective child* constitutes a "wrongful-life" suit. In bringing legal action, the defective child is in effect saying that life with defects is worse than no life at all. Such causes of action were virtually unheard of some fifteen years ago, and their continuing development largely reflects the recent advances in medical genetics.

The wrongful-life malpractice suits have largely arisen in clear circumstances. Most often, claims have been based on alleged negligence in genetic counseling. In such cases, the physician may have failed to recognize the presence of a genetic disorder in the parents or failed to advise them of their increased risk of having a defective child. Hence, failure to inform about the availability or advisability of either carrier-detection tests or prenatal diagnosis disallowed them the chance to avoid the birth of a seriously defective child.

A variation on the same theme concerns a physician's negligence to advise about the risks to the fetus from a medication prescribed for the mother during pregnancy. Other actions have been brought when physicians failed to diagnose a specific disease in pregnancy (such as German measles) that resulted in the birth of a seriously defective child. Finally, wrongful-life suits have also been brought on behalf of children born after a negligently performed abortion or sterilization operation.

Wrongful-life suits have met with considerable resistance in the courts. Professor Margery Shaw, director of the Health Law Program at the University of Texas Health Science Center in Houston, has summarized the judicial reasons for courts *denying* such claims: Courts are not in a position to assess life with defects compared to no life at all. The child is not in a position to sue the physician, since if no "wrong" had been done, the child would not even be alive to complain. Courts have recognized a reverence for life, holding that life is precious even if impaired. Some judges have held that the sanctity of life should take precedence over the quality of life. Other courts have maintained that to recognize a child's claim not to be born is really a matter of public policy that should be resolved by legislatures and not the judicial system. The point has been made that there is no legal right *not* to be born and, moreover, no fundamental right to be born without any defects. Courts have also made the point that a physician owed a duty of care to a mother but not to a fetus. In the opinion of some judges, the child's defect was not directly caused by the wrongful act of the physician.

Notwithstanding all these judicial reasons and opinions for denying claims made by children in wrongful-life suits, thus far the supreme courts of California, Washington, and New Jersey have recognized the validity of such claims. The courts have pointed out, in upholding such claims, that there is a conditional responsibility and duty to a future child. Moreover, these courts have recognized that the child both *exists* and *suffers*, and that there should be just compensation for legal wrongs. This trend in the acceptance of wrongful-life suits is likely to continue.

Some Instructive Cases

There are many instructive lawsuits involving genetic disorders, prenatal diagnosis, or both. Examining selected illustrative cases should be helpful in alerting prospective parents about pitfalls and problems that can be anticipated. The various cases I have chosen as examples have been tried in court, settled out of court, are in process, or have occurred without subsequent legal action having been taken. Many represent my direct involvement as an expert witness testifying either for a couple or for a physician.

FAILURE TO OFFER NECESSARY TESTS

J.L. was thirty-five years of age when she became pregnant for the third time. Her first two children were three and six years of age, respectively, and both were in perfect health. She lived in a small town in Florida and sought her obstetric care from an osteopathic physician. She went for her first prenatal-care visit when she was about ten weeks pregnant. At no stage of her prenatal care did her doctor inform her about the increased risk of chromosome defects as a function of advancing maternal age. Nor did he offer or recommend an amniocentesis for prenatal genetic studies.

After a perfectly normal pregnancy, J.L. had an easy labor and an uncomplicated delivery. A few moments after birth, a clinical diagnosis of Down syndrome was suspected and later confirmed by blood chromosome analysis.

J.L. and her husband claimed negligence by her physician and the case was ultimately settled in their favor out of court.

Comment: All women have the right to be informed about any increased risk they may have for bearing a child with a serious birth defect or genetic disorder. Physicians have the duty and responsibility to provide the necessary information or to refer the patient elsewhere for more expert consultation, if necessary.

Jim and Sonya were excited about their first pregnancy. Both were in excellent health and both were in their midtwenties. They had not actually planned this pregnancy but were nevertheless happy that it had happened.

Sonya went for prenatal care when she was about eight weeks pregnant. Her obstetrician inquired about a family history of birth defects and Sonya mentioned her sister, who was born with spina bifida and was paraplegic and without bladder control. He reassured her that there was no reason to worry since "that condition was not hereditary."

The rest of Sonya's pregnancy was uneventful, but labor was prolonged and delivery difficult. At birth, the child had spina bifida and hydrocephalus, the latter accounting for the difficult labor because of the enlarged head.

Comment: Jim and Sonya claimed reckless negligence by their physician for not offering an amniocentesis for prenatal studies to detect spina bifida. The physician successfully claimed that the alpha-fetoprotein test had not yet been licensed by the Food and Drug Administration and was not yet the standard of care. After the first avenue failed, the parents tried a second approach and instituted a new claim: that the doctor had not counseled them about the approximately 1-percent risk they had for having a child with this type of defect. The matter is yet to be resolved.

FAILURES IN COMMUNICATING RISKS

A New York couple had a child with polycysic kidney disease. The child died five hours after birth. Following autopsy, the parents were reassured that their risks were not increased for having a similarly affected child in a future pregnancy. Litigation began after their second child was born with the same disorder and later succumbed at two years of age. Their physician had failed to recognize that this particular kidney disorder was inherited as an autosomal recessive condition and that their risk in *each* subsequent pregnancy was 25 percent!

A woman diagnosed as having multiple polyposis of the colon died as a consequence of cancer that developed in the polyps located in her large bowel. The malpractice suit was brought against the physician who had cared for her father — the father who had died some twenty-one years before, suffering from the same disorder — because the doctor failed to warn his patient's daughter about the hereditary nature of this disease. He also failed to provide her with the necessary genetic counseling, which would have indicated the 50-percent risk of this fatal disorder turning up in any of the man's children. Moreover, the physician failed to point out another option: the total surgical removal of the colon *before* cancer supervened. The affected daughter, not knowing about the autosomal-dominant inherited nature of this disorder, unwittingly had a child, who now also had a 50-percent risk of having this potentially fatal condition.

Comment: Couples should become fully informed about their family histories. They should specifically inquire of their physician whether a particular disorder is hereditary. If there is any uncertainty, they should telephone a clinical geneticist in a medical school–affiliated hospital. They should simply ask if the disease in a family member is hereditary. If the answer is in the affirmative, they should seek an appointment for genetic counseling.

Following the birth of two normal children, a couple's third child was found to have multiple birth defects, including six digits on each hand and foot, hydrocephalus, mental retardation, and other abnormalities. In addition to the child's own physicians, a leading clinical geneticist was also consulted. Neither the expert physician nor the child's doctor were able to recognize a specific genetic syndrome from the presenting anomalous features. The parents were so informed and were counseled that there was "little chance" that such a disorder would recur again in any future pregnancy. Subsequently, the couple's fourth child was even more severely affected, but this time with somewhat more characteristic features diagnostic of a recognizable genetic syndrome.

Comment: Only in retrospect was it possible in this case to recognize the specific multiple-congenital-anomaly syndrome inherited as an autosomal recessive disorder. It had not been possible, after the birth of the first affected child, to recognize this disorder and its 25-percent risk of recurrence.

A woman aged thirty-nine and in her fourth pregnancy asked her obstetrician whether she should have an amniocentesis for prenatal genetic studies. She had read a magazine article that pointed out the increased risk of chromosomal abnormalities in the children of "older" mothers. Her obstetrician told her that abortion was amoral and that anyway, in his experience, "amniocentesis kills babies." She had great faith in her obstetrician, who had delivered all her other children, and she felt disinclined to seek any other advice. The pregnancy ended in the birth of a child with Down syndrome.

Comment: All physicians have the right to practice their religious beliefs. Their overriding *duty*, however, is to refer for consultation or tests those individuals who could benefit from a second opinion or treatment by others. No physician has the right to visit upon a patient religious beliefs, moral precepts, racial dictates, or other matters of conscience. The physician's responsibility, besides providing medical care, is to be certain that each patient is fully informed about the relevant options. In this case, the physician not only fabricated the risks of amniocentesis in a highly unethical manner but failed to inform his patient accurately about the standard recommendation given pregnant women of her age. Some physicians may also hold religious beliefs that would have them save the life of the baby, rather than the mother, when faced with that very rare choice.

WAS THE BABY DAMAGED DURING DELIVERY OR DEFECTIVE?

When the fetus has a serious birth defect, it is not unusual to find problems or complications during labor or delivery. Prolonged labor, cesarean section, breech delivery, excessive or insufficient amniotic fluid — all occur more frequently when the fetus has a major birth defect. Not unexpectedly, such concomitant occurrences generate considerable upset and thoughts of litigation.

Maria's pregnancy had been uneventful. Labor, however, was prolonged and difficult. After many hours and with good obstetric management, the baby was eventually extracted. At birth, the well-above-average size of the head prompted suspicion of hydrocephalus — a condition that was rapidly confirmed.

Studies showed that the interconnecting canal between internal spaces in the brain (called ventricles) was so narrow as to effect a blockage to the flow of cerebrospinal fluid within the brain. Evidence of brain hemorrhage was found in the spinal fluid obtained soon after delivery.

During the ensuing months, it became clear that the child had serious mental retardation. Maria and her husband, angered by the long and traumatic labor, sued their obstetrician for negligence. They contended that a cesarean section would have secured normal brain function for their son.

Comment: Unfortunately, no one referred the child to a clinical geneticist for further consultation, and neither the child's physicians nor his parents realized that the internal brain obstruction, called aqueductal stenosis, was a recognizable genetic disorder of the sex-linked type. In other words, there was a strong likelihood that the condition was directly inherited from Maria, who, though unaffected, carried the harmful gene for this disorder, which results in hydrocephalus. Since tests revealed no specific cause for the obstruction, the other possibility was that a spontaneous mutation in a gene caused the disorder. Either way, the condition, with its associated brain malfunction that leads to mental retardation, was present during fetal development and was in no way caused by the obstetrician. No significant disproportion existed between the baby's head and the mother's pelvis, and cesarean section would not have secured normal brain function.

Two important lessons emerge from this case. First, couples who plan to pursue legal action should first obtain expert medical opinion — for example, from a clinical geneticist — before embarking down this long and arduous path. Second, a consultation with a clinical geneticist is probably wise after the birth of a child with any serious birth defect. A clear understanding of the cause, subsequent risks in future pregnancy, and potential options for prevention could possibly then be discovered.

GENETIC DISORDERS AND OBSTETRIC MALPRACTICE

Just as we have learned extremely valuable lessons about normal human development from studies of very rare biochemical genetic disorders, lessons from cases of obstetric negligence, which are not as rare, can also help us achieve better care for all in the future. A few cases illustrate important lessons:

L.M., who was blind, came to her obstetrician for prenatal care. He provided routine care and treatment, pregnancy progressed without incident, and labor and delivery were uneventful. The child when born appeared healthy and was discharged from the hospital with the mother.

When the child was about seven months of age, her father observed a peculiar whitish appearance behind the pupil of one eye. Upon consulting an ophthalmologist, the parents discovered that the child had a malignant eye tumor called retinoblastoma — a defect that was shortly thereafter also found in the baby's other eye. Despite treatment efforts with anticancer drugs and X rays, the child's eyes had to be removed to save her life.

Comment: If only L.M.'s obstetrician had been careful and empathetic and had inquired about the cause of her blindness, he would have learned that her eyes had also been surgically removed when she was an infant. She knew that she had had eye tumors, but not the name of the disorder, nor that it was a genetic condition. The gene that causes malignant retinoblastoma in both eyes is transmitted as an autosomal dominant disorder. L.M.'s risk was therefore 50 percent for bearing a child with this condition.

L.M. had been adopted as a child and had no knowledge of her family history. Her physician, however, could easily have asked an eye specialist about what tumor of the eyes leads to their removal, if he lacked that knowledge. Once informed, the physician could have referred L.M. for genetic counseling and she could have learned her risks and realized her option to terminate the pregnancy. L.M. had, in fact, previously elected to abort two pregnancies.

The pediatrician who took the child under care after birth also failed to inquire about the cause of this woman's blindness. He immediately should have recognized the high likelihood of retinoblastoma and its genetic basis, and should have initiated a systematic and repeated examination of the child's eyes under anesthesia, at least throughout the first two years of life. In this way, an early diagnosis would have been made and prompt treatment initiated. There would also have been about a 90-percent possibility of not only saving the child's eyes but also ensuring reasonably good vision!

No physician, regardless of specialty, can be expected to know everything in medicine. There is, however, a duty to *confer with* and *refer to* other specialists. For this to occur, a physician must remain alert, informed, and caring enough to seek the best for each patient. Any lesser standard is unacceptable.

Carol had epilepsy with major seizures that were controlled by phenobarbital and Dilantin. She was very excited about her first pregnancy and went early for prenatal care. Carol continued to take her medications and her pregnancy was uneventful. She had no seizures throughout its course and her baby was born following a normal labor and delivery. A cleft lip and palate, a somewhat unusual-looking face, and very underdeveloped fingers and toenails were all evident immediately. By the end of the first year of life, it was apparent that the child was mentally and physically retarded.

The legal claim against Carol's physician was that he was negligent in failing to advise her about the risk of birth defects due to her epilepsy and associated drug treatment.

Comment: Women with epilepsy who take major drugs for seizure control have risks between 7 and 10 percent of bearing a child with a major birth defect, mental retardation, or a genetic disorder. This information has been known for over a decade and should have been shared with this patient. Indeed, she should have been made aware of these hazards even before becoming pregnant. Once pregnant, she would have wanted to consider this information in deciding whether to continue the pregnancy or not. Notwithstanding the failure of this obstetrician to inform his patient about the above risks or to refer her to a clinical geneticist who could, all women should take upon themselves the responsibility of determining the safety of *any* drug they *have to* take, especially during early pregnancy.

A couple who had two children with neurofibromatosis ("the Elephant Man disease") decided to have no further children. (This disorder has multiple and variable signs, including coffee-stain-like birthmarks, tissue nodules or lumps in skin or elsewhere, curved spine or other bone problems, possible deafness, and even mental retardation.) The husband was similarly affected and he elected to have a vasectomy.

About seven months after the man's vasectomy, his wife, concerned that she was again pregnant, contacted her obstetrician. After her suspicion was confirmed, the woman elected to terminate the pregnancy during the first trimester. She underwent an abortion by D and C (dilatation and curettage). Six months later, however, she delivered a child who also had neurofibromatosis!

Comment: In this case, there were major questions on a number of issues. First, was adequate advice provided to the father about the chances of a failed vasectomy, and were tests on his semen adequately performed to determine the presence of sperm some months after surgery? Since the couple had a 50-percent risk of having another affected child, did the obstetrician pay sufficient attention to the D and C he performed, and to the subsequent follow-up care in which he failed to diagnose a persisting pregnancy? Did the pathologist who received the tissues removed at the time of abortion fail to realize that products of conception were not present?

Negligence concerning the performance of the D and C and the failure to diagnose a persisting pregnancy is self-evident. A failed abortion procedure is very rare. The husband should have been counseled before the vasectomy and properly tested after it. Certainly, evidence of fetal tissues should have been seen by the pathologist if the D and C had been performed properly. The case was settled in favor of the couple after nine years of legal wrangling.

All couples should be aware that vasectomy does not necessarily result in 100-percent sterility and that testing of the semen is necessary a few times after the procedure. Recanalization of the tubes that carry the sperm from the testicles may occur in a very small percentage of cases, but not as soon as a few months.

ERRORS IN SPECIAL LABORATORY TESTS

In most cases, special reference laboratories are used for critically important genetic tests. These include tests to determine the presence of a genetic disorder in a fetus or to discover whether one or both parents carry the harmful gene for a particular hereditary disorder. One important case involved BioScience — a major commercial laboratory in California.

> A young couple of Ashkenazic Jewish ancestry had blood samples drawn to determine whether they were carriers of the gene for Tay-Sachs disease. The report provided indicated that neither was a carrier. In due course, they had a child who was eventually diagnosed as having the fatal Tay-Sachs disease.

Comment: This laboratory had previously been warned about the inaccuracy of its testing procedures. Subsequently, of course, properly executed tests revealed that both parents were carriers of the gene for Tay-Sachs disease. Couples who have *special* laboratory tests (those not performed in a routine hospital laboratory) should inquire about where their samples are being sent for study. They should also request information attesting to the experience and record of accuracy of that facility. The noncommercial medical school/hospital genetics laboratories are invariably able and willing to provide information about their accuracy and experience.

Erroneous prenatal diagnoses have occurred (see chapter 22), mainly as a problem inherent to such testing. For example, the result of prenatal genetic studies following amniocentesis may incorrectly indicate that the fetus is a chromosomally normal female. About once in 400 tests, accidentally admixed *maternal* cells, which can grow in the cell culture instead of the amniotic-fluid cells from the fetus, lead to a misleading result. Rare instances in which a male fetus with Down syndrome was missed are probably the result of this problem of maternal-cell admixture. Rarely, errors have been made in the laboratory interpretation of the chromosomal complement or of a biochemical test.

Mislabeling of a slide or tube mix-up has accounted for some errors in prenatal diagnosis. In one such case, a tube mix-up by a laboratory technician led to a test result being provided to the wrong patient, a normal fetus being aborted, and a defective child being born. Generally, however, prenatal genetic studies are remarkably accurate. The overwhelming problems arise almost invariably when physicians fail to provide or refer for genetic counseling or specific genetic testing when indicated.

ERRORS IN COMMUNICATION

Some people might be surprised to learn that disastrous errors have occurred in the process of communicating laboratory results. In one successful malpractice suit, a woman claimed that the results from prenatal genetic tests she had undergone had not been properly explained to her. As a consequence, she delivered a child with spina bifida and hydrocephalus, despite the fact that she had undergone the particular tests for this condition. Sufficient care had not been taken to explain carefully the implications of the laboratory and ultrasound study results. If fully informed, she argued, she would have chosen an elective abortion.

In yet another case, the amniotic-fluid cells had not grown well in the laboratory. Despite repeated phone inquiries by the parents, the laboratory technician told them, "Don't call us — we'll call you." Eventually, when some results were available, it was too late to offer an elective abortion. The patient delivered a child with

Down syndrome. At our Center for Human Genetics, standard laboratory practice is to inform patients that if no result is obtained by 23 weeks of gestation, they retain the option to terminate the pregnancy simply on the basis of risks, since on occasion no result may be forthcoming because of poor-growing cells. Generally, in prenatal studies that are done in a timely fashion, obtaining absolutely no results is rare.

Communication failure in another case involved a request made by a laboratory for a second sample of amniotic fluid following failed cell growth in the first specimen. The request was made to one physician in a group practice, and it was thought that the message had been communicated to the patient by a partner while the first physician was away on vacation. This turned out not to be the case. By the time it was realized that the recommendation for a second amniocentesis had not been communicated to the patient, it was too late for a repeat study. This patient, who had been particularly anxious about Down syndrome even though she was not yet thirty-five, duly delivered a baby with that disorder. The child also had the typical associated defects found in Down syndrome, including a severe malformation of the heart and intestinal obstruction (duodenal atresia), which required surgery soon after birth. The parents, in this case, refused to see or accept the child.

Sean and Laurie had a daughter with profound mental retardation and microcephaly. Very concerned about the risks for recurrence, they went to a medical school–affiliated hospital for genetic counseling. There they learned that the cause of their baby's birth defect was unknown. Care was taken, however, to point out that even though they had no family history of this disorder, it could still be hereditary. If so, the gene could be transmitted equally by both parents at the same time, giving them a 25-percent risk of having an affected child in every subsequent pregnancy. Such a hereditary disorder transmitted as an autosomal recessive trait is known for microcephaly and mental retardation.

Sean and Laurie carefully considered their options, and since no cause had been recognized, they decided on another pregnancy. They consulted their obstetrician and received reassurances that they had little to worry about and that he would do everything to be sure that they had a healthy baby.

About one year later, under this obstetrician's care, they delivered a child who was as severely affected as their first. Review of the medical records revealed that the obstetrician had taken no steps to determine whether the fetal head was developing normally. He had not ordered an ultrasound examination for such measurements to be made until 23 weeks of pregnancy. The study he ordered was simply to assess the duration of pregnancy, no mention being made about the previous microcephaly on the ultrasound requisition slip. The actual report indicated a discrepancy between fetal body size and diameter of the head. The body-size measurements indicated a pregnancy of 28 weeks while the head size was consonant with fetal development of 23 weeks.

Comment: Given that there was a potential 25-percent risk of microcephaly and mental retardation in this pregnancy, a normal standard of obstetric care would have required sophisticated, serial ultrasound measurements — for example, at 16, 19, and 23 weeks of pregnancy — in which accurate measurements of the fetal head, and other body parts for comparison, would have been made. This was a patent case of obstetric malpractice. In the legal deposition, Laurie told of her meeting with this obstetrician one Sunday in church. She tearfully approached him to inquire why he had let this happen. He responded that since he had seen her in church, he knew that she would understand!

Once again, no physician has the right to visit upon his or her patients personal dictates of race or religion. If a physician determines that a woman may elect to terminate a pregnancy in view of fetal abnormality and such an action runs contrary to his or her own religious or other beliefs, as in this case, the clear and recognized duty is to refer that patient elsewhere.

DIAGNOSTIC FAILURES IN CHILDHOOD

The now famous California case of Hope and Joy serves as a prime example of wrongful-life suits resulting from diagnostic failure.

Joy was the first child of a young couple. The new parents became concerned about her hearing and took the baby to an audiologist for a hearing test, which was reported as normal. Reassured that all was well, they proceeded with another pregnancy and duly delivered another child, Hope, who was later found to be profoundly deaf. Retesting of Joy later confirmed that she, too, had been born deaf.

Comment: This type of deafness results from a corresponding harmful gene being carried equally by each parent, who thus have a 25-percent risk in each pregnancy of having a deaf child. There was obvious negligence by the audiologist, who failed to detect Joy's deafness. The court, in awarding damages for Hope for extraordinary medical care, held that it would be "illogical and anomalous to permit only the parents, and not the child, to recover for the cost of the child's own medical care."

Similar legal cases of diagnostic failure after birth (involving cystic fibrosis and Duchenne muscular dystrophy) have occurred. Delayed diagnoses prevented couples from learning about their high risks before embarking upon a second pregnancy and the subsequent delivery of a second affected child.

Prevention or avoidance is better than coping with a condition for which there is no cure. This is especially relevant in regard to genetic disorders. Being fully aware of your family history and well informed about the implications of familial or specific ethnic disorders is critically important. Seeking legal redress in the courts never corrects a defect, never changes a tort. Couples should plan ahead, seek consultations, and — given the ongoing dramatic progress in genetics — keep in touch with a geneticist if carrier detection or prenatal diagnostic tests are not yet possible for a particular disorder that has occurred in their family.

27

Should the Defective Newborn Be Allowed to Die?

Excitement in the labor ward was running high. The mother had been in labor for seven hours and delivery was nigh. Two doctors were in attendance, since twins were expected. The father, also a doctor, stood lovingly at the head of the bed, encouraging and supporting his wife.

The first of the twins — bruised and blue, with the umbilical cord wound around the neck — appeared at the entrance to the birth canal. Suction and oxygen were immediately provided. Moments later, both twins were delivered and shock transcended the labor room. They were joined at the waist and had three legs. A nurse heard the obstetrician say: "Don't resuscitate, let's just cover the babies." The father seemed to agree by gestures. The order in the medical record read: "Do not feed in accordance with parents' wishes." The Siamese twins were then taken to the nursery to die.

In the nursery, the nurses were very disturbed. At least one fed the babies a few times, and several nurses complained to the doctor. One week later, an anonymous telephone call brought the authorities in, and following a hearing at which several nurses testified, the judge decided that the infants had been neglected and awarded temporary custody to Family Services. The twins were then moved from Danville, Illinois, to a Chicago hospital for further evaluation and treatment.

Some five weeks after birth, the local district attorney charged the parents and their doctor with conspiracy to commit murder. At the preliminary court hearings held to determine whether there was probable cause that the defendants had committed the crime, none of the nurses were willing or able to testify against the parents and their doctor. Owing to lack of evidence, the judge dismissed the charges.

Four and a half months after birth, in September 1981, the court ruled that the twins could be returned to their parents.

The baby was born with Down syndrome at the Johns Hopkins Hospital in Baltimore. Within hours after birth, it became apparent that an intestinal complication was present. A diagnosis of duodenal atresia — a birth defect that involves complete obstruction of the passageway between the stomach and the intestine — was made, and the parents were advised that an operation was necessary to correct the condition.

The parents elected not to have surgery on the child, who, if operated upon, would almost certainly have survived, albeit with moderate to severe mental retardation. The untreated baby slowly starved to death over a two-week period. That was in 1971.

Sadly, the prolonged anguish of the parents of these babies and the distress of the doctors and nurses who were involved are not unique. Several other cases, from among many, illustrate the nature and complexity of the issues that arise after the birth of a seriously defective newborn:

In England, in 1981, Dr. Leonard Arthur, a pediatrician of impeccable professional integrity, was charged with the murder of a baby with Down syndrome. His note in the medical record read: "Parents do not wish it to survive. Nursing care only." He prescribed water feeds and codeine for any distress. The baby died sixty-nine hours later of pneumonia. Unbeknown to Dr. Arthur at the time, the baby had an additional brain abnormality, heart disease, and lung abnormalities. The jury acquitted Dr. Arthur after only two hours of deliberation.

At the time, a highly respected organization in England conducted a public-opinion poll concerning issues raised by the Arthur case. Interviewers asked, If a doctor, following parental wishes, "sees to it" that a *severely handicapped* baby dies, should he be found guilty of murder? Of those polled, 86 percent said no and only 7 percent said yes.

In 1982, a baby with Down syndrome complicated by esophageal atresia and tracheo-esophageal fistula was born in Bloomington, Indiana. In this anomaly, the esophagus ends blindly so that no fluid or food can reach the stomach and there is a connection between the windpipe and the remaining lower esophagus. Major, and hazardous, surgery is required to remedy the defect. The parents decided to forgo surgery and allow the infant to die, which it did, six days after birth. Over the course of those six days, a national medical, legal, and ethical debate raged. An Indiana court, responding to hospital administrators, sustained the parents' right to have medical treatment withheld from their child.

The parents of a newborn in Maine refused surgery for their baby with tracheo-esophageal fistula and only one eye and one ear. The physicians successfully sought a court order forcing surgery. Before the operation could occur, the child's condition deteriorated, with periods of no breathing (apnea), seizures, and evidence of brain damage. At this point, the physicians also felt that no surgery should be done. The court had ruled that a newborn is "entitled to the fullest protection of the law" and that every human being has the right to life itself. In so ruling, the court negated parental rights and ignored any future suffering of the child.

A Massachusetts court was petitioned by the parents of a baby born with congenital rubella syndrome to allow them to forgo treatment of the child's life-threatening heart defect. In addition, the baby had cataracts, appeared to be deaf, and had a high probability of mental retardation. Once again, the court ordered surgery.

These and many other similar experiences have been devastating for all concerned. Some of the examples occurred in renowned medical centers that are noted for the highest possible standards of care. The cases described reflect the medical, legal, and ethical tensions that face all parties involved with a baby who is born with serious or severe birth defects. A problem of *much* greater magnitude now confronts an increasing number of parents — those who have a very-low-birth-weight premature baby. For many, many years, both these infants and defective babies have been allowed to die in newborn-intensive-care units. A report from the Yale–New Haven Hospital in Connecticut (which appeared in *The New England Journal of Medicine* in 1973) indicated that 43 of 299 infants (14 percent) admitted to the newborn-intensive-care unit in one year with irreparable damage or defects were allowed to die.

In newborn-intensive-care nurseries the world over, I have seen the tragic situations doctors face: babies with open spine defects, who will never be able to walk or have bladder and bowel control;

babies with severe hydrocephalus, who will be severely mentally retarded and blind; babies who, through brain damage from lack of oxygen, hemorrhage, or other causes, will be spastic, totally paralyzed, or unable even to turn over in bed; babies with severe genetic defects that herald, without a doubt, profound mental retardation and vegetative existence, as well as pain and suffering.

Just a few years ago, parents, doctors, and society did not have to confront the situations we face so often today. Advances in medical technology now enable the genetically or otherwise defective child to be saved (albeit defective) or to be kept alive with a respirator or by other artificial means. The difficulties are compounded further: while in certain cases it is possible to be absolutely sure of a disastrous outcome, in others a degree of uncertainty in the prognosis cannot be denied. This is especially the case with very tiny premature babies.

Remarkable advances in our lifesaving technology now make it possible to save 40 to 70 percent of infants born between 24 and 28 weeks of pregnancy. However, 40 to 60 percent of infants who survive have bleeding in or around the brain. Of these, 30 to 40 percent have some sort of permanent brain damage. In 70 percent of premature infants who are born before 28 weeks of pregnancy and require oxygen, an eye disorder called retrolental fibroplasia, probably caused by the oxygen, will develop and result in blindness or serious visual impairment. Indeed, almost all the lifesaving and life-supporting techniques cause either pain or injury that may prove fatal to the preterm infant.

The crises that erupt following the birth of a defective or damaged newborn involve not only the parents and child, but their family, the doctors and nurses, and the protective interests of the state. Profound implications, questions, and judgments, of a medical, ethical, moral, and legal nature, bedevil any attempt to find simple solutions. While it is beyond the scope of this book to explore these issues in depth (see *Genetics and the Law*, cited earlier), insight can be gained by briefly considering the anguish and dilemmas of those involved.

The Defective Child

Consideration should first be given to the defective infant. He or she may survive — tortured for endless weeks, months, or years by all manner of tubes, needles, catheters, and so on, as well as by the associated pain, suffering, and indignity. A question may justifiably be asked: What if that child could demand the continuation of artificial life-support systems? While the child has a right to life, some courts have already indicated that that right extends to life with a normal and healthy mind and body. Deferring to a higher being does not release you or the doctors or nurses from considering the grave difficulties encountered in caring for such defective infants. Indeed, the Roman Catholic church accepts that judgment may be employed when extraordinary measures are required to preserve and extend life. Most Protestant and Jewish theologians would concur with this view — drawing attention, however, to the difficulty in distinguishing between ordinary and extraordinary measures.

There are genetic defects so profound that life for those affected is meaningless. Anencephaly, chromosome defects such as trisomy 13 or 18, *profound* hydrocephalus, and forebrain disorders with or without cyclopia (single eye) are among the most tragic examples. "Right-to-life" groups wish to force mothers to deliver (instead of abort) such offspring; they then try to use the power of the law to keep them alive. Wouldn't compassion and mercy for child and parents dictate a humane policy allowing such infants to die?

Public policy when there is a less certain prognosis is harder to establish, and should best be left to doctors in consultation with parents and a hospital-appointed child advocate. Here I have in mind less catastrophic birth defects and the very common tiny premature baby whose risks of normality approach zero. The heartrending saga of the suffering of baby Andrew told by his parents should be read by all who doubt that parents should have rights. In *The Long Dying of Baby Andrew* (Little, Brown, 1983), Robert and Peggy Stinson recount the intolerably prolonged dying of their 24½-week, 800-gram (1¾-pound) premature baby. Despite multiple organ failures and repeated medical disasters (some caused by doctors), the Stinsons' heartfelt pleas to let their baby die a humane and natural death were ignored. The infant was hooked up to a respirator without the parents' consent and against their wishes. He survived five long months — all on the respirator — until he was allowed to die.

Something is gravely wrong with a system that removes parental rights, places a threat of criminal sanctions on doctors for complying with parental wishes, fails to recognize the need for mature care with dignity and humane medicine, and impoverishes already grief-stricken families.

Torture for the Parents

Armchair critics abound in this as well as in other painful controversies. Consider yourself the parent of a newborn infant with grave defects that will allow the child only a vegetative existence. What would your reactions be?

The doctor caring for the infant, by law, probably would have to inform you and your spouse that you may be criminally liable for making a nontreatment decision — that even if you do not wish to keep the child alive, you are obliged to, in the face of the law, at least until formal alternative arrangements have been worked out.

Even if you and your spouse decide to risk prosecution, the doctor is legally obliged to take steps aimed at saving the child. He might have to report the case to a child-welfare officer or follow a child-abuse-law reporting scheme; or the hospital itself might have to initiate neglect proceedings. The parents may feel even more entrapped when they realize that by discharging the doctor from the case, they do not necessarily avoid criminal liability: if the doctor is aware of parental wishes that conflict with the law, he or she may be obligated to communicate with the authorities.

For at least fifty years now, courts have been whittling away at the rights of parents to make decisions about their children. They have ordered lifesaving blood transfusions, limited choice of cancer treatment, and ordered surgery on a child with severe birth defects against parental wishes. And, finally, there is the tragedy of the Siamese twins in Danville. Parents' behavior in all these cases has been construed as child neglect.

I am very concerned about this erosion of parental rights, especially in the context of decision-making about defective newborns. On the one hand, judges have no training or insight about the prognosis and the many implications of serious birth defects. Their decisions reflect interpretations of the law and the rights of the *living*. Sadly, judicial wisdom that would temper the force of law with the poignant realities is seldom seen for fear that precedent would be abrogated. On the other hand, in the absence of definite guidelines, medical training alone seems insufficient when one deals with these life-and-death decisions. I have seen a whole range of opinions by physicians discussing how they would respond to parental wishes after the birth of a defective child. Often, however, such decisions are made in the crisis atmosphere immediately after delivery, in the absence of careful and deliberate consideration of *all* options.

Physicians who make decisions and judges who order treatment against parental wishes go home to their families and may discuss their "difficult day." The parents, meanwhile, may be left with their lives in chaos, financially ruined, and irrevocably bound to a retarded/defective child. Antiabortion groups, having successfully fought to keep defective children alive, have also spurred the government into cutting federal aid to the very handicapped people they "saved"! Parents of defective children have been left powerless, angry, frustrated, burdened, financially stricken, or bankrupt — all because they do not constitute a recognizable political voting bloc. (I have, however, rarely met parents of a defective child who were working on behalf of other parents *before* they confronted tragedy themselves.)

The Family

Consideration of the family of the defective infant, especially any other living siblings, is crucial. In chapter 1, I discussed in detail the consequences to a family faced with a seriously defective or mentally retarded child. A study of families with open-spine defects showed that 62 percent were judged to have made poor adjustments and that 43 percent of the parents were divorced or separated. And, of course, there is no measure of the sadness and emotional chaos. However, other studies on similarly affected families found fewer cases of broken marriages and indications that many families were willing to make great efforts and sacrifices for the defective child's sake.

Only rarely during the decision-making crisis in the newborn-intensive-care nursery do parents consider the effects that taking home a defective baby will have on their other children. Without belaboring the issue, I have long witnessed the possible effects: physical and emotional exhaustion, financial strain, and relative neglect of the normal children. When all of your energy is spent on caring for a defective child, how much is left over for your other children, who truly would benefit from your attention? Endless consultations with adults who grew up with a defective brother or sister have provided moving testimony about such normal siblings' feelings. Invariably, they come seeking any possible way to avoid having a child with the same defect that influenced their childhood.

Certainly, it can be argued that families may give up their obligations to care for a defective child. Odious as that possibility would appear to be, it might for some be less traumatic than withholding vital treatment. It would seem, however, that consideration of the siblings is frequently less compelling than concern for the welfare of the defective child.

Doctors and Nurses

The physician represents more than just "an interested party." Cognizant of the likely future and concerned about the general health and welfare of the family, the physician will realize the degree of devastation in store for both the defective child and the family. Although predisposed to the idea of death with dignity, and committed to avoiding the torture to all concerned, the doctor is nevertheless prevented by law from following a parent's wishes that he or she abandon extraordinary or heroic measures to save the life of a defective child.

The decisions to be faced are difficult, and pediatricians have widely varying opinions about "proper care" and "extraordinary care" and how long it should continue. For example, a survey study of pediatricians in the mid-1970s noted that only half felt that babies with Down syndrome and duodenal atresia should be treated surgically; the others maintained that the child should be allowed to die. Opinions about care for a tiny premature baby or one with other types of defects differ even more.

Physicians' attitudes toward advocating or withholding surgery have been seen to depend upon their religious persuasion and activity, age, and degree of specialization. Small wonder, then, that great distress has been engendered when duodenal atresia, which can be repaired by extremely simple and safe surgery, is present in a newborn. The operation has always been done *except* when a baby has Down syndrome. Advocates for the handicapped clearly and correctly see this as discrimination against their kind. To make matters worse, doctors, in the past, have decided with parents not to operate, and then have callously left the child to starve to death in the nurses' care.

It seems to have taken doctors a long time to learn that nurses are a critical part of the decision-making team. Neglecting the opinions and feelings of nurses — who dispense tender, loving care in these very difficult circumstances, and who are ordered to watch defective babies starve to death — has had predictable consequences. Calls to the authorities in most of the celebrated cases that have reached the courts, were, I believe, made mainly by nurses.

Achieving consensus about care among parents, doctors, nurses, and an advocate for the child is easier when there is a hopeless prognosis. It is with the much more common problem of an uncertain medical future that serious dilemmas are spawned.

The State

Until the Danville case, an atmosphere of trust and concern pervaded decision making by parents and doctors confronted by a defective newborn. That experience in 1981 and the Bloomington case in 1982 radically changed the way physicians could empathize and follow the wishes of distraught parents. In 1983, the Reagan administration, motivated by antiabortion groups, construed non-treatment of defective newborns as discrimination against the handicapped and issued regulations precluding any such activities. The government further threatened to withhold all federal funds from any institutions found guilty.

Fortunately, in 1986 the Supreme Court invalidated these regulations and ruled that withholding treatment from handicapped infants in accordance with parental wishes does not violate laws intended to protect handicapped persons. Before this court decision, however, Congress amended the Child Abuse and Treatment Act to provide general guidelines that prohibit withholding medical care from handicapped infants.

Society has a vested interest in decisions concerning the disposition of the defective newborn. There is a general empathy felt for all individuals who are subjected to pain and suffering. Hence, society could make available facilities for the care of the gravely defective infant and for the institutionalization that so often follows. Resources for long-term care, however, are scarce the world over; objectively speaking, certain other health priorities have to be recognized. Moreover, institutions for the severely deformed or defective are in poor repute in some countries.

When it is realized that the projected cost of lifetime institutional care for one child born with Down syndrome exceeds a million dollars, the argument is advanced that instead of spending literally billions of dollars for the care of seriously defective offspring, those financial resources should be put to better use, such as for the prevention of such defects. Sadly enough, if such a priority should take over, many others — for example, the elderly or the chronically sick — may get hurt in the process. Hence, the state becomes inexorably involved in making comparative assessments as it tries not to deviate from its fundamental commitment to the safety and sanctity of all persons.

Who Shall Die? Who Shall Live? Who Will Decide?

In the United States and elsewhere, laws designed for the protection of human life clearly extend to defective newborns, as well as young or old adults. The legal constraints on medical procedures are backed up with threats of punishment for noncompliance. Hence, it is clear that all participants caught in the dilemma of trying to decide the fate of a defective newborn place themselves in danger. Parents, doctors, nurses, hospital administrators, and various other members of the hospital staff, despite their good intentions, may be liable for crimes that range from murder to manslaughter, including child abuse, child neglect, or even conspiracy to withhold medical care. So explicit is the law that even advising or making recommendations to parents may expose the physician to legal jeopardy — for example, on the basis of conspiracy to commit homicide by withholding treatment. (The case in Danville represents the first time in the United States that parents and their doctor have been prosecuted for withholding care from a defective newborn.)

The time when anguished parents huddled with their doctor to decide, in the best interests of their gravely ill or defective child, what best to do, has passed — at least for now. Government, prodded by right-to-life groups, has stripped parents of their right to make these decisions and has left them not only with a potential lifelong burden, but facing financial ruin.

In April 1985, the U.S. government published rules under its "Child Abuse and Neglect Prevention and Treatment Program." These regulations do not make physicians secure or confident about all decisions concerning the care of gravely ill newborns.

The key intent of the rules is to prevent "the withholding of medically indicated treatment from a disabled infant with a life-threatening condition" by making such withholding equal to medical neglect. This makes it necessary, if a treatment exists, to provide such care. Three exceptions are recognized: (1) when the infant is in an irreversible coma; (2) when treatment would simply prolong dying or be futile in securing survival; and (3) when "the treatment itself under such circumstances would be inhumane."

The way the federal rules read, a treatment that may be "inhumane" may be withheld only if it is also virtually futile. It also remains possible for physicians or parents to be charged under state child-abuse and child-neglect statutes. Surely, then, doctors are bound to treat aggressively all but the most hopeless cases to avoid any possible connotation of neglect. This is hardly in the best interests of child and family. While a life-support system could theoretically be turned off, the regulations explicitly forbid withholding of fluids, food, or appropriate medication.

Decisions to allow a defective baby to die can now be made only pursuant to consultations among parents, doctors, and infant-care review committees in hospitals. Child protective services may frequently have a role in such decisions, and judges stand ready to decide who shall die when other mortals fail.

Pursuit of this policy, in the words of Father John Paris of Holy Cross College, means that "life is the ultimate value, and something that is to be preserved regardless of prognosis, regardless of cost, and regardless of social considerations." Doctors J. C. Moskop and L. Saldanha, writing in the *Hastings Center Report* (April 1986), summarized the essential thrust of the government's regulations: "The policy assumes . . . that noncomatose, nonterminal life is always preferable to nonexistence; it expressly prohibits consideration of the future quality of life of the infant." They argue, and most would agree, that "there are conditions other than irreversible coma or death in the near future in which people would overwhelmingly choose a shorter span of life over a longer life of very poor quality."

The pain and suffering of a newborn with grave birth defects, the devastation caused in the family, and the implications for society as a whole call for a definite policy in regard to decision-making, and one that is legally safe. Of the various options presented, it is clear that the present situation, in which parents and doctors are under continual threat of prosecution, is most unsatisfactory. The third-party technique, or death committee, is no better than giving the government the power of arbitrary euthanasia. It would seem best for parents and doctors to determine, case by case, what should be done or not done. They are best situated to evaluate the suffering of the child and to allow the infant to die with dignity in the most humane way. I believe, together with many other doctors, that such decisions should be made independent of the law. If selecting nontreatment can violate the law, then the law should be changed.

Inheritance of Common Diseases

Treatment of Genetic Disorders

28

Heart Defects and Disease, Hypertension, and Heredity

Will you have a heart attack? Why? When? Does it matter if you do or do not have a family history of heart disease? What can you do about heart disease?

Diseases of the heart and blood vessels affect virtually everyone. Both hereditary and environmental factors acting separately or, more commonly, together cause the vast majority of these disorders. I will focus on the hereditary aspects of diseases of the heart and blood vessels, drawing attention to environmental influences that may interact with a hereditary predisposition.

Disorders of the heart and blood vessels may take on many different forms (only the most important are discussed here). You or your child may have been born with a structural defect of the heart, such as a hole in the heart or a narrowed valve. Again, birth defects of the heart may be of genetic or environmental origin, or can result from an interaction between the two factors. Indeed, the common type of heart attack also results from such a combination, whereas a significant number of persons with coronary-artery disease suffer from a directly inherited condition. Genetic mechanisms also have much to do with hypertension (high blood pressure), as well as with defects of heart structure and function — common conditions such as mitral-valve prolapse and disturbance of heart rhythm. In your middle years of life, you may have or may develop high blood pressure or coronary-artery disease, a clogging of the main blood vessels that supply the heart. Let us briefly consider the important genetic aspects of heart defects and diseases.

Birth Defects and Genetic Disorders of the Heart

Structural defects of the heart are remarkably common: over 1 in 100 children are born with one of these congenital defects. Hereditary and environmental factors, acting separately or, frequently, together, cause these heart defects. It is a salutary lesson, however, to realize that less than 10 percent of all cardiac defects are of direct genetic origin.

Among environmental factors, viruses (such as German measles) are important, but a causal factor cannot be identified in about 80 percent of cases. Alcohol taken in the first eight weeks of pregnancy may result in the fetal alcohol syndrome, with cardiac defects being present in 45 percent of affected infants. Anticonvulsant drugs for the treatment of epilepsy are another important environmental factor, and lithium, used in the treatment of manic-depressive illness, is strongly suspected as also causing a specific type of heart defect.

Illnesses in mothers during pregnancy may also pose serious threats to normal fetal-heart development. For example, diabetic mothers who are treated with insulin have a higher risk of bearing a child with a congenital heart defect. In mothers with phenylketonuria, chemical susbstances may accumulate in the blood and damage the newly forming heart in 25 to 50 percent of the offspring. Mothers with the connective-tissue disorder systemic lupus erythematosus have a 20- to 40-percent risk of having a child with complete heart block due to interference in the conduction system of the heart by antibodies that pass from the mother into the fetus. Considerable controversy has existed about the role of sex hormones — usually taken inadvertently during the first eight weeks of pregnancy — in causing heart and other congenital defects. The overall current consensus is that these hormones probably do not cause cardiac defects.

The fact that in identical twins a heart defect, if present, is likely to affect only one of the pair, strongly suggests an environmental factor as the main or controlling cause.

The heart is formed during the first eight weeks of pregnancy. Hence, it is not unusual for a prospective mother, not even knowing that she is pregnant, to have already completed eight weeks of pregnancy and be carrying a fetus whose heart is already formed. She may unwittingly have been drinking alcohol or taking certain drugs, or may have been exposed to certain infections that could cause birth defects of the heart. Heart defects due to

German measles may include a hole between the two lower chambers of the organ — a condition called *ventricular septal defect*; or an overly narrow valve outlet in the artery that connects the heart to the lung — a defect called *pulmonary stenosis*; or the persistence of a so-called communication between the aorta (the large blood vessel through which blood leaves the heart to supply the rest of the body) and the large pulmonary artery that supplies the lung — a disorder called *patent ductus arteriosus* (PDA).

It is known that virtually all these structural abnormalities of the heart, such as narrow valves or communicating holes between chambers, can also arise through hereditary mechanisms. Children with chromosome defects frequently also have cardiac anomalies. For example, over a third of those born with Down syndrome have congenital heart defects (most often ventricular septal defect). Between 35 and 50 percent of girls born with Turner syndrome also have such defects, the most common being *coarctation of the aorta* — a dramatic narrowing of a short segment of the body's main artery.

Different single genes are known to cause a host of disorders in which structural heart defects are important, if not fatal, features. These conditions are either uncommon or rare, and may affect any structure of the heart, including the valves, the intervening walls, the muscle, the nerve conduction system, and, of course, the arteries. Modes of inheritance are mainly autosomal dominant and recessive, only rarely sex-linked. In certain inherited disorders of muscle — such as Duchenne muscular dystrophy and myotonic muscular dystrophy — the heart muscle itself is eventually involved, and this may even be the cause of death.

Frequently, it is virtually impossible to separate out those cases in which the cause has been environmental rather than hereditary. Indeed, interaction between both modes is the most common explanation for heart defects. Some occur more frequently in births during the fall and winter, which suggests the action of an environmental factor such as a virus.

When it is clearly difficult to separate out environmental from hereditary factors, it is safer to assume a hereditary factor, with risks of recurrence of 2 to 5 percent; if a couple has had two children affected, the risks of recurrence rises to 8 to 12 percent; and if three are affected, the recurrence risk approximates 25 percent. If one of the parents was born with one of these heart defects, then the risk in a first pregnancy of having a child with a cardiac defect may be 1.5 to 3.0 percent if the father is affected and

2.5 to 18.0 percent if the mother is affected, depending on what the defect is; of these, the identical defect will be present in only 30 to 60 percent of cases. The heart defects that fall into these risk categories include the aforementioned hole between the two lower chambers of the heart, a hole between the upper chambers (*atrial septal defect*), patent ductus arteriosus, and pulmonary stenosis, to name only the major types.

So far as prevention of congenital heart defects is concerned, avoiding known causes is obviously critical. Mothers should be sure to have rubella immunization before becoming pregnant. Diabetes should be extremely well controlled before pregnancy is initiated. Alcoholic beverages during the first four months of pregnancy should be scrupulously avoided. Anticonvulsant medications should be reviewed in consultation with a neurologist and the least hazardous of them (probably phenobarbital) used when possible without risk to the mother. Every effort should also be made to avoid other strong medications (such as lithium) if, after consultation, it is clear that their discontinuance can be achieved without hazard to a mother. The rare mother with phenylketonuria should initiate pregnancy on a strict special diet to minimize the effects of the chemical disturbance in her body that disrupts normal fetal development.

Diseases of the heart and blood vessels are the most common cause of death by far in the United States, and probably in most Western industrialized societies. The vast majority of these diseases are due to thinning and hardening of the walls of arteries, a process called *arteriosclerosis*. The most important type of arteriosclerosis is called *atherosclerosis*, which by itself is the greatest single cause of death in the United States.

Coronary-artery diseases in the Western world arise mostly from genetic-environmental interactions — which also explains the considerable differences in their frequency and in the precipitating factors seen in and among various countries. Directly inherited genetic heart disease accounts for fewer cases, and will be discussed after the common multifactorial, or *polygenic*, type. Either way, regardless of the cause, the final common pathway is atherosclerosis.

Polygenic Heart Disease

One in every 3 men in the United States develops major heart disease before the age of sixty. Women's odds are 1 in 10.

Specifically, for coronary-artery disease in men before sixty, the odds are 1 in 5. Over 43 million Americans — about 19 percent of the U.S. population — have heart disease or hypertension; two-thirds of these individuals are under sixty-five years of age. Heart disease accounts for one-half of all deaths in the United States, and the cost to the economy reaches to a staggering $97 billion.

ATHEROSCLEROSIS

Atherosclerosis is not a new disease, although it has become a major scourge of modern civilization. It has been recognized in Egyptian mummies and is described in the ancient writings of the Greeks. In atherosclerosis, fatty and fibrous substances are deposited and form streaks, patches, and nodules mainly on the inner walls of the largest arteries, including the three coronary arteries that supply the heart. These fatty and fibrous substances increase in size, become associated with blood clots, and eventually can obstruct a coronary artery to cause a heart attack. Evidence that atherosclerosis begins in childhood is clear. Children who died before puberty have been found with such deposits in their major arteries. Indeed, U.S. soldiers under twenty years of age who died in battle in the Korean War, frequently had well-established atherosclerosis.

A heart attack basically results from a lack of blood — and therefore oxygen — reaching the heart muscle. Such heart attacks are the main cause of death in males after thirty-five years of age and in all persons over age forty-five. Heart-attack deaths before fifty occur predominantly in men, and about one-third of all deaths from these heart attacks in males occur before age sixty-five.

Between thirty-five and fifty-five years of age, the death rate in the United States from heart attacks is five times higher in white men as opposed to white women. This difference becomes less significant after the menopause. Women with high blood pressure, diabetes, or an early menopause share the same risks as the male.

The six nations that have the highest death rates from heart attacks for white men between forty-five and sixty-four years of age are, in descending order, South Africa, the United States, England, Scotland, Canada, and Australia. Rates of death from heart attacks in these age brackets are much lower in most Western European countries and Latin America. In Japan, the heart-disease-death rates in this age group are about one-fifth of those in the United States. However, Japanese in Hawaii and

California have about twice the rate of coronary disease as do their peers in Japan.

Other factors that produce variation are socioeconomic: the higher the income and the better the standard of living, the more frequent the occurrence of premature heart attacks. Elements that might bring on more attacks among the affluent groups are the higher fat content of the diet and the decreased amount of physical activity. Generally, however, significant differences in the frequency of heart attacks among different cultures are not easily explained. For example, Scots and North Americans have a death rate from premature heart attacks that is more than twice that of Swedes. While premature heart disease is decidedly common among the whites of South Africa, it is extremely unusual in blacks living in the same locality. Differences in the occurrence of heart disease associated with high blood-fat levels (cholesterol or triglycerides) among different groups in the same country, or among national groups, probably reflect a fundamental variation in hereditary susceptibility to the particular environmental factors operative in different cultures.

Genetic studies comparing coronary-artery disease in siblings and others, based on twins, clearly show that an inherited predisposition is important. Indeed, data indicate that genetic mechanisms have much to do with the variation and control of cholesterol and other critically important fats that are carried by so-called lipoproteins. We already know that raised levels of certain lipoproteins (such as low-density lipoprotein cholesterol and apolipoprotein B) are strongly correlated with coronary disease. Even better prediction seems possible by finding *lower* levels of apolipoprotein fractions A-I and A-II. Most recently, using DNA analysis of the genes that code for the apolipoproteins, even greater, more basic insight into the genetic causes of heart disease is emerging.

FACTORS THAT CAUSE, AGGRAVATE, OR PRECIPITATE HEART ATTACKS
Notwithstanding the possibility of genetic predisposition, other recognized factors are known either to cause, aggravate, or precipitate premature heart attacks. Whether you are genetically predisposed or not, attention to the following hazards may help you prevent premature heart disease:

1. High blood-fat levels (cholesterol or triglycerides) caused by a variety of factors

2. High blood pressure
3. Cigarette smoking
4. Sugar diabetes
5. Lack of physical exercise
6. Obesity
7. Gout
8. Type-A (coronary-prone) personality
9. Emotional stress
10. Family history of premature heart attacks in near relatives, especially parents or brothers or sisters
11. Oral contraceptives

Women under forty years of age who take oral contraceptives have a negligible risk of heart disease provided that they have none of the risk factors just mentioned. In particular, oral contraceptives should be used with great caution — or perhaps better, not at all — in women over forty years of age, those with a family history of coronary disease, and those who smoke more than twenty cigarettes a day.

Women who take oral contraceptives have also been shown to have increased blood-fat levels and to have an associated increase in risk of disease of the blood vessels supplying the brain. This could mean clotting within those blood vessels and might lead to strokes. Certainly, the evidence suggests that heart attacks in women before menopause occur more frequently among those who use oral contraceptives. It is not yet clear whether women after menopause increase their risk of heart attack by using female hormones. Men who smoke more than one pack of cigarettes a day have 70 percent more heart attacks than nonsmoking males. The increase in the death rate is clearly proportional to the amount smoked. The relation between smoking and premature heart attacks also applies to women, but not as severely as to men.

THE HEART-ATTACK-PRONE PERSONALITY

Almost thirty years have passed since a certain behavior pattern was related to a greater likelihood of a heart attack. So-called type-A individuals have characteristic personalities. They exhibit overly aggressive, competitive, and ambitious behavior. These men or women are obsessed with deadlines, are workaholics, are chronically impatient, and usually have a strong sense of time urgency. These coronary-prone individuals have about twice the risk of heart attacks as do others.

HEART DISEASE AND STRESS

The risk factors for heart disease mentioned above make it clear that all that is familial is not necessarily hereditary. One other example is too good to omit. The Swedes studied identical twins, only one of whom had coronary disease. They focused on the psychological aspects in the lives of the twins. Their observations showed that the higher the degree of dissatisfaction with life, the greater was the severity of heart disease.

THE CORONARY-ARTERY BLUEPRINT

By now, you will not be surprised to hear that even the anatomy — the layout — of the three main coronary arteries and their interconnections is genetically determined. Particular differences in the way the heart muscle is supplied by blood may, on occasion, make certain areas more vulnerable to the effects of increasing blockage of the coronary arteries by atherosclerosis. Moreover, genetic influences affect the actual construction of the coronary arteries themselves. For example, Ashkenazic Jews in Israel, who have a higher prevalence of coronary-artery disease than Yemenite Jews or Bedouins in Israel, have a greater development of the inner lining (intima) and muscle-elastic layer of the coronary arteries than the other two groups. These structural features may enhance the formation or deposition of cholesterol, or both — that is, the process of atherosclerosis.

Hereditary High-Risk Heart Disease

Most people who develop coronary-artery disease do so through the interaction of a hereditary predisposition and one or more of the risk factors listed earlier. However, a very important 20 percent of those who have heart attacks before age sixty instead suffer from the effects of a single gene that causes atherosclerosis.

Studies have shown that between 50 and 85 percent of all individuals who contract heart disease before they are fifty have high blood-cholesterol levels, high blood-triglyceride levels, or both. There are four hereditary disorders that can almost invariably be found responsible for this condition.

FAMILIAL HYPERCHOLESTEROLEMIA

In this very common disorder, high blood-cholesterol levels are associated with coronary-artery disease. Familial hypercholesterolemia is one of the most widespread of all genetic diseases that afflict people. Approximately 1 in every 500 of the population possesses the gene, which means that familial hypercholesterol-emia is much more prevalent than cystic fibrosis or sickle-cell anemia. In the United States, this disorder accounts for about 5 percent of all persons who have a heart attack.

Familial hypercholesterolemia is transmitted through families via a dominant form of inheritance. This means that if you are affected, there is a 50-percent chance that each of your children will develop the same disorder. Should it happen that both parents carry the same abnormal (mutant) gene for familial hypercholesterolemia — a 1 in a million chance — the risk of their having an affected child rises to 75 percent in each pregnancy. Those children who receive the gene from both parents will be severely affected. Their blood-cholesterol levels will be so high, because of the double dose of the abnormal gene, that they will show fatty accumulations that can be seen on the skin (around the eyes or tendons); moreover, they will develop coronary-artery disease and have heart attacks in adolescence or as early as eighteen months of age! Children who have received an abnormal gene for familial hypercholesterolemia from each parent only rarely survive beyond thirty years of age.

Familial hypercholesterolemia may be detected at birth by analyzing blood samples from the umbilical cord, or, of course, by checking blood samples taken later in childhood. You should note that a single "normal" cholesterol reading does not entirely rule out the possibility that you may have the disorder, especially if you are at risk, since some overlap of cholesterol values does occur between those affected and normal family members. If you are at risk, then, in addition to repeatedly testing blood samples for cholesterol levels, it is also important to have more sophisticated tests to measure low-density lipoproteins. Always remember that blood-cholesterol testing should be done after an overnight fast if the result is to be reliable.

About 16 percent of affected males with this disorder will have some kind of symptoms of coronary-artery disease by the age of forty. By age sixty the figure rises to at least 52 percent for males and 33 percent for females. The risk of coronary-artery disease for women who carry the gene is somewhat lower than for males but is nevertheless greater than that for the general female population.

The available treatment to reduce high cholesterol levels includes a diet that is low in cholesterol and saturated fats and high in polyunsaturated fats. Certain drugs, such as cholestyramine resin or colestipol, may be used to further decrease the blood cholesterol by as much as 30 percent. Addition of nicotinic acid is also useful to help block the body's increased production of cholesterol. The unfortunate person who has received the abnormal gene from both affected parents requires drastic measures that include major surgical shunting of blood, exchange plasma transfusions, and other procedures not free of danger themselves.

Recently, cells grown from the skin (cultured fibroblasts) have been shown to manifest the presence of the abnormal gene for familial hypercholesterolemia. In addition, gene-analysis techniques (see chapter 8) have revealed that the gene for familial hypercholesterolemia is long: it has some 75,000 base pairs. Moreover, a mutation that causes a small deletion has been detected in affected individuals of French-Canadian ancestry but not in others with this disease. Clearly, given the length of this gene, different mutations are possible resulting in the same disease. These advances now make it possible to make the prenatal diagnosis of this disorder early in fetal life. In the rare situation in which both parents know they have the gene, they may select the option to terminate a pregnancy when the fetus can be shown to have the severest form of the disease, which will cause heart attacks and death by adolescence or earlier.

FAMILIAL HYPERTRIGLYCERIDEMIA
Another hereditary disorder, *familial hypertriglyceridemia*, is almost twice as common as familial hypercholesterolemia. Its frequency in the population has been estimated at close to 1 in 300 people. It is also associated with premature coronary-artery disease, but appreciably less so than in familial hypercholesterolemia. As opposed to the latter condition, this disorder is characterized by high levels of triglycerides, or blood fats, and is *not* usually associated with high blood-cholesterol levels. About 5 percent of all those who have heart attacks have familial hypertriglyceridemia.

This disorder is also transmitted as a dominant trait; that is, if you have the disorder, there is a 50-percent chance that each of your children will have the same problem. In actuality, and despite what was theoretically expected, only about 10 to 20 percent of children and adolescents of such affected parents appear to show signs of it. Different mutations in the responsible gene can each lead to hypertriglyceridemia.

For reasons that are not yet clear, diabetes, hypertension, gout, obesity, and resistance to insulin seem to occur more frequently in individuals with high blood-triglyceride levels. Very high blood-fat levels are found especially in persons with this disorder when their diabetes has gone untreated, if they develop hypothyroidism (an underactive thyroid gland), or after excessive alcohol intake, as well as after taking oral contraceptives or estrogen hormones. Obviously, then, if you are affected by such a disorder, diabetes should be watched for and excess alcohol and the use of female hormones containing estrogen should be avoided. In addition, if there is obesity, weight reduction is very important. The drug Atromid-S (clofibrate) is used to lower the blood-fat levels in this condition.

FAMILIAL COMBINED HYPERLIPIDEMIA

Familial combined hyperlipidemia, a third type of hereditary disorder, is even more common than either of the first two described. It affects close to 1 in 200 of the U.S. population. Again, in this disorder, there is a close association with a high frequency of heart attacks. The disorder may not become apparent until sometime after puberty. In about one-third of cases, both the blood-cholesterol and blood-fat, or triglyceride, levels are raised; in roughly another third, there is only high blood cholesterol; and the remaining third shows only high blood-fat levels. Combined hyperlipidemia accounts for about 10 percent of all those who have heart attacks.

Again, the hereditary transmission here is by a dominant trait: if you have it, there is a 50-percent chance you will pass it on to each of your children. Treatment similar to that given the other high-blood-fat disorders is usually recommended.

FAMILIAL DYSBETALIPOPROTEINEMIA

In *familial dysbetalipoproteinemia,* severe atherosclerosis involves not only the coronary arteries but also the other major arteries, including those in the limbs. Thankfully, the disorder is rare; about 1 in 40,000 people are affected. Its mode of inheritance is autosomal recessive. The matter seems even more complex, however, since it appears that in addition to the defective gene, one of the three other inherited disorders discussed above also needs to be present.

Familial dysbetalipoproteinemia in some persons will become obvious if diabetes or hypothyroidism occurs. Indeed, these complications, together with obesity and disease of the limb and other arteries (called peripheral vascular disease), are characteristic of the disorder. People with this condition, which usually first shows itself in individuals who are twenty or older, typically have fat deposits (which look like orange-yellow discoloration) on their palms and finger creases, or as small lumps around their eyes, elbows, and knees. Both cholesterol and triglyceride levels are high. Treatment, which is usually successful, is as described above for familial hypertriglyceridemia.

At least one of the four hereditary disorders just discussed affects 1 in every 100 individuals in the United States, and probably in other Western nations. The enormity of this figure needs no embellishment. Heart attacks kill a large number of people in whom one of these disorders has never been recognized. Specific diagnosis is not simple, as family-history analysis and other studies may be necessary to differentiate the four diseases from the more common polygenic heart disease discussed earlier in this chapter.

What Should You Do?

If you already know that you are affected by one of these hereditary high-blood-fat diseases, then I would hope that you have already stopped smoking and are on a schedule of treatment, including weight reduction, careful control of diabetes, and exclusion of oral contraceptives, estrogen hormones, and alcohol. The drugs mentioned above or others recommended by your physician may also be helpful.

It is more likely that as you read this section, you may not be aware whether you are in fact affected or not. If you have, or had, a parent, a brother, or a sister who had a heart attack before fifty years of age, then you should have a general medical checkup that includes blood-cholesterol and triglyceride studies performed after fasting. Treatment is available, and you would be doing yourself and your family harm by not taking adequate care of yourself. The United States Coronary Drug Project did show, however, that drug treatment to reduce blood fats was of much less importance than dietary modification and attention to the high-risk factors mentioned earlier. Incidentally, the blood-cholesterol and fat tests may need to be repeated two or three times, especially if there is reason to suspect one of these disorders. Other family members who have not yet reached thirty years of age should also be retested later, since children and adolescents frequently may not show the characteristic high blood-cholesterol or fat levels.

To Screen or Not to Screen?

Because of the serious impact heart disease has on all of us, due consideration has been given to testing blood samples early in life. The idea, of course, is that early diagnosis would lead to the instigation of preventive measures. Consequently, tests have been run on blood collected from the baby's umbilical cord right at birth, while other studies have focused on blood samples taken during the first year of life. Neither approach has really succeeded, mainly because the predictive value of a high or even a normal blood-cholesterol (or triglyceride) level has been unsatisfactory. For example, some children without a family history of heart attacks have been found to have high blood-cholesterol levels at birth. Others have had normal cholesterol values at birth and have been found to have the genetic condition at one year of age. Finally, genetic disorders with high blood-fat levels may not manifest themselves until adolescence or later.

At present, instead of screening every child born, it would seem wisest to examine samples only from the children of a parent with one of the four genetic disorders considered above.

Hypertension

THE PROBLEM

About 1 in 3 persons in the United States between twenty-five and seventy-four years of age has high blood pressure. When all ages are considered, 1 in 4 has hypertension — about 60 million people! Hypertension accounts for about 250,000 deaths in the United States each year by causing or contributing to heart failure, strokes, and kidney failure. The higher the blood pressure, the greater the risk of these eventualities.

Hypertension varies with age, sex, physical activity, stress, and a whole host of other factors. Given such variability, a casual measurement taken at any time gives a less reliable idea of a person's blood pressure than when repeated, regular measurements are made under controlled circumstances — for example, in bed at rest, in the morning, before rising. Some individuals have sustained hypertension, while others find that their levels go up and down with stress or other conditions. For example, blood pressure has been found to be higher on a first visit to the doctor than on subsequent visits. Some people normally have higher blood pressures than others and remain perfectly healthy. Others, on the other hand, may develop heart attacks, kidney disease, or stroke even at pressures that do not necessarily cause problems in others.

The frequency of hypertension varies in different parts of the world. For example, some primitive populations have virtually no hypertension. In contrast, when such individuals migrate or change their food and living habits, a significant increase in hypertension occurs. The black Zulu and Xhosa South African tribes show no increase in their blood pressures with age while living in rural areas but do when they become city dwellers. Similar changes have been noted in the migration of Polynesians from Tokelau Island to New Zealand. While considerable weight gain may account for some of these changes in those who have immigrated, excessive salt intake may also be very important. However, even among such groups, some individuals may be more susceptible to salt than others. Indeed, some individuals have been recognized as being more "salt-sensitive" than others, who are regarded as "salt-insensitive." (Maybe this latter group have among them those individuals who seem to reach out instinctively for the table salt and heavily douse their food even before having tasted it.)

It is known that high blood pressure is more common among blacks at all ages, of both sexes, than among whites. (High blood pressure among Japanese is also very common.) About 1 in 4 blacks ultimately develops hypertension. While this observation may clearly implicate genetic mechanisms, it has also been effectively argued that blacks have different cooking habits from whites. In particular, they are known to consume more salt than whites do. Salt, of course, is known to aggravate hypertension. The matter becomes more complicated when you recognize that calcium in the diet, especially in milk and milk products, may act to dampen the effect of salt on the blood pressure. It is a fact that many more blacks than whites are intolerant of milk — actually, to a sugar in milk called lactose — and that this intolerance is genetically determined. Hence, it has been argued, in a rather circuituous way, that there are hereditary factors that influence the blood pressure of blacks. Many others undoubtedly exist.

THE CAUSES OF HIGH BLOOD PRESSURE

The actual cause of hypertension is recognized in only about 10 percent of all cases. The most common known causes are kidney disease or disease of the blood vessels that supply the kidneys. Only in a few percent of the recognized causes of high blood pressure are there opportunities through surgery to correct the defect and cure the hypertension.

Women who use oral contraceptives have between a two and six times greater likelihood of developing hypertension than non-users. The risk of developing such hypertension may be about 5 percent. The condition may be mild to severe, and it will not necessarily disappear when the oral contraceptives are discontinued. If you are taking oral contraceptives and have been found to have high blood pressure, it would be wise to consult with your physician and perhaps to discontinue such medications for at least six months in order to assess the impact of these pills on your blood pressure.

In the vast majority of individuals, an obvious cause is not apparent and the term *essential hypertension* is applied. Genetic mechanisms contribute to this common form of hypertension, which we will consider next.

HEREDITARY ASPECTS

While the cause of hypertension remains undiscovered about 90 percent of the time, there is a great deal of evidence pointing to hereditary elements. Although it is well recognized that hypertension tends to aggregate within families, it is not at all clear how it is transmitted from parent to child. Adopted children have been compared to their adoptive parents and their biologic parents and comparisons made between siblings. These studies have concluded that whatever factors are shared in the home, they are not of particular importance in finally determining the level of blood pressure. We do know that a person develops a characteristic blood-pressure pattern in infancy that tends to remain for life. So, if your blood pressure is on the high side at one month of age, it is likely to remain on the high side throughout your life (and, similarly, to stay low if it was low initially). Those who begin with higher levels appear to be more likely to develop essential hypertension in adulthood. Since blacks tend to start with higher blood pressures, their pressures, not surprisingly, rise faster as they age than do those of whites. Certainly, the frequency of strokes due to hypertension is highest among people who as children had the highest blood pressures.

Hypertension within families is a common observation. We know that the blood pressures of related individuals have clearly been shown to resemble one another more closely than those of unrelated individuals living in the same house. For example, the blood pressure of parents has been shown to be much more closely related to their natural than to their adopted children. In twin studies, the genetic contribution to the cause of hypertension has been estimated to be between 30 and 60 percent. Attesting further to likely genetic factors in hypertension is the high concordance found between identical twins (both affected) compared to nonidentical pairs.

The genetic mechanism that results in hypertension remains unproven. Salt is viewed as a key suspect in this mystery. It has been noted that children of hypertensive parents showed a higher blood-pressure response to psychological stress, which was further accentuated after they ingested excess salt for two weeks beforehand. This suggests a genetically determined heightened response to both stress and salt. In fact, studies of sodium (salt) and potassium movement in and out of red blood cells show abnormal results in half the children when one parent is hypertensive and in three-quarters when both parents are hypertensive. Such a finding jibes with autosomal dominant inheritance. But the matter is very complex and nowhere near being settled. The current consensus is that essential hypertension results from multifactorial inheritance: it reflects an interaction between certain environmental factors (stress and salt) and a genetic predisposition.

Hypertension may also occur in association with other uncommon or rare genetic diseases, including adult polycystic kidney disease, neurofibromatosis, Fabry disease, porphyria, and others.

PREVENTION AND TREATMENT

Until there is clear knowledge and understanding of the basic mechanisms that cause hypertension, the best recommendations will have to depend on first principles. The advice therefore is to cut down on salt intake, avoid becoming fat, limit alcohol consumption, and exercise regularly. In consultation with their doctor, women should carefully consider whether to exchange oral contraceptives for another method.

Medical treatment is usually successful, especially if hypertension is diagnosed early on. The main problem, however, is that most people do not have their blood pressure checked regularly, which is why the early onset of hypertension is often overlooked. When did you last have your blood pressure checked?

Disorders of Heart Rhythm

The nerve impulses that speed through the conduction system of the heart causing it to beat regularly and rhythmically may be disturbed by either a structural or functional abnormality in that track. Not surprisingly, just like everything else in the body, the conduction pathway is inherited. We know, for example, that there is a much greater heart-rate correlation between identical twins than there is between nonidentical ones. Inherited differences between identical and nonidentical twins can also be seen on the electrocardiogram, which reflects the speed and pattern of an electrical nerve impulse as it moves through the heart.

The vast majority of disturbed heart rhythms result from acquired rather than inherited heart disease. Fewer than twenty abnormalities of heart rhythm — some uncommon, others rare — are due to clearly recognizable inherited disorders. Most disorders that result in rhythm disturbances are inherited as autosomal dominant traits or are due to a gene mutation. Only two or so have been clearly shown to be inherited as autosomal recessive traits.

Those conditions resulting in very fast heart rates that are regular in rhythm are called *tachycardias,* while those fast rates with irregular beats are called *tachyrhythmias.* A very slow heart rate due to blocked passage of the nerve impulse through the heart's conduction system is termed a *bradycardia.* Symptoms from either tachycardia or bradycardia mostly result from inadequate blood flow to the brain, which can lead to sudden weakness, fainting, and even sudden death. One of those dominantly inherited conditions is associated with deafness from birth.

Total blockage of nerve impulses in the conduction system may be detectable not only at birth but even in the fetus during the last three months of pregnancy. This condition, called *congenital heart block,* is not usually inherited. However, when a fetus or child is diagnosed as having congenital heart block, lupus erythematosus or rheumatoid disease should be sought for in the mother, since the association between these conditions is well known. It appears that heart antibodies from the mother may cross into the fetus and interrupt the conduction system of the child's heart. By the same token, pregnant women known to have lupus or rheumatoid arthritis should have the fetus or newborn carefully checked for congenital heart block, since this is a treatable condition.

Mitral-Valve Prolapse

Flow of blood from the upper left chamber of the heart (the left atrium) to the lower left chamber (the left ventricle) passes through the *mitral valve.* Many acquired diseases may affect the function of this valve, especially rheumatic fever. In the absence of any recognizable disease, one or both valve leaflets may be structurally defective and slowly become stretched and thin, eventually resulting in leakage or other problems. Defects in the structural backbone of these leaflets result in their ballooning, or *prolapsing,* into the left atrium when the left ventricle contracts. When this happens, a click, or a heart murmur, might be heard with a stethoscope. Other than palpitations, most patients with mitral-valve prolapse do not experience particular symptoms. Only a tiny fraction of such patients (around 1 percent) experience serious complications, which involve major valve leakage, infection of the valve, and even sudden death.

Mitral-valve prolapse is remarkably common. It occurs in approximately 4 to 7 percent of the general population, which makes it the most common abnormality of the heart valves in humans. The disorder appears to be somewhat more common among females. The most accurate diagnostic method is by ultrasound testing of the heart (echocardiography).

In the absence of any other associated genetic disorder, mitral-valve prolapse is likely to be inherited as an autosomal dominant disorder. In keeping with this form of inheritance, manifestations may be highly variable even within the same family. Not surprisingly, since the connective tissue that makes up the backbone of the valve leaflets is defective in this condition, it has been noted that as many as half of those with mitral-valve prolapse have associated abnormalities of the shape of their chest or breastbone, or have a curved spine, a high-arched palate, or long, thin fingers. All these features suggest that such individuals have a generalized connective-tissue disorder that is dominantly inherited.

Mitral-valve prolapse occurs frequently as part of certain well-known hereditary connective-tissue disorders, including the Marfan syndrome, the Ehlers-Danlos syndrome (dominantly inherited double-jointedness and fragile, easily bruisable skin), osteogenesis imperfecta (dominantly or recessively inherited fragile bones that fracture repeatedly), and even Duchenne muscular dystrophy.

There is an intriguing association of mitral prolapse with a condition called panic disorder. An individual with panic disorder has multiple episodes of overwhelming anxiety without obvious reason, associated with a fast heart rate, chest pain, and breathlessness. Panic disorder is also remarkably common, occurring in up to 6 percent of the population, appears to be inherited as an autosomal dominant condition, and is also more frequent in females. Some studies show that as many as half of the individuals with panic disorder also have mitral-valve prolapse.

Infection can settle on damaged valves and therefore people with mitral-valve prolapse should be conscious of situations in which significant showers of bacteria may enter their blood — for example, during routine dental work. Antibiotic coverage, beginning the day before, might be advisable, so affected persons are advised to check with their doctors.

Heart disease can be a depressing subject, since so many of us will eventually succumb to it. Early detection, coupled with vigorous efforts to prevent or reverse aggravating factors, has made a big difference in both the survival rate and quality of life of those affected.

29

Diabetes and Obesity

The most common type of adult diabetes is closely associated with obesity. While these two conditions are interrelated, useful clarification will be achieved by discussing them separately.

Diabetes: An Overview

Most every adult has heard of *diabetes mellitus,* or *sugar diabetes.* And small wonder, since there are some 200 million diabetics in the world.

Most people know that diabetics have a high level of *glucose,* or sugar, in the blood that spills over into the urine, and that they have a relative or absolute lack of *insulin,* the hormone secreted by the pancreas. The insulin deficiency creates two major problems. The first is a biochemical disturbance in body chemistry that creates a high blood-glucose level that changes the body's fat and protein chemistry. These changes may lead to diabetic coma and death if untreated. The second major danger approaches slowly over many years in the large and small blood vessels, particularly affecting the eyes, kidneys, and nerves, and brings on blindness, kidney failure, nerve damage, and possibly death. The longer the duration of certain types of diabetes, the more likely it is that these blood-vessel changes will occur. As a consequence, coronary-artery disease in diabetic males is two to three times more common than in nondiabetics, while diabetic females have a rate twenty times that of nondiabetic females. As if that is not enough, the risk of stroke is doubled; and there is a fiftyfold increase in the frequency of obstruction of small blood vessels (especially in the feet).

HISTORY

Knowledge about diabetes dates back thousands of years. Around 1500 B.C.E., the Egyptian Papyrus Ebers described an illness associated with excessive urination. Early Chinese medical writings also described a condition that was almost certainly diabetes. Physicians in India around 400 B.C.E. observed the sweetness of the urine as well as the association between obesity and diabetes and the familial nature of the disorder.

The disease was named about the beginning of the common era by the Romans Aretaeus and Celsus, who called it *diabetes* ("siphon") and *mellitus* or *melli* ("honey" or "sweet"). This was about 70 C.E. That the sweetness in the urine was due to sugar was only recognized late in the eighteenth century. In 1859, the great French physiologist Claude Bernard demonstrated the increased glucose content of blood in the diabetic patient. About ten years later, the areas of the pancreas that we today know secrete insulin were noted to be abnormal in patients with the disease. This led to experiments in 1889 that first produced diabetes in dogs by removing the pancreas.

These latter observations gave tremendous impetus to efforts to prepare extracts from the pancreas, which were then used to attempt to correct the deficiency associated with the disease. Continuous success using extracts prepared from dog pancreas to reduce the blood-glucose level was achieved in 1921 in Toronto by a young Canadian surgeon, Frederick Banting, and his graduate-student assistant, Charles Best.

Long-acting insulin was introduced in about 1936. Extracts of pancreas, of course, contain insulin. The chemical structure of human insulin was established only in 1960. In 1967, in Chicago, proinsulin, a large molecule from which insulin is derived, was detected in the blood. Not unexpectedly, proinsulin, too, is under genetic control.

WHEN IS DIABETES, DIABETES?

It is a simple matter to diagnose overt diabetes in an individual who complains of typical symptoms and has sugar in the urine plus an elevated blood-sugar level. There are various states of diabetes, and it is not at all inevitable for one state to progress to another. The first state is, as mentioned, *overt diabetes*.

Second, an individual who has no symptoms of diabetes may be found to have elevated blood-glucose levels only after a meal or after a glucose-tolerance test. (This test involves giving a person a "load" of sugar, either by mouth or intravenously, then measuring how high the blood-sugar level goes and how long it takes to return to normal levels over a three- to five-hour period.) This second state is called *asymptomatic* or *chemical diabetes.*

The third state, called *latent,* or *stress diabetes,* occurs in a person who has a normal glucose-tolerance test, but who was diabetic at some time in the past — for example, during a period of stress, during pregnancy, during serious infection, when very much overweight, or in association with a heart attack or serious burns.

The fourth state is one of, truly, *potential diabetes* — signaled, for example, by a medical history of a female who has given birth to a baby weighing more than ten pounds. This state cannot easily be diagnosed with certainty, nor can accurate predictions be made that such a woman would necessarily become diabetic.

These various states of diabetes provide no clue to the basic cause in any particular case. Moreover, some patients may be insulin-deficient while others have high levels of (ineffective) circulating insulin.

HOW LARGE A PROBLEM?

Between 11 and 12 million people in the United States have some form of diabetes. Non-insulin-dependent diabetes mellitus (NIDDM) accounts for over 90 percent of these cases and constitutes one of the most common chronic diseases in the United States. About 2.5 percent of the population — over 6 million individuals — have this type of diabetes. A further 4 million to 5 million people between twenty and seventy-four years of age have mild undiagnosed diabetes. The National Diabetes Data Group of the National Institutes of Health (from whom all these figures originate), further estimates that between 7 million and 18 million more persons in this age group have impaired glucose tolerance. Insulin-dependent diabetes mellitus (IDDM), which accounts for only 5 to 10 percent of all diabetes, occurs in about 1 in 600 school children.

While diabetes occurs in all countries, its frequency varies considerably, undoubtedly reflecting environmental, cultural, and, probably, genetic differences. It is rare in Eskimos, but is incredibly common among certain American Indians, such as the Pima in Arizona. Over 50 percent of the Pima population may develop NIDDM, although no IDDM has been found among these people. The incidence of diabetes in the Pima Indians is rivaled by that of Asian Indians who emigrated to South Africa. Asian Indian immigrants in many other places around the world have rates of diabetes that are higher than those of Asian Indians who did not leave their homeland. Japanese immigrants in Hawaii have about twice as much NIDDM when compared to Japanese who reside in Hiroshima. For many immigrants, changes in life-style are usually reflected by a higher standard of living. A diet with higher carbohydrate and fat content; less exercise; more stress, coupled with inherited predispositions — all probably account for these findings.

The frequency of diabetes is increasing for a number of reasons. The population is growing and people are living longer. (About 4 of every 5 diabetics are over forty-five years of age.) Carefully treated diabetics are living longer and, as a consequence, having children, something they may not have been able to do so often many years ago. Many of these children will inherit the gene for, or susceptibility to, diabetes. Finally, obesity, which appears to precipitate or aggravate the disease, is also increasing, thereby making more potential diabetics clinically obvious. At least 85 percent of diabetics are, or were at one time, overweight.

WHO SHOULD BE TESTED?

Since it is not feasible to do complete blood tests on all adults in the population, it is worthwhile to focus on individuals who have a higher risk of diabetes. Blood-glucose tests should be done at least yearly on the relatives of diabetics, who have a higher risk, on overweight individuals at every annual checkup, especially if they are over forty, and on mothers who have had babies with high birth weight (over ten pounds). Such women clearly have a higher risk of developing diabetes. Everyone having an operation or being admitted to a hospital, or having a preemployment physical or an annual checkup, should have a routine test for diabetes, even if there is no known family history of it.

THE CAUSES OF DIABETES

In 1922, when insulin was first detected, it seemed so clear: if your body failed to produce enough insulin, you got diabetes. Any process that interfered with insulin production — such as destruction of the pancreas because of cancer, serious infection, inflammation, or other disease — could therefore result in diabetes.

Certain diseases of other hormone-producing glands, such as the pituitary, might affect insulin secretion or actions and precipitate diabetes, as might certain drugs — for example, cortisone and some diuretics (pills to rid the body of water).

Only in the last twenty years has it become clear that a diabetic person rarely is absolutely insulin-deficient. In fact, quite the opposite situation exists: a majority of diabetics have *increased* levels of insulin!

We now know that while insulin is still the key, the mechanisms resulting in diabetes are incredibly complex, and reflect genetic and environmental influences. A clearer understanding is possible if we simply follow what may happen to insulin in the body. The insulin gene, located on chromosome 11, is responsible for producing structurally normal insulin, which functions properly by keeping the blood sugar under control. Autoantibodies to insulin or to cells that produce it are known to exist, and their destructive action may result in diabetes.

Insulin molecules in blood circulate and fit like a jigsaw puzzle into a "receptor" molecule on the surface of all cells, enabling insulin to enter and do its chemical work. Autoantibodies may literally block the receptor sites, depriving cells of the needed insulin. Alternatively, the receptor structure may be abnormal, or the number of receptors may be reduced or even absent. At least for all the "autoantibody action," we know that close interrelationships with HLA groups exist (see below). Failures to transfer chemical messages, or defective chemistry within the cell *after* insulin enters, may also result in a diabetic state.

Many other factors may interfere with the production or action of insulin. The first gene products are molecules, called proinsulin, which require further modification by the body before they are changed by enzymes into insulin. Defects in this conversion result in diabetes with high levels of proinsulin — a type of the disease genetically transmitted as an autosomal dominant disorder.

The insulin gene may produce abnormally structured insulin (three variants are known). These variant insulins caused by a mutation in the insulin gene function ineffectively, causing diabetes with high levels of insulin in the blood. Again, genetic transmission is of the autosomal dominant type.

Another type of diabetes could result from a defect in the release of insulin from cells in the pancreas. Affected individuals might develop diabetes in childhood and may or may not require insulin. Once again, the mode of inheritance is autosomal dominant. The mechanism responsible for childhood diabetes is uncertain. Antibodies may grab insulin molecules and render them ineffective, thus leading to relative insulin-deficiency and hence diabetes. Or the body may make antibodies to its own organs — for example, autoantibodies to cells in the pancreas, the thyroid gland, and so on.

RISK FACTORS

Singly or together, many factors other than heredity may precipitate the emergence of diabetes.

Obesity: The prevalence of obesity increases the incidence of diabetes. It is well established that glucose intolerance increases with body weight. One study of this correlation involved the brothers and sisters of diabetics who developed the disease mainly between their third and fifth decades. The siblings who were not obese were five times less likely to become diabetic than those who were obese. (On the other hand, among the South African Bantus, the most obese women have the lowest rate of diabetes.)

It is still not clear whether obesity itself or something in the diet is the most important factor. The best evidence implicates the fats in our diet. Evidence that high sugar intake is equally guilty has not been generally accepted. Curiously, obese individuals with heavy waists have higher rates of diabetes than those with heavy hips. Meanwhile, it appears that the influence of obesity on diabetes is greater in those presumed to carry a genetic susceptibility to this disease. It is noteworthy that the incidence of diabetes fell off considerably in Japan during the Second World War and the years of food privation that followed.

Studies of immigrant populations have yielded interesting results along these lines. One of the best was done in Israel. The frequency of diabetes in Yemenite Jews who had recently arrived was compared to that of Jews who had lived in Israel for twenty-five years or more. The latter were found to have approximately forty times as many diabetics among them as the newest immigrants. The Indians of South Africa are another example. They are reported to have a frequency of diabetes some ten times higher than that found in the Punjab, where their ancestors once lived. Similarly, a four- to twentyfold increase in diabetes was noted in Polynesian peoples in the Pacific compared to those living on the New Zealand mainland.

Reflecting either dietary influences or genetic susceptibilities, high blood-fat levels (cholesterol and triglycerides) or high uric-acid levels (which cause gout) are known to be associated with diabetes.

Race: The highest rates of NIDDM are found in the Pima Indians and the natives of the Central Pacific island of Nauru. After forty years of age, 1 in 12 Pima Indians and 1 in 36 islanders have diabetes. Nauru inhabitants developed one of the highest per-capita incomes in the world and the second-highest frequency of diabetes.

In the United States, between 1966 and 1981, the number of diabetics among blacks increased twice as fast as among whites. Mexican-Americans in San Antonio, Texas, have also been found to have higher rates of diabetes than their Caucasian neighbors.

Gender: For reasons that are not entirely certain, the relative frequency of diabetes in males versus females has varied over the years. For example, an excess of males diagnosed as diabetic was replaced by an excess of diabetic females at the beginning of the twentieth century. During the past thirty years, however, the direction has shifted back to an excess of men. Possibly, changing fashions in family size and variations in the frequency of obesity contribute to change in sex ratios.

Age: As we age, the body's ability to handle sugar decreases. This results in a steady increase in the incidence of diabetes as people get older.

Infection: The idea that diabetes could result from an infection might surprise you. It is, however, well known that in animals diabetes can be induced by viruses such as the Coxsackie B4 virus or the hand-foot-and-mouth virus. It is hard to prove that diabetes in a human has been induced by a virus, as it can always be argued that the virus acted as a stress phenomenon that precipitated diabetes in someone so predisposed. However, there are some data that show a seasonal variation in the incidence of new cases of insulin-dependent diabetes in individuals under thirty years of age; the peaks come in September and December, which strongly suggests the machination of a virus. Coxsackie B4 virus has definitely been implicated in these seasonal peaks. High levels of antibodies against that particular virus have been unexpectedly prevalent in juvenile-diabetes patients. Also, children damaged in the womb by German measles have a higher risk of developing diabetes later in life.

In diabetes with an onset in childhood, antibodies to the child's own pancreas have been repeatedly detected in the blood. It is not known whether infection of the pancreas initiated the process; but it is known that in certain diseases, the body responds by making antibodies to its own organs, such as the thyroid or adrenal glands, and that such conditions are frequently associated with diabetes.

Simultaneous development of the disease in brothers and sisters is another clue that suggests an infectious agent. There is one family I know of that suffered three children developing diabetes within three months of each other. In spite of such evidence, however, the matter becomes increasingly complicated when you recognize that some of us may be genetically predisposed to certain infections and, in that way, have a higher risk of developing diabetes. At this point, it might be useful to consider the known genetic aspects.

The Genetics of Diabetes

The many complex factors that are known to interfere with insulin production and function reflect the latest advances in the understanding of diabetes. Most of our knowledge to date simply reflects our understanding of five major clinical classes of diabetes: (1) type-1 insulin-dependent diabetes mellitus (IDDM); (2) type-2 non-insulin-dependent diabetes mellitus (NIDDM) — in nonobese or obese individuals; (3) diabetes mellitus associated with other disorders; (4) impaired glucose tolerance (IGT) — in nonobese or obese individuals, and in individuals with associated disorders; and (5) diabetes in pregnancy, or gestational diabetes mellitus (GDM).

Most of our knowledge about the inheritance of diabetes mellitus is based on these five major classes and not on the highly sophisticated new technologies that are splitting up the causes of diabetes into a larger number of recognizable genetic entities. It will still take years, using the latest analytic techniques, to determine the mode of inheritance for each newly separated class of diabetes. Meanwhile, for the main two types of diabetes, there is much we know that can provide useful guidance.

TYPE-1 INSULIN-DEPENDENT DIABETES MELLITUS (IDDM)

Sudden onset is typical in type-1 diabetes, which affects about 1 in 500 Westerners. At least 70 percent have been diagnosed within four weeks of noticing symptoms, which mainly consist of excessive urination, excessive thirst, a markedly increased appetite, paradoxical weight loss, and often fatigue, irritability, and moodiness. Symptoms may, however, be present for months, and sometimes even for many years, before overt diabetes is recognized.

IDDM is regarded as an autoimmune disease that is influenced by genetic factors. In this type of diabetes there is progressive destruction of the so-called beta cells in the pancreas that make insulin. Whatever factor — viruses have been implicated — causes these key cells to disintegrate, the body follows by producing substances to fight molecules released by infected or damaged cells. These autoantibodies home in to the beta cells, further destroying their function and thereby rendering the individual insulin-deficient and therefore diabetic.

There is only a modest familial association for IDDM: some 5 to 10 percent of close relatives develop the disease. In general, males are affected about 10 percent more often than females. Curiously, there appears to be a paternal influence in inheritance; children of fathers with IDDM seem to be about five times more likely to develop the same type of diabetes than are children of diabetic mothers.

Specific HLA groups (see chapter 9), DR3 and DR4, occur with high frequency in those with IDDM. Some studies show that 95 percent of those with IDDM have either HLA-DR3 or HLA-DR4, compared to only 50 percent who possess one of these antigens in the general population. About 40 percent of those with IDDM have both DR3 and DR4, in contrast to only 3 percent who have both in the population at large. In one report on 100 affected sibling pairs, 60 had identical HLA types, 35 had one HLA group in common, while only 5 shared no HLA group. Others have confirmed these rates and also shown that for the population at large only 25 percent have these identical HLA groups, 50 percent have one HLA group in common, and 25 percent have none of these HLA groups in common.

Individuals with IDDM do not directly inherit this disorder but rather have an inherited susceptibility, probably to a viral infection, which in turn either *directly* damages the insulin-producing beta cells of the pancreas or stimulates the body to produce damaging antibodies, thereby causing diabetes. If a parent has one child with IDDM, there is about a 2- to 3-percent risk of having another affected child. Where one identical twin has IDDM, there is about a 68-percent likelihood that the co-twin will also develop this disorder. This suggests that other, still poorly understood environmental factors are at play.

HLA antigens are involved in the immune process mounted by the body to fight infection. Calculations based on which HLA group an individual inherits make it possible to estimate the likelihood of IDDM developing in that individual. Those who have HLA-DR3 and HLA-DR4 have an approximate risk of 1 in 40 of developing IDDM. A brother or sister of someone with IDDM who has the identical HLA groups has about a 1-in-7 risk of developing IDDM. When those HLA-identical siblings have the HLA-DR3 and HLA-DR4 groups, their risk of developing IDDM rises to about 1 in 4.

The matter gets much more complicated, but suffice it to say that for IDDM, susceptibility to infection is the real problem. The hope for the future is to identify children with these specific HLA groups, which make them susceptible, recognize which viruses directly or indirectly threaten the pancreas, and prevent the infection by using a special vaccine in very early childhood. Treating or blocking the antibody process is a goal in which some success has already been chalked up.

TYPE-2 NON-INSULIN-DEPENDENT DIABETES MELLITUS (NIDDM)
Unlike type-1 diabetes, NIDDM cannot always be explained by a single defect. Affected individuals may have abnormalities in insulin structure, secretion, and action. In contrast to the sudden onset of type-1 diabetes, NIDDM begins imperceptibly and develops slowly. An astonishing 5 to 6 percent of the U.S. population have NIDDM, and about half do not even know it! Only about one-quarter of those with NIDDM have typical symptoms reflective of high blood sugar, including excessive urination and excessive thirst with weight loss. A remarkable 85 to 90 percent are obese. One ten-year study of individuals with impaired glucose tolerance showed that 15 percent became overtly diabetic (with sugar in the urine and symptoms as described), 22.8 percent remained glucose-intolerant, while the majority rapidly recovered or were only temporarily intolerant. The single most serious risk factor leading to NIDDM was obesity.

Type-2 diabetes, common among the middle-aged and elderly,

is usually mild in comparison to type-1 diabetes. In type 2, however, genetic factors are extremely important. If, for example, an identical twin develops NIDDM, there is over a 90-percent likelihood that the other will do likewise. The risk of an individual developing NIDDM depends upon his or her age, and upon whether both or only one or neither parent has this disease. The risks are summarized in table 18, which combines type-1 and type-2 diabetes, with over 90 percent being type 2. It can be seen from the table, for instance, that when both parents are diabetic and their offspring is twenty to thirty-four years of age, there is a 5.8 percent chance that that child will have diabetes. If only the mother is diabetic, the chance is 2.1 percent; and the risk is virtually identical if only the father is diabetic. If neither parent is diabetic, however, the risk of developing diabetes between the ages of twenty and thirty-four is just 0.7 percent. Current estimates suggest that the siblings of those with NIDDM have nearly a 40-percent risk of developing this form of diabetes, especially if they are obese.

A subtype of NIDDM has been recognized in which insulin is not required for treatment, although onset occurs in childhood or young adulthood. This type of maturity-onset diabetes of the young is transmitted as an autosomal dominant disorder, and the offspring and siblings of such individuals have a 50-percent risk of being affected. Luckily, however, this type of diabetes appears to be milder than the IDDM type, and it is associated with fewer complications.

Table 18. Chances of Diabetes in Children of Diabetics versus Nondiabetics

Age of Offspring (in years)	Both Parents Diabetic	Only Mother Diabetic	Only Father Diabetic	Neither Parent Diabetic
20–34	5.8%	2.1%	2.0%	0.7%
35–44	7.3	3.6	3.2	1.1
45–54	10.5	5.8	8.0	2.9
55–64	25.5	9.6	13.2	3.7
65–74	21.5	11.3	11.7	4.6

SOURCE: Data from M. I. Harris, National Diabetes Data Group, National Center for Health Statistics.

Ongoing research has already clarified that multiple subtypes exist among individuals with type-2 diabetes. Dominant inheritance clearly accounts for diabetes that is due to abnormally constructed insulin (three types have been recognized thus far), and, as mentioned, to the maturity-onset type of diabetes of youth. Another subtype with onset approximately between twenty-five and forty years of age may be caused by one or more genes. The siblings of affected individuals have a 69-percent risk rate of glucose intolerance. Further studies are necessary to determine the risks of overt diabetes in the offspring of such individuals.

DIABETES MELLITUS ASSOCIATED WITH OTHER DISORDERS
The vast majority of people with diabetes have one of the above type-1 or type-2 forms of the disease. Not infrequently, however, diabetes may develop in association with other genetic or nongenetic conditions. These secondary forms of diabetes emerge, for example, in association with diseases of the pancreas (such as cystic fibrosis), pancreatitis (inflammation of the pancreas), following certain medications (such as cortisone or certain diuretics), and in conjunction with a surprisingly long list of hereditary disorders.

The risk of occurrence or recurrence of diabetes in these situations will be related directly to the nature of the disease with which it is associated.

IMPAIRED GLUCOSE TOLERANCE (IGT)
Persons found to be intolerant of glucose clearly have a higher risk of developing diabetes than those with normal glucose tolerance. Such persons have mostly been described as chemical, borderline, or latent diabetics. The risk each would have of becoming overtly diabetic would depend upon the genetic or other causes of the particular type of diabetes. Long-term studies of individuals with IGT show that a relatively small percentage actually develop full-blown diabetes. Moreover, there is evidence that there is improvement in some. The single most important piece of advice for a person with IGT is to achieve a normal weight.

DIABETES OF PREGNANCY
This type — also called gestational diabetes mellitus, or GDM — refers to impaired glucose tolerance or diabetes that begins during pregnancy. GDM is not uncommon: it occurs in 2.5 to 5 percent of all pregnancies. Factors that increase the likelihood of a woman

with GDM developing overt diabetes after pregnancy are advancing maternal age, the previous birth of a heavy baby, a previous stillbirth or newborn death, obesity, and a family history of diabetes.

Dr. John O. Sullivan of the Boston University School of Medicine reported on a remarkable study of 615 patients with GDM who were followed for twenty-two to twenty-eight years. Diabetes developed in 1 in 4 — sixteen times the rate of women who had not had GDM.

As noted elsewhere, women with IDDM have about a two- to threefold risk compared to nondiabetics of having a child with a serious birth defect. Mounting evidence also suggests that the children of women with GDM are at an increased risk for obesity and glucose intolerance during childhood and later life. Once again, the best advice for such women is to ensure right from the earliest weeks of life that their children do not become obese.

Obesity

Obesity runs in families. A person is regarded as obese if his or her ideal body weight, as calculated from averages in the general population for persons of various heights, is exceeded by 20 percent. Obesity is a common and serious problem in Western societies. United States Public Health Service figures indicate that 14 percent of men and 23.8 percent of women twenty years and older are obese.

The seriousness of obesity is reflected by its association with multiple life-threatening diseases. More than any other factor, obesity is associated — in over 85 percent of cases — with non-insulin-dependent diabetes. High blood pressure, high blood-cholesterol levels, lung and gallbladder diseases, as well as certain cancers — all occur more often in the obese than the nonobese. In the famous Framingham Heart Study, persons who were 20 percent or more overweight had hypertension ten times more often than normal-weight individuals.

While there is no argument that fatness is familial, discovering whether this reflects an inherited trait or family eating and other habits has been difficult. Let us examine the current facts.

GENETICS OR ENVIRONMENT?

We know that about 80 percent of the children of two obese parents become obese adults, as compared with no more than 14 percent of the children of two parents of normal weight. If one child in a family is obese, the likelihood of a brother or sister being obese is 40 to 80 percent. However, these correlations are similar for all children in such households, *including* those who have been adopted! Furthermore, genetic explanations cannot explain correlations showing that when one spouse is obese it is highly likely that the other is similarly overweight. The essential role of the environment is no better shown than by a study that demonstrated the similarity in fatness between *pets* and their owners!

Obesity is also thought to occur more frequently among children whose parents have been separated or divorced. Single-child families have the highest prevalence of obesity, and the number declines with increasing size of the family. It is indisputable that family eating, exercise, and other habits are key factors in the development of obesity. But what of the genetic contribution?

Not surprisingly, studies of twins have shown a strong genetic component to obesity. While twin studies clearly point to the significant contribution of heredity in the development of obesity, an abiding criticism is that identical twins share identical home environments as well as the same womb. Adoption studies, which would considerably enlighten the question, have yielded conflicting results in the past. A recent study published in *The New England Journal of Medicine* was based on the remarkable and outstanding *Danish Adoption Register*. This register contains the official records of every nonfamilial adoption granted in Denmark between 1924 and 1947. The study, which compared adopted individuals with their biological parents as well as with their adoptive ones, showed that there was a clear relationship between adoptees and their biologic parents in terms of both thinness and fatness. The analysis was consistent with a multifactorial type of inheritance, but did not exclude the possibility of a contribution by single or multiple genes. These researchers also showed that obesity in children was more strongly correlated with the mother's rather than the father's size.

Notwithstanding the conclusions of this excellent study, certain criticisms continue. For example, offspring who receive increased amounts of glucose while in the womb may ultimately be different from those not so exposed. The children of women who were diabetic *during* pregnancy are different from those of women who subsequently became diabetic after pregnancy, in that they had an increased incidence of obesity and impaired glucose tolerance when compared with the offspring of women without diabetes.

Moreover, women with diabetes *during* pregnancy are themselves more likely to be the offspring of women with a similar type of diabetes.

We can conclude that heredity plays a significant role in obesity, together with highly significant dietary and other environmental factors. There is no doubt that even with a hereditary predisposition to fatness, detailed attention to limiting calorie intake and to exercising adequately will enable satisfactory weight control.

While these efforts may be difficult, every effort should be made to prevent the development of obesity (which is so hard to reverse) — in order to improve the quality of life, decrease the frequency of associated diseases, and prevent premature death.

30

Mental Illness and Heredity

Many find mental illness in the family much harder to deal with than physical problems. Perhaps this is in part due to fear, age-old stigmas and taboos, or to the frustration created by the invariably chronic nature of such an affliction. In recent years, however, treatment with drugs has made a significant difference.

There is a clear distinction between neuroses and much more serious psychoses. The term *neurosis* implies some kind of deranged bodily function in the absence of an actual "physical" disease; the abnormality is dependent, in a way usually unknown to the afflicted person, on some kind of mental disturbance. Neurotic individuals — and we probably are all neurotic at least at some periods in our lives — may complain about an astounding variety of different symptoms that are all very real: headaches and other aches and pains in various parts of the body, palpitations, breathlessness, loss of appetite, weakness, lethargy or fatigue, a feeling of pins and needles in the fingers, vomiting, diarrhea, constipation, excessive sweating, or a feeling of a lump in the throat. Mental disturbances may produce fear of heights, sounds, or open or closed spaces; or there may be anxiety because of troublesome thoughts, inability to sleep sufficiently well, or compulsively repeated acts. There is no evidence that neuroses have a genetic basis. Through a shared environment, it could be anticipated that severely neurotic parents will be likely to have neurotic children.

This chapter will concentrate on the more severe types of mental illness classified as *psychoses* or as *affective disorders*. Individuals who are psychotic have signs or symptoms that separate them from reality either constantly or intermittently. A number of such disorders have been recognized, but good evidence for a genetic basis has emerged only for schizophrenia, major depressive illness, manic depression, and, in some cases, infantile autism.

Schizophrenia

The term *schizophrenia* was first introduced about seventy-five years ago to describe a splitting between thought and emotional responses — giving rise to the idea of "split personality." Today, schizophrenia is considered to be a disorder of thought processes and expresses itself as a disorder of feeling, through inappropriate behavior and an increasing withdrawal from interpersonal contacts. There are various typical symptoms. A person may believe he is God, dispensing blessings, charity, or punishment. Patients hallucinate, and walk around intently listening to voices they maintain they are hearing. These voices, they say, are giving them instructions to perform certain acts that may even include murder. Some schizophrenics become totally withdrawn, hardly moving a limb. They may stand with an arm in the air or in some other unusual posture for hours at a time — a so-called *catatonic state*. Other patients may become paranoid, insisting that someone or many people are out to kill them or to persecute them, and so on.

Because a wide range of schizophrenialike disorders or schizoid personalities exists, formal criteria are used to diagnose schizophrenia. The five cardinal points are

1. delusions or hallucinations, or marked thought disorder associated with blunted responsiveness, delusions, hallucinations, or very disorganized behavior;
2. deteriorated function at work or in social relations and personal care;
3. a length of illness of at least six months that includes at least one week of symptoms such as delusions, hallucinations, or disorganized behavior;
4. the onset of symptoms before forty-five years of age; and
5. the absence of other medical disorders that cause altered brain function or mental retardation.

Schizophrenia is common and is thought to occur in about 1 in every 100 individuals in the United States. The first episode may occur sometime between adolescence and fifty years of age. Most patients are diagnosed in their twenties. Males and females are equally affected, although an earlier onset is typical in males. With the advent of the newer medications, the outlook for schizophrenics is better than it used to be. About 20 percent of schizophrenics experience a full and permanent recovery; in about 35 percent, full recovery is interrupted by psychotic episodes; and in about 35 percent, long-lasting milder psychoses persist, with patients mainly being able to live at home or in shelters. Sadly, about 10 percent remain severely psychotic and require permanent hospital care.

IS SCHIZOPHRENIA INHERITED?

Schizophrenia is indeed common. In the United States alone, a staggering number of persons are affected — over 2.5 million! Of course, many of these individuals are affected only periodically, with mild symptoms, and are easily treated. Those who are severely affected come to the attention of psychiatrists or society fairly quickly.

A number of risk factors for schizophrenia have been recognized that raise "environment" at least as a contributor to its origins. Schizophrenics have a higher rate of winter births in both the Northern and the Southern hemispheres. They also experience a higher rate of stressful life-events. Completely unexplained is their sixfold *lower* rate of rheumatoid arthritis. The facts that more schizophrenics are in the poor social classes and that more are single are probably explained by their illness, even if not yet obvious, hindering attainment of commonly achieved goals.

The most important advances, however, have been in recognizing the genetic contribution and interaction with environmental factors that result in schizophrenia. You will not be surprised, then, at how often people come for genetic counseling concerned because of schizophrenia in their family, wondering whether either they or their offspring may ultimately develop it. There have been seemingly endless studies to determine whether schizophrenia is hereditary or not. Three groups of studies have provided the best data: those on families of schizophrenics, those on twins, and those on adopted children.

SCHIZOPHRENIA IN THE FAMILY

Reliable data on familial schizophrenia took a long time to formulate since definitions of schizophrenia have varied appreciably.

Information available today clearly indicates that there is an increased likelihood of schizophrenia occurring when someone else in the family is affected. The estimated risks of such occurrences are summarized in table 19. Parents of schizophrenics, for instance, are affected themselves about 5 percent of the time, while a brother or sister of a schizophrenic has an average risk of about 8.7 percent for developing schizophrenia. Curiously, and for unknown reasons, the frequency of schizophrenia is significantly higher than normal among Irish residents of Ireland and residents of the Istrian peninsula of Yugoslavia.

SCHIZOPHRENIA IN TWINS

Enormous efforts have been made to study twins with all kinds of genetic disease (see chapter 12). The idea, of course, is that if the disorder is genetic, there is a high likelihood that both identical twins will have or will develop the disorder. Recent data on identical twins have shown that both twins have been affected with schizophrenia in 33 to 52 percent of cases. These figures are significantly higher than the 10- to 12-percent occurrence rate noted for nonidentical twins. Many critics have maintained that identical twins reared in the same home may be subject to the same kinds of pressures and environmental variables and have therefore questioned the significance of those studies. Notwith-

Table 19. Frequency of Schizophrenia among the Relatives of Schizophrenics

Relationship to Schizophrenic	Estimated Risk of Having Schizophrenia
Parent	5.1%
Brother or sister	8.7
Identical twin	33–52
Nonidentical twin	10–12
Child (one parent affected)	10–15
Child (both parents affected)	39.0
Uncle or aunt	2.0
Nephew or niece	2.2
First cousin	2.9
Unrelated (general population)	0.9

standing such criticism, an authoritative critical review in 1983 concluded that twin studies have clearly established that schizophrenia is partly inherited.

SCHIZOPHRENIA IN ADOPTED CHILDREN

Since the family environment could potentially affect the upbringing of identical twins, thereby possibly influencing the development of schizophrenia, researchers devised elaborate studies to examine adopted children and determine the frequency of schizophrenia in children removed early in life from their home and family. These adoption studies were done in Denmark by American and Danish psychiatrists in collaboration. Denmark was selected because the records of adoptions and of episodes of schizophrenia (and other vital records) are so magnificently collected and kept.

The goal of the first study was to determine the frequency of schizophrenia in the adopted children of parents one or both of whom were schizophrenic. These figures were compared to figures indicating the frequency of schizophrenia in the natural offspring of the families of adopting parents. The researchers found that schizophrenia among the adopted offspring of schizophrenics was three times that found in the natural offspring of adopting parents. Later studies found the same frequency of schizophrenia in children of schizophrenic parents who were themselves adopted early in life as in those who were raised by schizophrenic parents themselves. Moreover, schizophrenia was found to occur among the adopted offspring of normal parents (those without schizophrenia) at about the same rate as noted in the general population.

To exclude the effects of the environment in the womb or influences surrounding birth or early mothering, another study was devised to determine the frequency of schizophrenia in adopted offspring who had the same fathers but different mothers. This study concluded that significantly more offspring of the same father with schizophrenia had the disorder than did matched control groups.

The adoption studies were meant to separate out the hereditary effects on child rearing from the environmental ones. One unexpected observation was the slightly higher level of mental illness in those parents who adopted children who later became schizophrenic.

CONFOUNDING FACTORS

The inexorable conclusion from these remarkable, painstaking family studies is that genetic influences are involved in schizophrenia. It is equally obvious that schizophrenia, like mental retardation, represents a variety of different disorders and causes; consequently, different patterns of inheritance can be expected. As for environmental influences, some evidence suggests that childhood schizophrenia develops more often in those children who have had complications during and immediately after the birth process than in their brothers or sisters who had normal births. Moreover, the offspring of parents of whom at least one is a schizophrenic have been noted to have a higher frequency of fetal death or death in the newborn period.

A curious observation was made by researchers who noted the sex of children born to women hospitalized for schizophrenia. Those mothers whose symptoms of schizophrenia had occurred within one month of conception of their infant all had female babies. The researchers further noted that if the schizophrenic episode began within two or three months before or even after conception, there were significantly fewer males among the babies born, as well as an increased frequency of stillbirths, malformations, and birth defects.

Observations made as long ago as 1910 have implicated the left side of the brain in some 90 percent of schizophrenics. The absolutely latest in sophisticated technologies, using not only X-ray brain scans but also computer-controlled brain-electrical-activity mapping, also point repeatedly to left-brain involvement. It is also remarkable that the children of schizophrenics have a higher rate of both psychosis and left-handedness than their siblings.

It has long been noted that certain drugs that cause hallucinations, such as LSD, may precipitate schizophrenia — perhaps in the predisposed individual. Some psychiatrists have separated out their psychotic patients into those in whom no cause whatsoever can be suggested or found — called *nonorganic* psychosis — and those in whom some suggestion or sign of abnormality in brain function, such as mild mental retardation, limited speech, and so on, can be determined — called *organic* psychosis. Evidence has shown that children or adolescents with deviant behavior gross enough to be called psychotic probably include a number with dysfunction of the brain.

This insight shed some light on a remarkable observation made some years ago. Dr. William Pollin noticed that if only one twin of an identical twin pair was psychotic, he or she was lighter in birth weight and was more likely to have been exposed to damage that interfered with brain development. He also found that many of the affected psychotic twins had some suggestion of brain dysfunction.

Two 1985 studies of the children of parents where one was schizophrenic compared children of normal parents and showed that between eight and fourteen years the children of the former had a higher frequency of brain dysfunction involving coordination, perception, and attention.

SUSCEPTIBILITY TO SCHIZOPHRENIA

There have been a number of studies aimed at determining the susceptibility of individuals to the development of schizophrenia. Using a computer to analyze patients' speech patterns, researchers have observed that schizophrenics have a sense of disorientation and chaotic thought in association with certain apparent defenses against confusion and discomfort. These speech-pattern analyses are being applied to relatives of schizophrenics who are not overtly affected.

Another approach has been to analyze eye movements, as the eyes track a moving pendulum. This is done by recording electrical signals through tiny electrodes near each eye, to discover how smoothly the eyes move in response to a slow-moving target. Deviations have been noted more frequently among schizophrenics (in 59 percent) and manic depressives (41 percent) than in healthy persons (8 percent). Of particular interest was the observation that "abnormal" eye movements were noted in at least one parent in 55 percent of couples with a schizophrenic child, compared to 17 percent of parents with a manic-depressive child. This dysfunction in eye tracking is unexplained, but may serve to identify those who are vulnerable to schizophrenia.

Many other tests to determine vulnerability to schizophrenia are being explored — tests that involve brain enzymes, HLA groups, and DNA analysis. Thus far there seems to be a link between HLA-A9 and paranoid schizophrenia. While more detailed studies are still in progress, it appears that HLA-A9 occurs about twice as often in schizophrenics than in nonschizophrenics.

Manic-Depressive Illness

Depression is a very common symptom that affects each of us at some time in our lives. Depression may, for example, occur perfectly normally after sad events such as the death of a loved one, personal illness, or family problems and not be considered a neurosis. When it occurs without any relationship to obvious problems, then it may well be part of anxiety neurosis. I will not discuss depression of this sort but rather concentrate on the very much more serious disorder of manic depression: a mood (affective) disorder associated with extreme elation or profound depression, either of which dominates the patient's entire mental life in cycles. A person may suffer intermittently from periods of severe depression (called the *unipolar type*) or from episodes of mania or depression (the *bipolar type*). The unipolar and bipolar disorders may represent the effects of one or more genes, which in addition may be influenced by environmental factors.

Very definite criteria are used to recognize major depression, including a mood of depression, despair, and irritability, loss of interest in usual activities, as well as at least a two-week period in which a minimum of four of the following eight associated symptoms are present:

1. A major increase or decrease in appetite and weight
2. Insomnia or excessive sleep
3. Impaired intellectual function and lack of movement or agitation
4. Decreased interest in sex or other usual activities
5. Lack of energy
6. Inappropriate expression of guilt
7. Poor concentration
8. Repeated thoughts of death or suicide

In contrast, individuals with manic depression, or bipolar illness, may have alternating episodes of mania or extreme, uncontrollable excitement and deep despair.

The patients with the manic phase of the disorder are characterized by an extremely elated though unstable mood, flights of ideas, and tremendous physical and mental activity. Specific criteria are used to diagnose manic depression. These include a period requiring hospitalization or lasting at least one week characterized by a excessively happy (euphoric), generous, or irritable mood and involving three or four of the following seven symptoms:

1. Increased activity or restlessness
2. Excessive talking
3. Racing thoughts or flights of ideas
4. Incredible generosity and grandiose actions
5. A reduced need for sleep
6. Easy distractability
7. Shopping sprees, reckless driving, or other risky activities

I have never forgotten a particular patient with mania that I saw thirty years ago while still a medical student. She had been a perfectly healthy young woman with a normal, happy disposition. One day, out of the blue, she suddenly developed manic excitement, became extremely restless, and started talking and screaming continuously, her elation reaching points of ecstasy at times. She expressed peculiar ideas, declaring that she was acting under orders of God, with whom she had spoken. At the time she was admitted to the hospital, she insisted that the doctor kiss her and she berated him when he refused. She could hardly finish a sentence, moving from a description of what God instructed her to do, to active efforts to embrace doctors in attendance, to exposing her body and inviting anyone walking by to make love to her. She became extremely excited and angry when such efforts failed. She then smashed crockery, tore her bedclothes, broke glass, and flung things at other patients. She ultimately recovered completely, but had amnesia and could not remember the particular period during which she was ill.

Profoundly depressed individuals have difficulty in thinking and may have associated depressed physical activity. They may be suicidal, have hallucinations, hear voices, feel they are being persecuted, and hardly talk at all. They may be careless in their dress and, in extreme cases, even soil their clothes with urine or feces.

IS MANIC-DEPRESSIVE ILLNESS INHERITED?

Close to 1 in 100 in the general population have or will develop manic-depressive illness. And surprisingly, at least 6 percent will develop serious depression at least once in their lifetime. The symptoms within a given family are usually similar; the same bipolar disorder, for example, will affect different members.

There has been evidence of a genetic basis in manic depression for many years. The familial nature of the disorder and the contrast between identical and nonidentical twins have been key points. The estimated risks of the disorder recurring within a family are summarized in table 20. Studies of adopted persons with manic depression showed that only 2 percent of their adoptive parents had the disorder, while 30 percent of their biological parents did.

Until recently, the mode of transmission has not been entirely clear. In 1987, researchers using DNA analysis techniques demonstrated for the first time the existence of a dominant gene — located on the tip of the short arm of chromosome number 11 — that causes manic depression. This study was done in the very cooperative Old Order Amish — a group of over 12,000 individuals of German descent who live in relative isolation in Lancaster County, Pennsylvania. This remarkable group, whose values could be a lesson to us all, originated from about fifty couples who came from Germany between 1720 and 1750. They use neither drugs nor alcohol, and there are virtually no crimes or acts of violence in the Amish community.

A fascinating observation in the Amish study was that only 60 to 70 percent of those who inherited the gene for manic depression actually developed this disorder. Work is in progress to discover why the gene does not have any effect in 30 to 40 percent who inherit it. One intriguing clue: close by to the gene for manic depression is another gene responsible for producing an enzyme (tyrosine hydroxylase) that plays a key role in transmitting brain messages!

Table 20. Frequency of Manic-Depression among the Relatives of Manic-Depressives

Relationship to Manic-Depressive	Estimated Risk of Having Manic-Depression
Child (one parent affected)	27%
Child (both parents affected)	50
Identical twin	75–80
Nonidentical twin	20
Unrelated (general population)	1–3

Researchers in England, France, and Iceland, investigating families with dominantly inherited manic depression, have been unable to detect any abnormal gene on chromosome 11. This means that more than one genetic type of manic depression exists and that we can look forward to further research into this painful puzzle.

Meanwhile, it should be realized that young children may also be affected by manic depression and that they will usually benefit from medical treatment. This focus is important since studies show that up to 40 percent of children of parents with manic-depressive illness suffer from depression or other psychiatric or behavior disorders. In addition, compared with children of mentally healthy parents, they seem to experience more complications during the labor-and-delivery and newborn period, as well as more convulsions, head injuries, operations, and suicide attempts. Careful observation of these children will facilitate early detection of problems and timely treatment.

Infantile Autism

The cause of infantile autism, a condition first described in 1943, for the most part remains obscure. A developmental brain dysfunction probably lies at the root of this intractable disorder. Notwithstanding difficulties in understanding the cause, the characteristic and devastating signs are only too well known.

Most cases are probably evident at birth, but the hints are so subtle they are frequently missed. Parents and doctor may fail to notice the newborn's lack of eye-to-eye contact with the mother when feeding, or the lack of anticipation of and responsiveness to being picked up. Other typical signs as the child grows older include extreme self-isolation, aloofness, withdrawal, apparent insensitivity to pain, treating people as objects (the child may stand on the head of a playmate), failure to develop language, repetitive body movement, rocking or handflapping, self-mutilation, head banging, and so on.

Various causal factors have been implicated. Brain injury — for example, due to lack of oxygen during and immediately after birth — has been associated with subsequent development of autism in some cases. Epilepsy, with onset during adolescence, in association with an organic brain disorder has also been linked with autism. Detailed X-ray studies (CAT scans) of the brains of autistic children have revealed structural abnormalities of the ventricles (inner spaces). Clearly, some retarded children with specific genetic disorders such as phenylketonuria and tuberous sclerosis may show autistic behavior. Recent advances, however, suggest that genetic factors may be important.

We now know that 2 to 3 percent of the siblings of autistic children are also affected. This is a rate fifty times in excess of that expected by chance. Further supporting a genetic basis for autism are studies of identical twins showing that both are affected in 40 percent of cases, in contrast to the 10-percent rate for nonidentical twins. Of special interest is the fact that learning difficulties and language problems also seem to occur more frequently in families with autistic children. In one study, some 25 percent of autistic children had a sibling or parent with a history of delayed speech or other language disorder. A 1981 study showed that 15 percent of siblings of autistic children had IQs in the retarded range, compared with the 3 percent of siblings of children with Down syndrome who had such low IQs. Most recently, following analysis of multicenter surveys that compiled information on a total of 614 autistic males, estimates were made that some 12.3 percent of fragile-X males (see chapter 6) are autistic.

So what, finally, can be said about the causes of autism? The best evidence currently points to brain-cell disorganization that interferes with the perception and reception of every kind of sensation or stimulus, from pain to speech. Clearly, different factors alone or together may result in the disorder. It is now apparent that there is a genetic component involved in causing autism. The nature and extent of this contribution is uncertain, however, and requires further study.

31

Heredity and Cancer

Genetic mechanisms lie at the root of all cancers. This is not to say that all cancers are hereditary. In fact, fewer than 10 percent are directly inherited. Whatever environmental cancer-causing agent (*carcinogen*) exists, it probably wreaks its harmful effects by damaging one or more genes. Once impaired, the gene's normal function of controlling or regulating cell growth goes awry and the cancer process is thus initiated. The mechanism may not be so direct. For example, gene damage may impair the body's protective immune system, leaving it open to attack by another carcinogen. Before tackling the vast subject of cancer causation, it would be helpful to explain some basic points and reflect on the magnitude of the problem.

What Is Cancer?

Before we go further, it should be understood that a *cancer* is a group of cells, probably originating from one cell, that begins to grow abnormally, invades the surrounding tissues, and spreads through the bloodstream and the lymphatic system to other parts of the body. It is then deemed a *malignant* tumor. If a tumor simply grows and remains in its tissue of origin without spreading, it is called *benign* and can be easily removed, leaving the person with a very low risk of recurrence. The distinguishing feature of a malignant tumor is therefore not so much the abnormal growth of certain cells, but rather a defect in the body mechanisms that normally set the territorial limits of cells and keep the abnormally growing cells from invading other tissues or spreading to other parts of the body.

While cancers may arise in any single tissue of the body, broadly speaking there are mainly three recognized types. The first are called *carcinomas*: they arise in sheets of cells that cover the outside skin of the body and that line the inside of all hollow organs, such as the stomach. The second are those that arise in the blood-forming cells of bone marrow, lymph glands, or spleen; these are called *leukemias* or *lymphomas*. The third and rarest type of cancer are those that originate in the bones and connective tissues of the body and are called *sarcomas*.

The Magnitude of the Problem

The probability that any of us in the United States will develop cancer at some time during our lifetime is about 1 in 3. Not surprisingly, after heart disease, cancer is the second-most-common cause of death in the United States. Some 20 percent of American deaths are due to cancer — about 483,000 alone in 1987. Worldwide, over 4 million deaths occur annually.

Cancer develops most frequently in association with aging, and this applies to all its varieties. It is known that the death rate from cancer of the large intestine increases about 1,000 times between the ages of twenty and eighty years, and most of this increase occurs after age sixty.

About half of all cancer deaths are caused by cancers of three organs: the lungs, the breast, and the large intestine. Trying to determine the frequency of a cancer can be quite difficult. For example, the usual estimate of cancer of the prostate in seventy-year-old men is about 200 cases per 100,000 men per year (0.2 percent). However, examination of the prostate glands at autopsies of seventy-year-old men who have died from other causes have shown cancer that is still microscopic in size in 15 to 20 percent of these cases. The incidence of prostatic cancer measured by this technique would therefore be 100 times greater than that observed in patients who have symptoms. Hence, it is possible that each of us may develop several groups of abnormally growing cells in various tissues. The crucial point is that we do not know what factors allow those abnormal cells to spread through tissues and lead to clinically obvious cancer.

The one thing that is certain is that more-easily-observed tissues allow more-careful surveillance. The stomach, for example, is harder to examine than the skin. In fact, a survey in a rural district of Tennessee revealed that about 4 percent of the adult population had some kind of skin cancer. Obviously, the better the surveillance, the quicker the diagnosis — and the more likely the cure.

The Immigrants

The occurrence of cancer in a few individuals in the same family living in the same state or country immediately raises a very complex question: Is heredity or environment at work here? There are a few clues provided by groups who have emigrated from one country to another, and many studies of such groups have been reported. Take, for example, the incidence of cancer of the stomach in Japan: the disease is much more common there than in the United States. (In contrast, cancer of the large intestine, the breast, and the prostate are much less common in Japan than in the United States.) It has been noted that when Japanese emigrate to the United States, these differences in the incidence of cancer are lost within a generation or two. Since, for the most part, Japanese immigrants and their children have tended to marry within their ethnic group, the change in incidence of cancer is probably more related to their changed environment than to genetic factors.

Another example: the Jews who emigrated to Israel from Europe or the United States. They have been found to have an incidence of cancer more typical of the country from which they originated. Their children born in Israel, however, were found to have a much lower incidence of almost all kinds of cancer — rates similar to the Jewish population already in the area. Since families are likely to stay in the same areas, or at least in the same countries, it is difficult to separate out the factors.

The problem is further bedeviled by the likelihood that certain environmental agents may have initiated the cancer mechanism a few decades before the patient actually comes to the doctor with an obvious tumor. The frequency of lung cancer is directly proportional not to the number of cigarettes smoked this year but rather to the number of cigarettes smoked about twenty years ago! In a similar way, cancers that occur as a result of exposure to certain industrial chemicals may take ten to twenty years or more to appear — and the victim may already be retired. One curious example of the long "incubation" period, which can be so deceptive, is cancer of the penis. This cancer is mainly seen in old men.

However, it is now well established that cancers of the penis can be prevented by circumcision in the first few days of life, but not if circumcision is delayed for a few years.

Genetic or environmental factors acting singly or together are responsible for all cancers. Dissection of the possible causes is often difficult because of the intricate overlapping roles of genes and other extraneous factors. The matter is further confounded by the realization that there is a common final pathway to cancer that involves gene damage, regardless of the original cause. Let us first consider the genetic and then the environmental causes of cancer, remembering the artificial nature of such categories.

The Genetics of Cancer

Fortunately, fewer than 10 percent of all cancers are of genetic origin. These situations often involve very high risk, however, so you should be very much aware of family genes that cause cancer. Whatever the basic mechanism, the evidence implicating genes is abundant:

1. There are well-recognized inherited types of cancer.
2. Cancer is more frequent in identical twins than in nonidentical twins.
3. Specific hereditary disorders exist in which cancer is the main feature or a common complication.
4. Cancer is a common problem when the body's protective immune system (which is under gene control) is affected.
5. Certain individuals are genetically susceptible to cancer.
6. Disorders in which gene-repair mechanisms go awry are associated with cancer.
7. Virtually all tumor cells show chromosome (and hence gene) defects.
8. Agents that cause gene damage (mutation) also cause cancer.
9. Certain families seem especially prone to develop cancers, often of different types.
10. Certain genes can undergo mutation and become cancer-causing genes.

Let us examine these ten lines of evidence that link genes to the genesis of cancer.

INHERITING CANCER

Celeste was born to a mother who had had both eyes surgically removed in infancy for malignant tumors. Her own parents had been killed in an automobile accident and no one was left to tell her that this eye cancer — called retinoblastoma — was of genetic origin and transmitted as an autosomal dominant disorder.

After her birth, Celeste's pediatrician correctly assumed what her mother's tumor was and, recognizing the 50-percent risk, instituted examination of her eyes under anesthesia every alternate month. At four months of age, the same kind of tumors were recognized in both Celeste's eyes, and chemotherapy and radiotherapy were initiated immediately. After prolonged treatment and follow-up, Celeste's eyes and vision were saved because of her alert doctors. She now has a 50-percent risk of passing the retinoblastoma gene to each of her children.

While retinoblastoma occurs only in 1 in 20,000 infants, we have learned much about the development of genetic cancer from studying the tumor. Let me simply summarize "the bottom lines" of a prodigious amount of work. The gene that causes retinoblastoma is on the long arm of chromosome number 13. In most cases, the gene occupying the corresponding site along the *other* chromosome 13 (you will remember that our chromosomes are paired) controls the function of the retinoblastoma gene. A mutation of this *controlling* gene results in the failure to regulate the cancer-causing effects (in the eye) of the retinoblastoma gene, which then leads to uncontrolled cell growth, called cancer.

The presence of the retinoblastoma gene is mostly unsuspected. In about 10 percent of cases, however, *detailed* chromosome studies reveal that a sliver of chromosome 13 is missing or abnormal at the very site of the retinoblastoma gene. Chromosome abnormalities — for example, deletions or translocations — should therefore be sought for in every person with this tumor. Prenatal diagnosis can be offered when one parent has such an abnormality: the risk for affected offspring is as high as 50 percent. Even without any obvious chromosome defect, if one parent has the tumor in *both* eyes, each offspring has a 50-percent risk. Prenatal diagnosis using DNA analysis (see chapter 8) is now available for such families. A parent with a single eye affected may also have a heritable retinoblastoma: about 10 to 12 percent of these unilateral cases are in this category. No family history is evident in about 75 percent of those with retinoblastoma.

Ominously, and for reasons unknown at present, individuals who survive retinoblastoma have a 10- to 15-percent risk of developing a second malignancy seven to fifteen years later. About two-thirds of these second cancers are of the head and neck, and they are thought to result from the original treatment of the retinoblastoma. One-third, however, occur at other sites, such as bones and bladder. A French study also showed an excess of cancer deaths (all types) in the grandparents of patients with retinoblastoma.

Another cancer of childhood — Wilms' tumor, which occurs in the kidneys — provides lessons similar to those yielded by retinoblastoma. Also inherited or transmitted as a dominant disorder, Wilms' tumors occur about 1 percent of the time in persons in whom the iris (the colored part of the eye) is missing — a tip-off suggesting the location of the causal gene. Some children with *both* Wilms' tumor and absent irises (aniridia) have a tiny sliver deleted from the short arm of chromosome number 11. The parents of such children should have their chromosomes examined. If any chromosome defect involving chromosome 11 is found, the risks of recurrence are high; but prenatal diagnosis could be offered in all future pregnancies.

Retinoblastoma and Wilms' tumor have taught us that cancer can be inherited as a dominant disorder or can occur sporadically and only thereafter be transmissible. This applies to the other cancerous tumors listed in table 21. Do you recognize any of the tumors mentioned there? If you or one of your parents have, or have had, one of these hereditary tumors (or others), you would be well advised to consult with a medical geneticist.

CANCER IN TWINS

One of the classic methods of determining the role of heredity in disease has been to examine identical and nonidentical twins (see also chapter 12). Observations for leukemia and for cancer of the breast, stomach, and intestine suggest a rate for identical twins that is more than one-and-a-half times that for nonidentical twins — which simply supports the evidence that hereditary factors are clearly operating, at least in these cancers. A Norwegian report documents an unexplained, much higher rate of cancer of the kidneys in twins of the same sex. A remarkable Connecticut study of over 32,000 twins born between 1930 and 1969 focused on X-ray exposures in the womb and subsequent cancer. Twice as many twins exposed to radiation in the womb developed leukemia or other cancers than did unexposed twins. While these malignancies appear to be radiation-induced, the ultimate mechanism is gene damage that results in cancer.

It was also observed, especially in Japanese studies, that the parents of children with leukemia are more frequently blood-related than are the parents of children without leukemia. There appears, furthermore, to be an increased susceptibility to cancer in the siblings of a child with cancer. Brothers and sisters of children with brain tumors may have as much as a tenfold chance of dying from tumors of the brain, bones, or soft tissues. Careful lifetime medical surveillance of such persons is obviously important.

HEREDITARY DISORDERS AND CANCER

I have been talking up till now about relatively rare cases. Mostly, cancer seems to reflect the interaction between genes and environmental agents, such as the correlation between lung cancer and cigarettes.

A whole host of inherited disorders are associated with the subsequent development of cancer. Over 161 such conditions are

Table 21. Examples of Inherited Malignant Tumors

Type of Tumor	Site of Malignancy
Acoustic neuroma	Inner ear[a]
Basal cell carcinoma	Skin
Keratoacanthoma	Skin
Medulloblastoma	Brain
Melanoma	Skin; eye
Multiple endocrine adenomatosis	Pancreas; thyroid, parathyroid, adrenal, and other endocrine glands
Neuroblastoma	Adrenal gland
Pheochromocytoma	Involuntary nervous system (may be associated with thyroid cancer)
Retinoblastoma	Eye[a]
Teratoma	Any organ
Wilms' tumor	Kidney[a]

[a] Especially, but not only, bilateral.

recognized; a few of the most common are listed in table 22. It is not absolutely clear, even now, whether some of these hereditary conditions predispose a person to develop cancer or whether they themselves and the growth of malignant tumors essentially reflect the same hereditary "fault." From the point of view of the individual with a hereditary disorder, the matter is academic, since what is needed is a way to *anticipate* malignant complications, to diagnose them early, and to effect permanent cures.

A number of well-known chromosomal disorders have a recognized association with subsequent development of various types of cancer. Take, for example, the most common chromosomal disorder, Down syndrome, in which a twentyfold increase in the frequency of acute leukemia is known to occur. When there is an extra number-13 chromosome — the condition called trisomy 13 — or an extra X chromosome in a male — Klinefelter syndrome — an increased incidence of leukemia has also been noted. Just as there is a higher frequency of sugar diabetes (diabetes mellitus) and thyroid disease in the parents of children with Down syndrome, a higher frequency of various cancers in these same families has been observed.

Table 22. Examples of Inherited Disorders Associated with Cancer

Genetic Disorder	Site of Malignancy
Agammaglobulinemia	Blood and lymphatic tissues
Albinism	Skin and eyes
Alpha$_1$-antitrypsin deficiency	Liver
Aniridia (absent iris)	Kidney (Wilms' tumor)
Ataxia telangiectasia	Blood
Bloom syndrome	Blood
Familial polyposis of the colon	Colon
Fanconi syndrome	Blood
Gardner syndrome	Intestinal polyps and multiple tumors elsewhere
Neurofibromatosis	Brain or nerves (any site)
Peutz-Jeghers syndrome[a]	Colon
Tuberous sclerosis	Brain and other organs
Xeroderma pigmentosum	Skin

[a] Also associated with abnormal pigmentation of the mouth, lips, face, fingers, and toes.

In Klinefelter syndrome, which causes males to develop feminine breasts after puberty, the frequency of cancer of the breast is about sixty-six times greater than in normal men — a frequency similar to the incidence of cancer of the breast in women. In the condition associated with an extra chromosome number 18 — trisomy 18 — a higher frequency of kidney tumors (Wilms' tumors) has occurred. And females with a male (Y) chromosome have a 20- to 30-percent risk of developing cancer in an ovary.

Certain chromosome translocations (see chapter 2) are characteristic in some cancers. One persistent translocation between chromosomes 9 and 22 is associated with myeloid leukemia. The abnormal 9/22 chromosome, which is present only in the blood-forming tissues, is referred to as the Philadelphia chromosome, having been first observed in that city. A host of other consistently found translocations are known in other types of leukemia.

In many other genetic conditions (such as those listed in table 22), cancer occurs more or less frequently. In familial polyposis of the colon, there is a 100-percent risk of developing cancer of the colon, rectum, or stomach — hence the need to avoid this problem by electively removing the entire colon! In contrast, individuals with neurofibromatosis, which occurs in about 1 in 3,000, have about a 10-percent risk of developing cancer. For tuberous sclerosis, the risk of cancer is only a few percent, but this is higher than in unaffected persons. Tumors of the nervous system — 9 of 10 involve the brain — occur in 1 in 1,000 persons.

While heredity is not of major importance in causing most brain tumors, for some types of tumor the opposite is true. Besides cases of neurofibromatosis and tuberous sclerosis, relatives of individuals with certain brain tumors called gliomas have a four- to tenfold increased risk of developing the same tumor, and the siblings of affected individuals have a nine-times-greater risk than chance alone. These sibs are also predisposed to cancers at other sites. As yet unexplained is the observation, in two-thirds of individuals with a brain tumor called meningioma, of a deletion of part of the long arm of chromosome number 22 in tumor cells.

While your genes may not predispose you specifically to skin cancer, they certainly do dictate the color of your skin and therefore your reaction to habitual exposure to sunlight. Furthermore, fair-skinned people who sunburn easily develop cancers almost exclusively confined to the exposed parts — the face, head, neck, arms, and hands. Blacks remain remarkably resistant to the development of skin cancer, as do Orientals. Not surprisingly, the frequency of skin cancer differs markedly with geography. Fair-skinned individuals have a higher chance of dying of skin cancer if they live in one of the southern states in the United States. Cancer of the skin is also much more prevalent among outdoor workers such as farmers or sailors.

Some correlation exists between ethnic origin and the development of skin cancer. In people of Celtic ancestry, skin cancer may occur earlier and more frequently than would otherwise be expected. Hence, fair-skinned people of English, Welsh, or Irish descent would perhaps be more prone to it than others. Albinos, with a hereditary defect in skin color, eye color, or both, may have almost no pigment in their skin. After prolonged exposure to sunlight, they invariably all develop cancer of the skin.

Cancers that develop in the pigmented cells of the skin and eyes, called melanomas, are hereditary in 3 to 11 percent of all cases. How this genetic predisposition is transmitted is uncertain, but, curiously, affected children are about twice as likely to have a mother rather than a father with melanoma.

Leukemias are mostly not genetic in origin. Many types occur — some with sudden, acute onset, others slow and insidious. Acute leukemia mainly occurs in children, and perhaps 5 percent of the time has a genetic basis — possibly dominant or polygenic. Chromosome abnormalities are found in the cancer cells in at least half of the cases. Occasionally, such an abnormality may even herald the coming of leukemia. In addition to Down syndrome, noted earlier, leukemia may complicate at least twenty different genetic disorders, including neurofibromatosis and disorders of the immune system.

Chronic leukemia of the myeloid type does not seem to have a genetic basis — in contrast to chronic lymphocytic leukemia, in which familial cases occur more frequently than expected by chance. Estimates are that 10 to 15 percent of individuals with this form of leukemia have a genetic basis for their disease — again, dominant or polygenic. An inherited disturbance in the body's protective immune system is the postulated cause. Such genetic perturbations may also apply to certain tumors (called lymphomas) of the immune system, including Hodgkin and non-Hodgkin disease. Both genetic and environmental factors appear to be involved in some of these cases, but much clarification is still needed. Meanwhile, since a close relation of someone with

Hodgkin disease has about a threefold higher risk of this lymphoma than other people, it would be advisable for such individuals to have physical exams at least annually.

There is a group of genetic conditions that are characterized by excessive breakage of chromosomes. Three well-known, though rare, disorders — Bloom syndrome, Fanconi syndrome, and ataxia telangiectasia — are associated with as much as a 100-times increased risk of cancer. This group of genetic conditions are all inherited in a recessive manner (see chapter 7) — that is, both parents must be carriers, and there is a 25-percent chance that each child will be affected.

Some of these chromosomal-breakage syndromes have been associated with a disturbance of the body's immune system, thus producing a greater susceptibility to infection. Indeed, there is a host of genetic disorders that spring from a disturbed immune system, and here again the increased incidence of cancer is well established.

CANCER AND THE BODY'S DEFENDERS

Our bodies are defended, so to speak, by a variety of mechanisms, including white blood cells, substances called antibodies, and a complex of other systems that all protect against invasion by "foreign" proteins, bacteria, viruses, and so on. The system that makes antibodies — which are protein molecules aimed at blocking the action of foreign proteins — has the ability to "recognize" an alien protein substance. The body, for example, recognizes its own cells by the specific protein that coats each cell. Proteins in the bloodstream, which are part of the immune system, may combine and inactivate foreign proteins.

When a cancer appears, its cells have a different coat of protein from that of the other body cells, and the immune system is thought to recognize the appearance of this coating, thereby activating the mechanism to destroy the cancer cells. It has long been believed that failure of the immune system to destroy a developing cancer is the reason why the cancer continues to grow and eventually to spread and cause death. That is, the surveillance by the body defense mechanism becomes inadequate, as in old age. Of course, the specific types of protein that coat each cell in the body are genetically determined, as are those that circulate in the bloodstream and are made in the lymph glands, spleen, and bone marrow.

Not unexpectedly, then, when hereditary disorders of the

immune system occur — and there are at least seven major such recognized immunodeficiency diseases — a higher frequency of cancer would be anticipated. This does, indeed, appear to be the case. Hence, if you or any of your relatives have been affected by this type of hereditary disease, the most careful and continual surveillance to detect cancer is advised.

Unfortunately, the problem has become even more complex. It turns out that patients who receive organ transplants, such as a kidney, require certain drugs to depress their immune systems so that the organ transplanted will not be rejected. Sadly, those individuals who receive organ transplants have a much greater frequency of malignancies of cells that make up the immune system. They do not, however, have a tendency to develop any of the other common cancers.

The body, does, indeed, recognize the abnormal proteins that coat the cancer cells and immediately makes antibodies aimed at destroying them, but the mechanism may go awry in that the antibodies act to protect the tumor instead of the person! Indeed, the notion has been advanced, again with supporting evidence, that these antibodies made to fight the cancer cells may even stimulate their growth!

AN INHERITED SUSCEPTIBILITY TO CANCER

A person may not have a specific genetic disorder yet may be genetically susceptible to cancer. It is likely that we know little about many such situations. Three we do know about concern individuals who *carry* the gene for but are not affected by either Fanconi syndrome, Bloom syndrome, or ataxia telangiectasia. About 1 in 100 persons is thought to carry the gene for ataxia telangiectasia. Carriers of the gene for any of these recessive disorders have an increased risk of cancer, especially cancers of the immune system, such as leukemias and lymphomas. One remarkable estimate is that among all persons who die of leukemia, 1 in 20 is thought to be a carrier of Fanconi anemia. Reliable tests to determine carriers are not yet available.

The basic mechanism that leads to cancer in persons with the three above-named disorders and in the gene carriers probably involves interference with gene structure and function, possibly at the sites of chromosome breakage. The genes involved are likely engaged in control, regulation, and function of the immune system. Malfunction in this system would make cancer-change more likely.

DEFECTIVE GENE-REPAIR AND CANCER

Our genes are regularly damaged by largely unknown mechanisms. These mutations may result in a remarkable array of diseases. Certain known physical agents — ultraviolet light, X rays, toxins, and so on — may also damage genes. Repair systems exist, and you will not be surprised to learn that even gene-repair itself is under genetic control. Certain enzymes are necessary to execute such repairs to damaged genes, and if deficiency of one enzyme occurs, one of the few known diseases in this category becomes obvious. The classic example is xeroderma pigmentosum — an autosomal recessive disease characterized by an abnormally high sensitivity to ultraviolet light, including sunlight, which results in skin changes that eventually become malignant.

TUMORS AND CHROMOSOMES

It would not be unexpected to find chromosome defects in wildly growing uncontrolled-tumor cells. Indeed, if carefully looked for, virtually all such cells will be found with one or more chromosome defects. If the same chromosome defect is persistently present, then clearly that chromosome (and its gene) are involved, one way or another. The example of the Philadelphia chromosome, noted earlier, is so characteristic of myeloid leukemia that finding it alone invariably means this disorder will soon become obvious. Solid tumors, too, have some typical associations. For instance, in certain lymphomas, translocations involving chromosomes 8, 14, 18, and 11 (in that order of frequency) are found consistently. There are many more examples.

DAMAGED GENES AND CANCER

Physical agents such as chemicals, toxins, X rays, and cosmic rays, and biologic agents such as viruses, can all damage our genes. This damage causes mutations in gene structure and may result in genetic disease, birth defects, or cancer. A common fear of nuclear explosions is related to radioactive fallout, which can cause cancer and other illness.

CANCER-PRONE FAMILIES

Cancer often aggregates in families. This fact alone cannot distinguish whether genes in common or the shared environment (or both together) cause cancer. A genetic explanation is more likely if the tumor type, the organ involved, and the age of onset are all the same. Cancers caused by single genes, as noted earlier, best fit this profile. However, quite distinct cancer-prone families have been recognized. The tumors these families suffer tend to be among the known common malignancies. For example, in a single extended family you might find a number of people with different cancers — say, colon, uterus, ovary, breast, prostate, stomach, pancreas, and skin cancer. Some unusual features tend to be obvious in these families. The cancer tends to occur earlier than usual; an individual may have more than one type of cancer; offspring of a cancer victim have more than a 25-percent risk of malignancy themselves; and the family cancer pattern is consistent with dominant inheritance (see chapter 7).

Studies from the Connecticut Tumor Registry showed a fivefold increased frequency of cancers in the relatives of children with brain tumors. Hence, since relatives of a child with a brain tumor have an increased risk of brain tumors themselves, as well as an increased risk of leukemia and of other tumors, immediate attention should be paid to persistent symptoms of any kind and medical attention should be sought immediately.

Breast cancer is familial in about 13 percent of cases. Half of these have a family history of breast cancer; the other half, a history of other cancers. Breast cancer accounts for more than 1 in 5 of all cancers in females. Indeed, the average lifetime probability of breast cancer in women is about 7 percent. Familial breast cancer tends to occur before the menopause and usually involves *both* breasts (sometimes not simultaneously). The daughter of a woman with bilateral breast cancer has about a 50-percent risk herself — consistent with dominant inheritance. Even males in such families have somewhat increased risks and, moreover, may transmit this disorder! But only 3 percent of all breast cancers occur via a recognized hereditary mechanism.

Postmenopausal cancer, in contrast, is more often caused by "environmental" factors. Typically, it is associated with first-time childbearing at a relatively late age, with never having been pregnant, with having had relatively few children, and with obesity.

CANCER-CAUSING GENES

A family of genes, now about forty strong, control or regulate how, if, or when normal cells grow. Current thinking suggests that perturbation of these so-called *proto-oncogenes* results in cancer, by one of these genes functioning in the wrong cell or functioning at the wrong time or in an uncontrolled way. A number of mechanisms exist by which the proto-oncogenes are activated to become *oncogenes* — that is, genes that code for proteins that make cells become cancerous. When a chromosome translocation occurs (as it does in various leukemias and lymphomas), a proto-oncogene may be misplaced and begin to function without the control exerted by its previous gene neighbors, or it may be influenced by its new gene neighbors. There is already ample evidence to show that proto-oncogenes are involved in some of the key translocations and other chromosome aberrations so typical of certain cancers.

In the cells of certain tumors, several hundred copies of a given proto-oncogene have been observed. In fact, for some malignancies, such as lung cancer, prognosis is directly related to the number of proto-oncogene copies per cell. Yet another mechanism involves a minute structural change — a mutation — in a single proto-oncogene. The effect is for the now-defective gene (the oncogene) to code for a protein that transforms that cell into an invasive cancer cell. While viruses can clearly cause such a mutation in animals, the evidence in humans is so far unconvincing. Chemical toxins, X rays, and other environmental "poisons" may well cause mutations in proto-oncogenes and result in permanent abnormal, uncontrollable stimulation of cell growth called cancer. New data show that "cooperation" between two oncogenes may be necessary before a specific cancer emerges. Moreover, evidence has recently emerged about the existence of genes that suppress tumor formation. Cancer could, therefore, be initiated if genes that "keep the lid on" undergo mutation by toxins or other agents.

The momentous discovery of oncogenes has great potential practical significance. These genes could be used very much like a homing device, as markers for detecting very early cancer and for monitoring treatment and providing surveillance after chemotherapy or surgery. Oncogenes should also prove valuable in establishing prognosis and even in predicting risk. Oncogenes used in this way may also have a role in preventing cancer.

Proto-oncogenes have been found in diverse species, from flies to humans — a fact that indicates their conservation for 600 million years! Their function therefore is regarded as fundamental to all life — probably in directing, regulating, and stimulating cell growth from the moment of conception. Not unexpectedly, any toxin or other factor that can cause a mutation in a proto-oncogene, turning it into an oncogene, may interfere with normal embryonic growth, cause birth defects, and, later, even cause cancer. In fact, viewing this borderland between normal cell growth, birth defects, and cancer, you will not be surprised to learn that a greater frequency of *minor* birth defects have been found in children with various cancers.

The evidence implicating genetic mechanisms in cancer is clearly overwhelming — whether it is a hereditary cancer or one spawned by an environmental factor. In a book devoted to genetic aspects, the reason to consider these environmental agents might not be obvious. We should, however, try to understand them — since, after all, they exert their bad effects through our genes.

Cancer and the Environment

Over 90 percent of all cancers are caused by environmental factors. Cancers are, therefore, very much preventable. Many tumor-causing agents, or carcinogens, are known; even more are suspected. A hereditary predisposition to an environmental factor probably exists — but in few cancers only. There are dozens of types and hundreds of subtypes of cancer, and a few predominate in each country. In both the United States and the United Kingdom, lung cancer predominates, followed by cancers of the breast, large intestine, and stomach. These four types account for more than half of all cancer deaths in these countries.

Many environmental factors are causally associated with cancer. *Cigarette smoking* heads the list and causes a staggering number of cancer deaths, not to mention heart and lung disease. Specific toxic components of cigarette smoke are responsible for disease — for instance, tar for cancer, and nicotine and carbon monoxide for heart disease.

Scores of *occupations* involve an excess risk of cancer. The worst involve asbestos dust, which accounts for as many as 2 percent of all cancer deaths in the United States. Other causative agents — used in all kinds of occupations — include arsenic, radon, polycyclic hydrocarbons, nickel, chromium, vinyl chloride, benzopyrenes, benzene, leather dust, wood dust, pitch, cadmium — and on and on, through an almost endless list. Overall, about 4 percent of all U.S. cancers are attributable to occupation. For manual workers in mining, agriculture, and industry, possibly 1 cancer in 5 may be due to exposure at work.

Alcohol is thought to account for about 3 percent of all cancer deaths, besides making victims of endless innocents. *Dietary factors* are also clearly implicated. Since 1967 we have known that the higher the fat content of food eaten, the greater the risk of breast cancer. In contrast, the higher the fiber intake, the lower the rate of cancer of the colon and rectum. When Seventh-day Adventists, who do not eat meat, were noted to have lower death rates from cancers than others, a vegetarian diet was thought to protect against cancer. Unexpectedly, however, nonvegetarian Mormons seem to have similar rates. Moreover, vegetarian nuns have rates of breast and colon cancer similar to those in the general population. So much for the protective benefit of this diet — at least for now.

Environmental *air pollution* is thought to account for about 1 percent of all cancer deaths, probably by enhancing the effects of cigarette smoke on lung cancer. *Sexual activity* has been known for many years to be correlated with the occurrence of cancer. Prostitutes, for example, have a very-much-higher rate than nuns of cancer of the cervix. (This, by the way, is probably due to one or more types of the human papillomavirus, rather than herpesvirus.) Cancer of the breast, however, is far more common in nuns than in prostitutes, since the earlier a woman becomes pregnant, the lower her risk of breast cancer.

The outbreak of AIDS (acquired immune deficiency syndrome) among homosexuals in the United States in 1981 was associated with a malignant tumor called Kaposi's sarcoma. The AIDS virus(es), alone or in association with other viruses, seems to lie at the root cause of this cancer. Cancer of the penis is also thought to be of viral origin (via the papillomavirus). (The wives of men with penile cancer have a higher rate of cancer of the cervix.) Other *viruses*, sometimes accompanied by an impairment of the body's immune system, are associated with various cancers. Primary liver cancer, certain lymphomas (tumors of lymph glands and associated cells), and tumors of the throat are good examples of other virus-linked cancers.

Radiation from sunlight, cosmic rays, minerals, and medical X rays and radiotherapy constitute the chief sources for human exposure to yet another environmental carcinogen. Radiation probably accounts for about 3 percent of all cancer deaths. One important study over a twenty-year period examined the frequency of cancer deaths among children who were exposed to X rays while still in the womb. A 50-percent increased risk of death from leukemia was observed for children aged ten and under. The risk of death from other cancers was increased by about 30 percent. (See chapter 14 for more on the effects of X rays.)

Given the vast experience in administering *sex hormones*, it comes as no surprise that we know that such drugs do influence the risks of cancer. The most notorious was diethylstilbestrol (DES), which was used in early pregnancy to prevent miscarriage. In 1971, it was shown that females exposed in the womb had an increased risk of vaginal cancer evident after puberty. The "DES disaster" has subsequently been studied carefully and the best estimate now indicates a vaginal-cancer risk between 0.1 and 1.4 percent for those exposed in the womb — a risk much smaller than previously feared. Males so exposed appear to have an increased risk of minor structural defects of the genitals such as undescended testicles; at present there are unconfirmed reports of a possible increased rate of tumors of the testicles. Mothers who were given DES are advised to have annual breast and vaginal examinations since there is a *suspicion* that they may be at increased risk of cancer of the breasts and genital tract.

The female sex hormone estrogen, when used *alone,* has been associated with an increased risk of uterine cancer. Combinations of estrogen and progestogen, used also as oral contraceptives, seem to protect against cancer of the uterus, cancer of the ovaries, and benign breast lumps. The "pill" may, however, increase the risk of cancers of both the breast and cervix.

Other hormones produced by the body itself may also stimulate specific organs — for example, the breast, prostate, ovary, uterus, testis, and thyroid — resulting in cancer. Such tumors account for about 10 percent of all cancers in men and about 40 percent in women. Other, nonhormonal, *drugs* may also cause cancer. Prolonged use of phenacetin, a commonly used painkiller, has been associated with kidney cancer. Anticancer drugs themselves, while killing cancer cells, may affect other cells, which later undergo malignant change. A few other drugs are known to be carcinogens, and many more are suspected to be.

It is entirely possible that some people are genetically susceptible to certain cancers that can be precipitated by drugs, hormones, viruses, X rays, toxins, and other environmental agents. Experts maintain, however, that over 75 percent of cancer deaths are avoidable!

Preventing Cancer

Here are some ways to avoid cancer or detect it early and institute lifesaving treatment:

1. Avoid smoking.
2. Limit alcohol consumption.
3. Avoid obesity.
4. Have an annual screening test ("pap smear") for cancer of the cervix.
5. Observe good genital hygiene.
6. Avoid unnecessary X rays.
7. Avoid unessential use of hormones.
8. Have an annual gynecological examination if you are a DES daughter.
9. Avoid excessive exposure to the sun, especially if you are fair-skinned.
10. Avoid exposure to carcinogens at work.
11. Consult your physician about prolonged ill health or other suspicious symptoms, especially these warning signs:
    - persistent, unexplained pain or lumps
    - hoarseness or difficulty swallowing
    - changes in warts or moles
    - unusual bleeding or discharge
    - shifts in bladder or bowel habits
    - a sore that will not heal
12. Learn how to examine your breasts for abnormalities if you are a woman and do so regularly.
13. Have a mammogram (X-ray examination of the breasts) annually if you are a woman over fifty.
14. Practice "safe sex" and take other steps to avoid AIDS.
15. Have a rectal examination annually if you are over fifty.
16. Eat a high-fiber diet.
17. Know your family history of cancer and birth defects.

The fear of having cancer works to your disadvantage if it makes you act as though you do not really want to know whether you have it. I cannot sufficiently emphasize how important it is for you to know your family history — to know your genes. And remember, most cancers, hereditary or nonhereditary, are curable if diagnosed early.

32

Genes, Aging, and Dementia

Why, you may ask, do we all die? We understand why death occurs when a person has cancer, heart disease, high blood pressure, stroke, or some other disease or disorder. But if no disease confounds our efforts to live on and on, why do we expire? Do we simply wear out?

And why do females statistically live longer than males? This characteristic, in fact, is not confined to the human species: female spiders, fish, water beetles, houseflies, chickens, and fruit flies all live longer than their male counterparts. (Yet in some bird species, especially pigeons, the males reputedly live longer than females.) Strangely, in humans, disproportionately more male than female newborn babies die.

Lessons about Aging from Animals?

It would seem that specifically circumscribed life spans are characteristic in the animal kingdom, as they are in humans. The life span of the mouse rarely exceeds two years; rats, four years; cats, thirty years; elephants, sixty years; and horses, forty years. Humans and virtually all animals age. Exceptions are those animals that continue to grow in size after reaching adulthood. For example, several fish and reptile species show no biologic age changes. They are not, however, immortal: invariably, they succumb to disease, predators, or accidents.

Experimentally, rats that have been fed diets that were sufficient in all constituents except calories showed retarded growth during the period of calorie restriction. After increasing the calorie intake, the rats proceeded to grow to adult size and eventually exceeded the normal expected life span for that strain of rats. They reached about twice the maximum age achieved by rats whose diets were not interdicted. Similar results have been noted in chickens, bees, silkworms, and other species. The effects on prolonging life were most pronounced when low-calorie diets were started soon after birth. Of particular interest was the observation that those rats who were initially on the calorie-restricted diet had an associated delay of onset of various tumors and chronic diseases related to aging.

Even the surrounding temperature in which we live has been questioned in its relationship to age. Fish raised at low temperatures have better growth and live longer. Rats, on the other hand, when reared at low room temperatures have a considerable decrease in their life span from all causes of death — including, for some strange reason, cancer!

Removal of the sex glands of salmon early in their development has been noted to prolong their life span.

Mice raised in a germ-free environment have been shown to have a longer mean life span, as have rats who have had their spleens removed early in life. The implication of both of these experiments is that the body defenses active against infection may clearly be implicated in the aging process.

The implication of different longevity in males and females, mentioned above, suggests the influence of the sex hormones. In this context, therefore, it is interesting to note that among cats, those who have been castrated have attained the highest recorded ages. No equivalent human data exist. Some female fruit flies that are virgins, born without ovaries, or that are sterilized live longer than their normal female contemporaries. While virgin mice live longer than spayed females, the oldest of all are castrated mice.

The Death Clocks

The death of cells is a normal accompaniment to development. In animals, whole tissues or organs are destined to self-destruct at a critical moment, and as part of normal development. This is the mechanism by which organs, useful only during the larval or embryonic stages of many animals, are eliminated. Degeneration and death of cells is crucial, too, in the development of limbs in certain animals and man, especially in modeling not only digits, but also the contours of whole limbs. Cell death is an intrinsic part of normal animal and human development. Whatever the magical biological drummer our cells "march" to, our bodies are controlled rhythmically. Puberty dawns at a certain time, and menstrual periods begin and, indeed, by this inner clock, cease at a given time (except for women who smoke, who have an earlier menopause). Tennyson's oft-quoted observation that "in the spring a young man's fancy lightly turns to thoughts of love" probably reflects the known seasonal changes in certain hormone levels. Evidence, at least from animal studies, indicates that our genes control even our internal cell-clocks or pacemakers. We know, from a host of hormone and other chemical studies, that with aging considerable rhythmicity is lost. Experiments show that serious disruption of body rhythms decreases the life span in animals. It remains to be proved whether desynchronization of our body clocks — and I do not simply mean jet lag — due to changes in gene expression as we get older causes the effects of aging on the body.

An intriguing and as yet unexplained phenomenon has been observed in both animals and man: the offspring of older mothers appear to have a shorter life expectancy — an effect that may extend through several generations!

Centenarians

In the Caucasus, a 1970 census counted 5,000 persons who had reached at least 100 years of age. Many centenarians live in the Soviet republic of Georgia. The reason for their aggregation there is uncertain. Very many live at high altitudes — 1,500 to 4,500 feet above the sea. Industrial centers appear to have the lowest percentage of centenarians. We do not know whether pollution, urban stress, and other factors hasten the aging process or not.

Even in Soviet Georgia, there is a preponderance of older women: of the total population over the age of ninety, about two-thirds are females. A similar preponderance of surviving females is evident in the United States. The difficulties of documenting exact ages in persons over 100 years in Russia and elsewhere are acknowledged by all authorities. A Russian male, who died in 1973, claimed to be 168 years old. The 1988 edition of *The Guinness Book of World Records* holds that the longest documented human life was 120 years, 237 days; that centenarian male was Japanese.

Why Do We Age?

The exact reason why we age is really not understood. Married men appear to live longer than unmarried men. I am sure you can think of a hundred reasons why this should be so. More — or less — sexual activity may be one of the thoughts that crossed your mind. Certainly, it is established in rats that old males live appreciably longer if a young female is provided to groom them. A number of studies have demonstrated that children of long-lived parents have a longer life span compared to the offspring of short-lived parents, though the difference is unimpressive. Meanwhile, we are still beset with theories or hypotheses about why we get old. Let us examine the latest theories that, singly or together, implicate malfunction in our genes, the immune system, the brain and hormones, and cells in general.

YOUR CELLS ARE PROGRAMMED BY YOUR GENES

Perhaps the most accepted theory of aging suggests that the blueprint for your total life span is "written in your genes" and is apparent from the moment you are conceived. The basis for this information is fascinating.

If you took a tiny pinch of skin and grew it in tissue culture in the laboratory, you would discover that its cells have a fairly fixed life span. These cells, or *fibroblasts,* have been grown for many persons. We know that each cell is able to grow and double itself about forty or sixty times over. After that, it simply dies. The same studies have been done for different animal species, and the same limited life span of cells has been demonstrated.

Moreover, it appears that the lifetime of each cell is directly proportional to the average life span of that species. Hence, for men and women, these fibroblast cells on average double about fifty times before dying and relate to the usual human life span of about seventy years. Cells from the chicken, in contrast, double some fifteen to thirty-five times, which is directly related to the average life span of thirty years. The cultivated cells of the long-lived Galápagos turtles also have a markedly extended life span. The evidence indicates that the number of times a cell can divide is fixed in our inborn genetic messages.

Scientists can take these fibroblast cells from any species, including man, and freeze them in liquid nitrogen at minus 196 degrees Centigrade. Dr. L. Hayflick of the University of Florida has, in fact, thawed and regrown human cells frozen for longer than twenty-three years. They simply resume dividing and doubling just where they left off! Their "memory" is clear and accurate: probably "programmed" in the genes.

Cells grown from tiny pieces of skin taken from an old person are found to have curtailed life spans when compared with cells taken from infants. Again the implication is that for each of us aging is basically fixed from the start by "aging genes" that program our cells for a certain span; at the end of that time, they cease to function normally, senile changes occur, and eventually we die. This theory is fortified by the inevitability of the menopause. Though not completely predictable, its exact timing is probably influenced or controlled by heredity — just as, some say, aging is.

Other evidence abounds to support the view that heredity plays a significant role in determining how long you will live. If there were a clear genetic effect on aging, then it should be evident in studying identical twins. After all, identical twins originate from a single fertilized egg, whereas nonidentical twins develop from two eggs. Any differences noted between identical twins, as discussed earlier, are presumably a reflection of environmental influences. Identical twins, on average, appear to die within five years of each other, although a much greater spread does occur in individual cases. Nonidentical twins die much further apart. It is interesting that male twins generally have shorter life spans than female twins.

As we get older — and this may be hard to believe — some of our cells tend to lose a sex chromosome. This is especially the case for women, who lose an X chromosome; males lose a Y less often. Occasionally, other chromosomes may be lost. The cause, and indeed the effects, if any, are unknown.

DNA — the genes themselves — as well as their immediate products, have been carefully investigated in the context of aging. An increased frequency of gene mutations has been found to occur with advancing age. Whatever the reason, the "errors" in subsequent products of genes may interfere sufficiently with cell function and yield an aging effect. Mounting evidence suggests that aging is associated with a generalized impairment of gene functions, including those having to do with duplication, sending messages, or gene repair.

THE MACHINE SIMPLY WEARS OUT

Another theory holds that there are no specific aging genes, but, rather, that our cells, in the course of living, dividing, and growing, are subject to environmental influences as well as to biochemical changes, as time goes on. The idea is that molecules within the cell become damaged in some way, causing breakdowns in the cell machinery and subsequent errors in function. Cells are simply worn out. This so-called rate-of-living theory relates the rate of chemical reactions within cells — the *metabolic rate* — to the life span. Heat, as is obvious, steps up the metabolic rate. Flies, for example, survive about three weeks in summer, but longer than six months in winter. Flies kept flightless but able to walk live two-and-a-half times longer than those allowed to fly — which implies that a slower rate-of-living is beneficial (at least to flies!). This theory does not, however, identify the causes of aging.

MALFUNCTION IN THE IMMUNE DEFENSE SYSTEM

Protection from infectious agents — bacteria, viruses, and so on — or foreign substances is afforded our bodies by the incredible ability to produce complex molecules called antibodies to entrap, "neutralize," or kill such invaders. This ability diminishes with age and even in healthy individuals eventually reaches a level only 5 to 10 percent of what it was at puberty. Moreover, as this immune responsiveness declines, the system becomes less able to discriminate between the body's own molecules and external invaders. This progressive failure to distinguish self from nonself not only increases our vulnerability to infection as we age, but also means that we increasingly make antibodies against our own organs! As a consequence of this malfunction, chronic autoimmune diseases — diabetes, rheumatoid arthritis, thyroid disease, and many others — emerge more often as we grow older. While these misled immune responses are basically directed by our genes (especially those on chromosomes number 6), we do not really understand why the body turns on itself.

Other, nongenetic factors interfere with the immune system. During the Second World War, studies performed during the development of atomic energy showed that various forms of X rays, when given in sublethal doses to young animals, such as mice, caused them to die of "old age" earlier than their contemporaries. The X-rayed animals not only looked older, but also died earlier of the same diseases their nonirradiated litter mates had experienced. It is certainly conceivable that agents such as X rays have an adverse effect on human longevity. Unfortunately, the immune system cannot by itself explain aging: many organisms that age *lack* an immune system!

IT'S IN THE MIND

Many of us have met people who, having suffered a bereavement, seemed to have "aged overnight." We now know that depression constrains the immune response, through ill-understood mechanisms. Older people who have died soon after the loss of a loved one may have succumbed from their increased vulnerability to infection, rather than simple "heartbreak." The adverse effect of grief on the immune system and the body generally is in all likelihood mediated by brain-manufactured hormones. Hormone levels vary with age and their control mechanisms are complex. Changes in certain hormones may predispose an individual to age-associated diseases. For example, steroid-hormone loss at menopause may set the stage for bone-thinning osteoporosis.

Why brain cells are systematically lost with age remains a mystery. We may not be accustomed to thinking of brain cells as hormone producers, but that is what they are. The hormones manufactured by the brain cells not only influence cell-to-cell communications, but also affect organs elsewhere in the body.

Any idea that brain hormones alone could affect the aging process is tempered by the realization that not all organisms that age have such systems. Moreover, it is still difficult to distinguish whether hormone changes are a result or cause of aging.

Premature Aging

It can no longer be disputed that some hereditary defects can bring on old age sooner, though the mechanism is still to be discovered. One distressing, rare condition is called *progeria*. It is characterized by an aging process so accelerated that children who have it resemble old people before they are ten years of age. They have stunted growth and serious disease of all the major blood vessels, including the coronaries; they may become bald and are often thin and wrinkled. Death comes early — almost always in the teens — usually from heart disease or pneumonia. All the evidence we have suggests that both parents of such an afflicted child are carriers of the gene, but the mechanism behind this tragic condition is unknown.

Another premature-aging disorder, *Werner syndrome,* is probably transmitted in the same way as progeria, by two parents who are carriers. The victims have stunted growth, develop cataracts early in life, become gray-haired or bald, and tend to develop diabetes as well as major disease of the blood vessels. Moreover, their bones become less dense (this is called osteoporosis), and there is a high frequency of cancer. Again, the causative mechanism is totally obscure.

As for normal aging, despite our technology, we can still only guess. The die may be cast the moment we are conceived; that single cell that comes into being when the sperm fuses with the egg may contain the blueprint — the plan — of a life that will last, if not cut off by disease or accident, for sixty-eight years. Or ninety years. And so on. At least now we know one thing for certain: in aging, genetic components interact with environmental factors. How the mechanisms work — or go awry — is still part of the mystery to be unraveled.

Alzheimer Disease and Other Dementias

We might become wiser with age but, unfortunately, our faculties in general do not improve. Beyond the normal range of aging symptoms and signs, there exist some sixty disorders known to cause *dementia*. Deterioration or loss of intellectual faculties, of memory, and of the ability to reason, associated with confusion, disorientation, disturbances in language and perception, and loss of ability to solve problems, think abstractly, or make judgments, all characterize a syndrome of dementia. In addition, there may be changes in personality, social withdrawal, and a general fearfulness. Some may be paranoid and think other people are talking about or plotting against them. Others may be irritable, agitated, and even verbally or physically aggressive toward family members or others. Delusions could occur and a person could think that he is someone else — Napoleon, perhaps. Some may become extremely apathetic and even lapse into a stupor.

Dementia is a common problem. Current estimates indicate that about 4.6 percent of people over sixty-five have severe dementia — they are so incapacitated that institutional care or full-time care in their own homes is necessary — while about 10 percent have a mild to moderate degree of dementia and are able to manage semi-independently. Based upon these estimates, there are over 1.2 million people with severe dementia and over 2.5 million with mild to moderate dementia in the United States. These numbers refer to all types of dementia but have not yet begun to reflect the latest recognized cause, AIDS (acquired immune deficiency syndrome).

Among the many causes of dementia are two that characteristically result in the onset of symptoms before the age of sixty-five. These two presenile dementias are called Alzheimer disease and Pick disease. While their symptoms (described above) are essentially indistinguishable, they can be differentiated by study of brain tissue. In both conditions, there is progressive mental disintegration caused by degeneration of brain cells.

ALZHEIMER DISEASE (AD)

The clinical features of dementia, as noted above, with onset before the age of sixty-five, typify this disease. Current estimates suggest that about 55 percent of all individuals with dementia have Alzheimer disease (AD), which means that about 2 million persons are affected in the United States alone! Indeed, the frequency of this extremely distressing disorder continues to increase with age, so that new cases of dementia between seventy-five and eighty-five years of age occur as frequently as heart attacks. Estimates are that 10 to 15 percent of individuals over sixty-five years and as many as 20 percent over eighty years have AD.

A clinical diagnosis of AD can only be made with about 90-percent accuracy. A recent discovery of an abnormal protein in the brain of victims with AD holds promise for the first time in providing a straightforward biochemical diagnostic test. This protein, called A-68, which also appears in the spinal fluid of individuals with this disease, clearly plays a role and appears to be unique to AD in that it has not been found in other brain disorders. However, a definitive diagnosis is at present only possible when brain tissue, obtained at biopsy or at autopsy, is examined under the microscope. Therefore, because of the genetic nature of this disease, an autopsy should always be sought if this diagnosis is suspected.

Extensive studies have been done for years aimed at determining the cause of AD. The familial nature of AD has focused attention on a genetic basis. In familial AD, the onset is usually much earlier — at around fifty years of age — than in the very-much-more-common late-onset sporadic form. There is, however, considerable variation in the time of onset, even in the inherited form. The early-onset familial disease is also more severe and progression is much more rapid. If just early-onset AD were genetic in origin, then heredity would account for about 1 percent of all AD patients. The evidence, however, suggests a much greater genetic component, even though the matter remains unclear. Certainly, in the 1 percent or so of early-onset familial cases, AD is transmitted or inherited as an autosomal dominant trait. The child of an affected parent would in such cases have a 50-percent risk of ultimately developing AD.

In other cases, although familial, other modes of inheritance might apply — namely, polygenic ones (see chapter 7). We know that 15 to 35 percent of persons with AD have affected siblings. While it has not yet been possible to discover how many of the children of an affected parent with a brain-tissue-proved diagnosis have developed AD, there are useful data about the risks to an individual when a brother or sister is affected. These risks vary according to the age of the person, whether or not a parent is affected, and the age of onset of AD in the affected sibling. Dr. L. Heston, in a Minnesota study, calculated the average risks shown in table 23. It can be seen from this table that, for example, a person fifty-five to sixty-four years of age whose brother or sister developed AD at age seventy or later has a 2.1-percent risk of developing this disease. If the sibling became ill before age seventy, however, the risk would be 5.6 percent if the parents are not affected; and if in addition to onset before age seventy, there is an affected parent, the risk would be 8.3 percent.

Studies of twins with senile dementia noted that both identical twins were affected five times more often than both twins in nonidentical pairs. Interplay of some other factors — unidentifiable

Table 23. Chances of Developing Alzheimer Disease for Persons Aged Fifty-five or Older with an Affected Sibling

Person's Age (in years)	If Sibling Was Affected after Age 69 and Parents Are Not Affected	If Sibling Was Affected before Age 70 and Parents Are Not Affected	If Sibling Was Affected before Age 70 and a Parent Is Also Affected
55–64	2.1%	5.6%	8.3%
65–74	4.2	9.6	20.6
75 or older	8.5	17.1	46.4

environmental or biochemical ones — is reflected in a report on identical twins in whom the first symptoms of dementia occurred thirteen years apart!

Curiously, there have been reports of an increased concentration of aluminum as well as silicon in the brain cells of patients with AD. The current consensus, however, is that neither of these elements causes this disease. Another inexplicable finding from two excellent studies was that 15 to 20 percent of persons with AD had suffered serious head injuries up to thirty-five years before the onset of the disorder, compared to only 5 percent in a carefully matched group of individuals without this disease. This observation is especially intriguing because examination of brain tissue from boxers who become punch-drunk and develop dementia show similar features in their brain cells, albeit in slightly different areas of the brain.

Still other "coincidences" are intriguing. The unaffected relatives of people with Down syndrome have a higher rate of various cancers. (Down-syndrome individuals themselves have a twenty-times-greater risk than others of developing leukemia.) Unaffected relatives of those with AD also have an increased risk of cancers — in this case, cancers involving the immune system, such as leukemias and lymphomas. Fingerprint patterns have been carefully examined in persons with AD and have been repeatedly found to be similar to those noted in individuals with Down syndrome.

Quite unexpectedly, relatives of patients with AD have been repeatedly observed to have an increased incidence of Down syndrome. This curious observation made little sense until it was quickly recognized that individuals with Down syndrome age prematurely (see figures 7–10 on page 30). (Actual further mental deterioration commonly and prematurely occurs in Down syndrome after thirty years of age.)

Moreover, microscopic examination of the brain in AD and in Down syndrome revealed one particular finding in common. In both disorders, deposits, or plaques, of a complex fibrous protein called amyloid appear within cells of the brain cortex — the area concerned with intellect, memory, emotion, and other functions.

Since Down syndrome was known to result from the presence of an extra number-21 chromosome (resulting in three copies in all), detailed investigations using gene-analysis techniques were launched to further explore the initial findings.

Sure enough, these analyses have shown that the gene responsible for making the amyloid protein is duplicated in persons with AD. Other studies aimed at detecting the gene that causes AD came up with the startling observation that either the amyloid gene itself or one located very close to it is the likely culprit. A whole host of questions arise. Why does gene duplication occur? Why is amyloid protein deposited? What regulates this gene? What, if any, environmental factors influence function of this gene? Since a few amyloid plaques are also seen in the brain cells of normal older people, can we discover the elements in disarray that cause not only dementia but also the normal aging process?

Studies thus far show that amyloid-gene duplication is present even in *nonfamilial* AD — which directly implicates heredity in *all* cases of AD. In fact, a 1988 New York study of 379 close relatives of 79 persons with AD concluded that even late-onset sporadic AD may be dominantly inherited. Pinpointing the exact location of the AD gene will someday make it possible to detect this dementing disorder decades before the disease becomes obvious in an individual — in fact, in the fetus as early as nine weeks of pregnancy! The wisdom of providing such early detection in the absence of any useful treatment is, of course, questionable. However, locating the gene for AD will facilitate research into methods aimed at delaying or completely preventing the onset of the condition. Once again, research on a genetic disease is likely to provide a major advance in our understanding of a normal body process: aging.

PICK DISEASE

As noted above, the symptoms of dementia in Pick disease are similar to those experienced in AD. While Pick disease is considerably less frequent, the dementia is equally problematic. Most described families show an autosomal dominant mode of inheritance, although there have been reports on a few families in which autosomal recessive inheritance seems to apply.

Some reports suggest that AD and perhaps Pick disease may sometimes result from an infectious agent such as a virus. This suspicion is based on experimental studies in which brain tissue from a patient with AD was injected into the brain of monkeys, who subsequently developed the tissue signs of this disease. It is still thought possible that "sporadic-case" families may have a hereditary predisposition to infection by a "slow-acting" virus. No such agent has ever been isolated, but research continues.

FAMILY AND FINANCE

The burden of AD and other dementias falls particularly hard on other family members, who may themselves be elderly. The sheer agony can only be told and understood by the nearest relatives. The U.S. Office of Technology Assessment in a 1987 report entitled *Losing a Million Minds* drew attention to the awesome costs of these tragedies. The report estimated *annual* direct and indirect costs of $40 billion and pointed to an expected doubling of the number affected by the year 2000 and a quintupling of the number by 2040! Vigorous attention must be paid to research priorities and health-care planning if financial disaster is to be avoided.

33

Treating Genetic Disease

How, you may ask, can a hereditary disease be treated? Is it not irrevocable and irreversible? Is actual *cure* of a genetic disorder possible?

The treatment of such disabilities is eminently feasible. Research has made leaps and bounds in this area of medicine. People can be helped tremendously; their lives can be saved by recognizing hazardous situations. Many people in this category live entirely normal life spans. The goal of treatment is either to prevent a genetic disorder from manifesting itself or to minimize its adverse effects.

What are the treatments possible for hereditary disorders? What progress has been made to cure genetic diseases?

Diet

The relationship of certain foods to aggravation of hereditary disorders is no longer in question. One example (mentioned earlier): Very many blacks are born lacking a certain intestinal enzyme, lactase, that is necessary for the digestion of milk and milk products. This deficiency may not become obvious until late childhood or even adulthood. These individuals suffer diarrhea after drinking milk or eating milk products. They can become symptom-free by avoiding these products.

For certain genetic disorders, some foods prove toxic or even lethal. Mental retardation and even death may be the result. There are specific approaches to various diseases. Let me take them up one by one.

Certain foods and liquids can be used to protect a patient against biochemical disorders. Some of the instructions are remarkably easy to follow. In sickle-cell anemia the blood cells become sickle-shaped when dehydration occurs. It is therefore very important for the patient to drink a fair amount of water. Water is also essential in the treatment of cystinuria, the genetic condition that produces kidney stones. Taking alkalis provides additional help in preventing stone formation in both kidney and bladder. A number of hereditary disorders involve low blood sugar, or hypoglycemia; simply eating fruit and other sweet foods may protect an affected person against the complications of the disease.

Some biochemical disorders require much more than simple dietary therapy. Certain sugars, such as glucose and galactose, can cause serious problems for persons with hereditary disorders of carbohydrate metabolism. For a baby with one such condition, galactosemia, the milk sugar galactose may cause brain damage, cirrhosis of the liver, cataracts, and even death soon after birth. All these are preventable by strictly excluding the offending sugars from the diet beginning with the very first days of life, and, even better, by also excluding these sugars from the diet of the prospective mother throughout pregnancy.

In a condition called fructose intolerance, the body is unable to stand the natural sugars in fruit. Dietary therapy is facilitated because the patient often develops an aversion to sweet-tasting foods.

Cutting out saturated fats and foods that contain cholesterol, such as eggs, is an important maneuver in treating some hereditary heart and blood disorders, as I mentioned earlier.

There is also a disorder involving a chemical related to chlorophyll that may cause serious nerve damage and blindness. The removal of all green fruits and vegetables from the diet may prevent or at least ameliorate the disease.

RESTRICTION AND SUBSTITUTION

There is a real problem when an essential food is toxic to a patient because of a hereditary disease. Deprivation of that essential food can lead to serious malnutrition, failure to grow, and even death. Substitution of synthetic products may become essential. The condition called phenylketonuria (PKU) is the best example. As I have said, the blood and urine of newborn infants is routinely tested for PKU; diagnosis during the first days of life, plus immediate initiation of a restricted diet, will easily allow such children to develop normally, free of the severe mental retardation, eczema, seizures, and other complications of PKU.

What has to be restricted is an essential amino acid called phenylalanine that is present in all high-grade proteins. The body needs phenylalanine in order to make all its own proteins. By carefully regulating dietary intake of this amino acid, and by frequently monitoring and substituting synthetic protein, it is possible to treat the affected child successfully. Treatment does, however, require very sophisticated care in a major medical center.

The severity of the diet restriction has, until recently, been eased up at about the time the child starts going to school. The latest reports indicate some behavior disturbances and a falling-off of IQ points after diets were discontinued in children five to eight years of age. Consequently, parents may wish to continue (or reinstitute) the special diet.

Many parents will attest to the enormous strain and emotional drain this therapy puts on the family. The unnatural diet deprives the child of many foods he or she wants to eat; and some of the synthetic substitutes have a definitely unpleasant taste. The continual attention by the doctor, the repetitive taking of blood samples, the special aura of caution that must surround the child — all take their toll. But there is no doubt that without such careful dietary treatment, almost all PKU children will develop severe brain damage and be untrainable or uneducable. Fortunately, the new DNA techniques described in chapter 8 have now made prenatal diagnosis and carrier detection possible.

Once the patient with PKU grows up, marries, and decides to have children, an additional dimension of care is needed. Formerly, such eventualities were rare indeed. Nowadays, if the diet has been properly controlled, persons with PKU will probably marry and bear offspring. In the past, virtually 100 percent of their children were mentally retarded — not from PKU, but from the accumulation of the toxic product phenylalanine in the mother's blood; such accumulations permanently damage the fetal brain and frequently cause miscarriage and birth defects. Now it is recognized that by putting a PKU woman on a restricted diet before she becomes pregnant and keeping her on it throughout pregnancy, doctors can ensure that the toxic phenylalanine in her bloodstream will not damage the brain of the fetus. It does appear clear that once damage has become evident in PKU, it is irreversible.

Wilson disease, which involves a defect in metabolism that leads to a tremendous overstorage of copper in the brain and liver, is another example of a disorder that can be treated through dietary restrictions. Possible signs of the disorder include neurological abnormalities such as tremors and lack of coordination, as well as cirrhosis and other liver malfunctions. All foods with a relatively high copper content, such as cherries, chocolate, or beef, must be cut out completely. (Drugs to rid the body of copper, as well as those to interfere with its absorption, must be taken as well.)

SUPPLEMENTATION AND REPLACEMENT

Specific dietary supplements may be lifesaving in the treatment of hereditary disease. There are a host of chemical disorders in which supplementation with amino acids or protein is crucial. Another type of supplementation, used in the treatment of diabetes insipidus, involves water plus a hormone. Diabetes insipidus, a sex-linked disorder carried by females but affecting only males, renders the kidney unable to concentrate urine. (It is a different sort of diabetes from sugar diabetes.) The patient passes excessive quantities of urine, becomes dehydrated, and, eventually, may die. Water supplementation temporarily saves the day. But a hormone replacement is also needed: the kidney's inability to deal with the urine is caused by a deficiency of a hormone made in the pituitary gland. Medical research has made it possible to obtain this hormone in powder form; the patient is given it to use as snuff. From the nose the powder is absorbed into the bloodstream and circulates to the kidney, where it enables that organ to concentrate the urine.

Vitamins added to the diet may be extremely important in treating hereditary disorders of metabolism. In diseases such as cystic fibrosis, for example, in which fat is not properly absorbed, the fat-soluble vitamins A, D, E, and K will also not be absorbed normally. Deficiency of vitamin K shows itself in excessive bleeding and bruising. In some complex biochemical disabilities, vitamin supplementation will stimulate compensatory chemical reactions in the body and overcome the basic derangement. Since vitamins are vital to certain body reactions, massive doses are sometimes needed.

A fetus suffering from the biochemical disorder methylmalonic aciduria was successfully treated by administering enormous amounts of vitamin B_{12} directly to the mother. This is the ideal we all seek: to avert the need for abortion by safe treatment of the fetus still in the womb. With this goal in mind, much work needs to be done to avoid the trauma of abortion and to ensure that wanted children are born normal and healthy.

In the case of cystic fibrosis, the most common genetic killer of children in the white population, much more than vitamin supplementation is needed. While the disorder is mainly characterized by recurrent lung infection, it is also associated with deficient secretion by the pancreas of enzymes necessary for digestion. An extract of the actual enzymes is usually administered by mouth to cystic-fibrosis patients; they usually take tablets with each meal for life. This does not cure the disease, but usually enables the patient to have close to a normal bowel habit.

Treatment with Drugs

A few examples will serve to illustrate the many kinds of drugs used to mitigate certain genetic diseases.

The thyroid gland may be extremely underactive at birth, and the causes, on occasion, can be hereditary. Simply initiating treatment with thyroid hormone will prevent the child from becoming mentally retarded. This treatment must be lifelong. In another hereditary disorder, the adrenal gland fails to manufacture cortisone for life. Certain forms of gout are hereditary, and the use of specific drugs will block the formation of uric acid, thereby preventing the excruciating attacks of pain. In Wilson disease, penicillamine, a drug related to penicillin, is used to flush excess copper out of the body through the urine. In hereditary conditions involving high blood cholesterol, various cholesterol-lowering drugs can be used. They work by interfering with cholesterol production.

Even in disorders in which there may be an extra sex chromosome or one too few, the affected person may be appreciably helped by drug treatment. For example, the short stature of individuals with Turner syndrome can be treated with a combination of growth hormone and a body-building steroid. Early results are positive, but what final height can be attained is still unknown.

Avoiding Harmful Factors

In earlier chapters, I mentioned genetic disorders that become especially problematic when the affected person unwittingly takes certain medications or undergoes certain exposures that may precipitate serious and even fatal complications. Table 24 is a compilation of genetic disorders that indicates what should be avoided or done to avert the potential serious or fatal consequences noted. (Use the index to find fuller discussions of most of these items.)

Modifying the Internal Environment

Modifying the body chemistry is another approach that fails to cure genetic disorders but may save life or tremendously improve its quality. A remarkably effective treatment was discovered in the early 1970s for a rare, usually fatal disease that affects mainly the skin and the bowel. The condition, which is inherited from both parents equally, is known by the long name acrodermatitis enteropathica. Before 1972, children with this disorder frequently died. The discovery was made in England that zinc taken by mouth remedies the problem in the bowel and the skin as long as the patient continues to take this trace element. Hence, a previously fatal genetic disease, while not cured, can be managed in such a way as to ensure the probability of a normal life expectancy — and this can be accomplished without even knowing the basic defect involved.

Some of the hereditary disorders that afflict newborns may be so severe as to cause immediate coma and death if not recognized in time. In these disorders, toxic biochemical products accumulate and rapidly kill the infant. Treatment may be astonishingly effective. It consists of changing the internal environment of the baby immediately — for example, by completely removing the newborn's blood and replacing it with blood from a donor. The toxic product can also be removed from the infant's body by a variety of other techniques, including kidney dialysis.

Treatment by a Replacement

Factor VIII, the blood-clotting factor that is missing in hemophilia and whose absence causes the victim to go on bleeding after receiving a small cut or bump, can be obtained and concentrated from donated blood plasma and put into blood products called cryoprecipitates. While they do not cure hemophilia, they make a tremendous difference to patients with the disorder. These costly

Table 24. Preventive Measures and Possible Consequences in Certain Genetic Conditions

Genetic Condition	What to Do	Possible Consequences If Advice Not Needed
Albinism	Avoid sunlight	Skin cancer
Alpha$_1$-antitrypsin deficiency	Avoid smoking and inhaled occupational pollutants	Lung disease
Familial hypercholesterolemia	Avoid smoking and fatty foods	Premature heart attacks
Galactosemia	Start special milk-free diet before becoming pregnant	Children with cataracts, liver disease, mental retardation, or other serious illnesses
Malignant hyperthermia	Avoid general anesthesia	Death during or immediately after surgery
Maternal phenylketonuria	Start special low-protein diet before becoming pregnant	Children with mental retardation and other serious birth defects
Multiple polyposis of the colon	Have the entire colon surgically removed	Colon cancer
Porphyria	Avoid barbiturate-type drugs	Abdominal pain severe enough to lead to unnecessary surgery
Pseudocholinesterase deficiency	Avoid the muscle relaxants suxa-methonium and succinylcholine, which are used in general anesthesia	Failure to start breathing again after anesthesia
Rh-negative blood in mothers	Be injected with immunoglobulin after miscarriage, amnio-centesis, or delivery if partner is Rh positive	Serious illness and even death in second and subsequent children
Sickle-cell anemia	Be vaccinated against pneumococcal infection; seek immediate treatment of any such bacterial infection	Severe, possibly fatal infection
Wilson disease	Take medicine to rid the body of copper; avoid foods with high copper content (cherries, beef, chocolate, etc.)	Irreversible brain and liver damage
Xeroderma pigmentosum	Avoid sunlight; take the oral drug isotretinoin	Skin cancer

cryoprecipitates, which use up enormous quantities of human blood plasma, are already being replaced by a genetically engineered product. Moreover, such a product will be free of the AIDS virus — the deadly contaminant that has caused so much tragedy and complicated the lives of all hemophiliacs who need repeated blood transfusions.

Replacement of proteins is important in the treatment of many genetic disorders. In one rare condition — again confined, like hemophilia, to males — the child may not be able to make the chemical called gamma globulin, which protects the body against infection. It therefore has to be administered for life.

Missing enzymes cause a large number of hereditary handicaps. The problem is illustrated in this diagram:

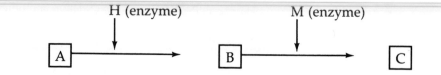

Say there is a sequence of clinical reactions in the body necessary for normal functions: a substance A taken in the diet is broken up by a specific enzyme H into substance B; B is normally converted by a special enzyme M into a chemical product called C. If the hereditary disease in question is characterized by a missing enzyme M, then the chemical substance B cannot be broken down or digested by the body and the entire sequence is disrupted. Meanwhile, substance B begins to accumulate excessively and is stored in organs such as the liver, brain, eyes, or heart, where it begins to cause serious malfunctions in those organs, slowly or rapidly leading to death. One exciting treatment has involved delivering missing enzymes to the body. In the example shown in the diagram, an effort would be made to provide enzyme M — either by mouth or through direct injection into the blood — in order to get the process of converting B to C working normally again. This, unfortunately, is easier said than done.

A slew of problems require funding and research if adequate progress is to continue in this field. First, the technology to isolate and obtain the specific enzyme has to be further developed. The enzyme so derived has to be active, very stable, sterile, and made in such a way as not to stimulate the body to reject it. Each of those steps poses tremendous challenges.

Once the enzyme is obtained or isolated, some method of delivery into the body has to be devised. The most appropriate route is by mouth — but that may not be possible, because in most cases, acid in the stomach is apt to inactivate the enzyme. While much progress has been made in efforts to establish what we call enzyme therapy in hereditary disease, major obstacles remain to be overcome. Suppose it were possible to administer the specific missing enzyme directly into the bloodstream. How could we ensure that the enzyme actually reaches and enters the tissue where the toxic products are accumulating? If that organ were the brain, the particular enzyme might simply not permeate the blood vessels into the brain cells where the real damage is occurring. This problem has proved to be a major obstacle. If solved, it may one day allow for the option we all seek: to diagnose a genetic disease in the fetus and to initiate treatment in the womb.

Repair and Replacement

The development of a large head, or hydrocephalus, associated with mental retardation because of brain damage, can be prevented by surgical techniques to decrease the pressure within the skull. Certain rerouting, or shunting, operations of major veins in the abdomen may be lifesaving in a few different kinds of hereditary disorders. Hereditary abnormalities of the face, the ears, the hands, the feet, and so on, may all be remedied by "cosmetic surgery." A defect involving the lack of one or both ears may be passed down from one parent to half the children; through plastic surgery, it is possible to reconstruct a missing ear.

New Organs for Malfunctioning Ones

In hereditary diseases, it is unusual to have only a single organ affected. Yet it is possible, when a particular organ is especially damaged in the disease process, to remove and even replace it. Indeed, this technique has been applied to diseased kidneys — the implication being that kidney removal and transplantation of one from a donor will prevent further complications of the hereditary disease and allow the patient a normal life. Some efforts have already been made in transplanting bone marrow, livers, spleens, corneas, and hearts, and in injecting laboratory-cultivated connective-tissue cells called fibroblasts, with only limited success.

Organs such as the bone marrow and spleen have been transplanted in genetic disorders that result in the storage of certain complex substances in the brain and other vital organs. The hope has been that the deficient enzymes that ordinarily chop up those complex substances could be made by the new organ. This approach has been only partially successful — for example, in metachromatic leukodystrophy and Gaucher disease — since newly made enzymes have entered the brain and other needy cells insufficiently.

One important experience materialized when an eight-year-old girl with sickle-cell disease developed leukemia. Luckily, her four-year-old brother was a close match and she successfully underwent bone-marrow transplantation. After several life-threatening complications, she recovered and was free of leukemia sixteen months later. Her new bone marrow functioned well, as evidenced by normal *male* cells in her blood, and her "bonus" was that she no longer suffered sickle-cell crises! Bone-marrow transplants are not, however, the effective answer for the treatment of sickle-cell disease. Among the complex problems that arise with transplantation are not only the rejection of donated organs but also the possibility that the hereditary disease may continue unabated and may even affect the donated organ.

Gene Therapy

The latest DNA techniques make it clear that we are now able to isolate, clone, and modify virtually any gene. On the face of it, this suggests that we are on the threshold of curing at least some genetic diseases. (See the discussion of gene therapy in chapter 8.) Sadly, much still needs to be done, even though we know that once isolated, a gene can be introduced into living cells and shown to produce its product.

One experiment in rats showed that the gene for producing growth hormone functioned excessively and a giant rat resulted. That experiment pointed out how difficult it is going to be to control an inserted functioning gene. We know precious little about the orchestration of gene control and regulation, and hence what may turn a gene on or off. Moreover, how would we safely insert needed genes into all cells, or at least into all cells of the organ in which they were needed?

Beginning with a single cell — say, the fertilized egg — we could successfully insert a gene, but how would we know that the embryo was deficient in that gene? Elegant techniques already in use in England make it possible to remove without injury one or two cells from an embryo under two weeks old and perform intricate diagnostic chemical and genetic tests.

Not long ago, researchers at the California Institute of Technology *cured* a genetic disease in a mouse! They were working on a remarkable genetic disease in which affected mice are born with uncontrollable shaking and shivering, convulsions, and short life spans. These shiverers have a defective gene critical to normal nervous-system function. The researchers cloned a normal gene and injected it into fertilized eggs of these shiverer mice. In about 1 in 150 treated embryos, the gene began to function at the right time and a normal mouse was born.

Much is still to be learned about how inserted genes function and how they could affect the activities of other genes. We already know that in one human genetic cancer — malignant retinoblastoma — a change in one gene sparks the uncontrolled cancer-causing activity of another gene. Clearly, we will want to be sure that no cancers are spawned in those whose genetic disease we are trying to cure.

There is sometimes a feeling of hopelessness when it comes to hereditary disease. The specific intent of this chapter was to provide just a brief insight into the many opportunities that exist for care and treatment, despite the absence of cure. Key to many of these approaches is the early initiation of treatment, which may prevent premature death or irreversible mental retardation. The astonishingly rapid technological progress in understanding the structure and function of human genes provides ample reason for optimism, for we can now begin to think of *cure* in the context of previously irremediable genetic disease.

Appendixes
Index

Constructing Your Family Pedigree

For the sake of your own health and that of your children and other family members, you should construct your family pedigree. Once drawn, the pedigree should be kept with your important papers and updated whenever family members have children — with or without birth defects — or develop serious illnesses. Adopted individuals should have the right to obtain information about their pedigree, and to be informed through social agencies about any newly discovered genetic disease in their biological family that may raise their personal risks or those of their children. You should record your ethnic origins as well.

The symbols conventionally used in constructing a pedigree are shown in figure 31. Using these symbols, you should be able to construct your own pedigree, basing it on the example shown in figure 32. Notice that in figure 32 each generation is labeled with a roman numeral on the left and that each person in each generation is numbered with an arabic numeral so that I can refer to the individuals more precisely. Instead of numbering, you should write the names of individuals below their symbols. This particular pedigree comes from a family I cared for some years ago. It might be helpful if I point out some of the facts I gleaned from this pedigree and communicated to "Janet," the individual IV-6 (marked with an arrow), who was the first person in this entire family to come for genetic counseling.

Janet's primary concern arose following her marriage, when she and her husband decided to have children. Her sister (IV-5) was born with spina bifida. I therefore advised Janet that her own risks of having a child with spina bifida or another type of neural-tube defect were about 1 percent and that prenatal genetic studies were recommended in all her future pregnancies.

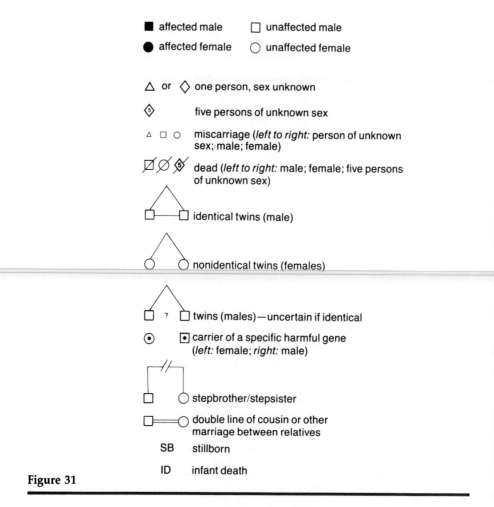

Figure 31

Pedigree symbols

She was also concerned about Alzheimer disease, which affected her maternal aunt (III-11), her grandfather (II-5), and her great-grandmother (I-2). Janet's own mother (III-10) was sixty years of age and in excellent health. All three family members who had Alzheimer disease became symptomatic around fifty years of age. Given the likely dominant form of inheritance for Alzheimer disease in this family, her own mother had a 50-50 risk of developing this disorder. It was good news that she had already reached sixty without any evidence of disease, but given the variability of dominant diseases, it was not possible to assure my patient that her mother could not develop this fatal dementia.

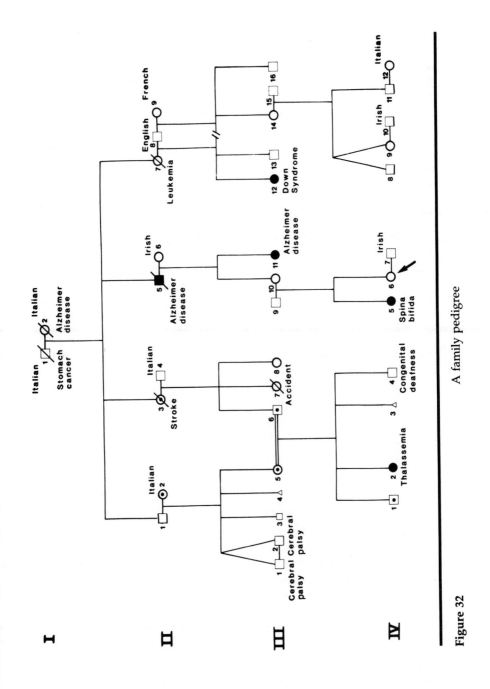

Figure 32

A family pedigree

Only if her mother developed this condition would Janet have a 50-percent risk of ultimately developing this disease herself — and, of course, of transmitting it to her offspring. I told her about the new gene studies that are now available (see chapter 8). I also pointed out that there is a known association within the same family pedigree between Alzheimer disease, leukemia, and Down syndrome (see chapter 32).

In addition, I advised Janet to inform her first cousins that their risk of having a child with a spina-bifida-type defect was slightly higher than average and that they could benefit from genetic counseling. Moreover, I offered to see two of her first cousins (IV-1 and IV-4) for genetic counseling since their sister (IV-2) was born with thalassemia — a potentially fatal hemolytic anemia (see chapter 7). Those appointments were subsequently made.

Examination of that side of the family's pedigree showed not only that individuals III-5 and III-6 represented a first-cousin marriage, but that Italian ancestry was present. Given the occurrence pattern of thalassemia, a disorder so typical in families of Mediterranean extraction, I obtained blood tests to determine whether IV-1 and IV-4 were carriers of the gene for thalassemia. Tests showed that IV-1 carried the harmful gene, while IV-4 did not. IV-1 was planning to marry an Italian woman and I advised the same carrier test for her. Only if both partners were carriers would they have a 25-percent risk in each pregnancy of having an affected child, I explained, and anyway, there were specific prenatal diagnostic tests available should she also be a carrier.

IV-4 was born deaf, almost certainly because of a gene carried equally by his parents — not a very rare consequence of first-cousin marriages. I advised that only if he married an individual carrying the same deafness gene would he have a 50-percent risk of having a deaf child in each pregnancy. Unfortunately, no specific tests are available yet to determine genes that result in deafness.

Expressed concerns about the identical-twin uncles (III-1 and III-2) who both had cerebral palsy were simply resolved. Both had been extremely-low-birth-weight babies who, unfortunately, had suffered brain damage that resulted in cerebral palsy.

Subsequently, I also saw couples IV-9/IV-10 and IV-11/IV-12, and I advised them concerning the trisomy-21-type Down-syndrome individual (III-12) to whom IV-9 and IV-11 were related only through their grandfather. Their exact risks of having a child with a chromosome defect were uncertain, I pointed out, but probably were not significantly greater than average — assuming that there was no Down syndrome in the family pedigrees of IV-10 and IV-12. Prenatal genetic studies were therefore not recommended but remained available as an option.

Genetics Resource Organizations

The National Center for Education in Maternal and Child Health developed this directory of voluntary organizations for health professionals, educators, administrators, and the general public. The organizations provide a variety of services and activities, including publishing educational materials, disseminating general information, making referrals, and furnishing support to assist in coping with specific needs. An asterisk (*) indicates that the organization provides a newsletter.

ACOUSTIC NEUROMA
(*See also* Deafness/Hearing
Impairment; Neurofibromatosis)

*Acoustic Neuroma Association
P.O. Box 398
Carlisle, Pennsylvania 17013
(717) 249-4783

ACQUIRED IMMUNE
DEFICIENCY SYNDROME (AIDS)
(*See* Immune Deficiency)

ADOPTION
*AASK (Aid to Adoption of
Special Kids) America
1540 Market Street, 21st Floor
San Francisco, California 94102
(415) 543-2275

*National Adoption Center
1218 Chestnut Street
Philadelphia, Pennsylvania 19107
(215) 925-0200

ADRENOLEUKODYSTROPHY
(*See also* Leukodystrophy)

ALD Project
c/o The JFK Institute for
Handicapped Children
707 North Broadway
Baltimore, Maryland 21205
(301) 522-5409

AIDS
(*See* Immune Deficiency)

ALBINISM
(*See* Skin Disorders)

ALZHEIMER DISEASE
*Alzheimer's Disease and Related
Disorders Association, Inc.
70 East Lake Street,
Suite 600
Chicago, Illinois 60601
(312) 853-3060

*National Ataxia Foundation
15500 Wayzata Boulevard,
  Suite 600
Wayzata, Minnesota 55391
(612) 473-7666

AUTISM
*National Society for Children and
  Adults with Autism
1234 Massachusetts Avenue,
  N.W., Suite 1017
Washington, D.C. 20005-4599
(202) 783-0125

BATTEN DISEASE
Children's Brain Diseases
  Foundation for Research
350 Parnassus, Suite 900
San Francisco, California 94117
(415) 566-5402
(415) 565-6259

BEREAVEMENT
AMEND (Aiding a Mother
  Experiencing Neonatal Death)
4324 Berrywick Terrace
Saint Louis, Missouri 63128
(314) 487-7582

*Compassionate Friends, Inc.
P.O. Box 3696
Oak Brook, Illinois 60522
(312) 990-0010

*HOPING (Helping Other Parents
  in Normal Grieving)
Edward W. Sparrow Hospital
1215 East Michigan Avenue
Lansing, Michigan 48909
(517) 483-3606

*Association of Birth Defect
  Children
3526 Emerywood Lane
Orlando, Florida 32812
(305) 859-2821

*March of Dimes Birth Defects
  Foundation
1275 Mamaroneck Avenue
White Plains, New York 10605
(914) 428-7100

*National Network to Prevent
  Birth Defects
530 Seventh Street, S.E.
Washington, D.C. 20003
(202) 543-5450

NORD (National Organization for
  Rare Disorders, Inc.)
Fairwood Professional Building
P.O. Box 8923
New Fairfield, Connecticut 06812
(203) 746-6518

BLINDNESS/
VISUAL IMPAIRMENT
*American Council of Blind
  Parents
Adelphia House, Mezzanine 5
1229 Chestnut Street
Philadelphia, Pennsylvania 19107
(215) 561-1079

*American Foundation for the
  Blind, Inc.
15 West 16th Street
New York, New York 10011
(212) 620-2000
(212) 620-2158 (TDD)

*Association for Macular Diseases,
  Inc.
210 East 64th Street
New York, New York 10021
(212) 605-3719

Jewish Guild for the Blind
15 West 65th Street
New York, New York 10023
(212) 769-6200

*National Association for the
  Visually Handicapped
22 West 21st Street
New York, New York 10010
(212) 889-3141

*RP Foundation Fighting
  Blindness
1401 Mount Royal Avenue
Baltimore, Maryland 21217
(301) 225-9400
(800) 638-2300

*Vision Foundation, Inc.
818 Mount Auburn Street
Watertown, Massachusetts 02172
(617) 926-4232
(800) 852-3029 (within MA)

CANCER
*American Cancer Society, Inc.
261 Madison Avenue
New York, New York 10016
(212) 599-3600

*Association for Brain Tumor
    Research
2910 West Montrose Avenue,
    Suite 200
Chicago, Illinois 60618
(312) 286-5571

*Candlelighters Childhood
    Cancer Foundation
1901 Pennsylvania Avenue,
    N.W., Suite 1001
Washington, D.C. 20006
(202) 659-5136

*Familial Polyposis Registry
Department of Colorectal Surgery
Cleveland Clinic Foundation
9500 Euclid Avenue
Cleveland, Ohio 44106
(216) 444-6470

*Leukemia Society of America, Inc.
733 Third Avenue
New York, New York 10017
(212) 573-8484

CARDIOVASCULAR DISORDERS
*American Heart Association
7320 Greenville Avenue
Dallas, Texas 75231
(214) 750-5300

*American Lung Association
1740 Broadway
New York, New York 10019
(212) 315-8700

CATAPLEXY
(See Narcolepsy)

CELIAC-SPRUE DISEASE
American Celiac Society
45 Gifford Avenue
Jersey City, New Jersey 07304
(201) 432-1207

*Celiac-Sprue Association
2313 Rocklyn Drive, Suite 1
Des Moines, Iowa 50322
(515) 270-9689

CEREBRAL PALSY
*National Easter Seal Society
2023 West Ogden Avenue
Chicago, Illinois 60612
(312) 243-8400

*United Cerebral Palsy
    Associations, Inc.
66 East 34th Street
New York, New York 10016
(212) 481-6300

CHARCOT-MARIE-TOOTH
DISEASE
(See Peroneal Muscular Atrophy)

CHILDREN'S HEALTH
*Association for the Care of
    Children's Health
3615 Wisconsin Avenue, N.W.
Washington, D.C. 20016
(202) 244-1801
(202) 244-8922

*Children in Hospitals, Inc.
*Federation for Children with
    Special Needs
312 Stuart Street
Boston, Massachusetts 02116
(617) 482-2915

*Children's Defense Fund
122 C Street, N.W., Suite 400
Washington, D.C. 20001
(202) 628-8787

*Children's Hospice International
1101 King Street, Suite 131
Alexandria, Virginia 22314
(703) 684-0330

*Coordinating Council for
  Handicapped Children
20 East Jackson Boulevard,
  Room 900
Chicago, Illinois 60604
(312) 939-3513

Foundation for Child
  Development
345 East 46th Street
New York, New York 10017
(212) 697-3150

*Make a Wish Foundation of
  America
2600 North Central Avenue,
  Suite 936
Phoenix, Arizona 85004
(602) 240-6600

*Make Today Count, Inc.
101½ South Union Street
Alexandria, Virginia 22314
(703) 548-9674

*National Easter Seal Society
2023 West Ogden Avenue
Chicago, Illinois 60612
(312) 243-8400

*SKIP (Sick Kids [need] Involved
  People), Inc.
216 Newport Drive
Severna Park, Maryland 21146
(301) 647-0164

CLEFT LIP/PALATE
(See also Craniofacial Disorders)

*American Cleft Palate Educational
  Foundation
National Office
313 Salk Hall
University of Pittsburgh
Pittsburgh, Pennsylvania 15261
(412) 624-0625

*Prescription Parents, Inc.
P.O. Box 426
Quincy, Massachusetts 02269
(617) 479-2463

COCKAYNE SYNDROME
*Share and Care
124 S Street
North Valley Stream, New York
  11580
(516) 829-6768
(516) 825-2284

COLITIS
(See Ileitis and Colitis)

CONNECTIVE-TISSUE
DISORDERS
*American Brittle Bone Society
1256 Merrill Drive, Marshallton
West Chester, Pennsylvania
  19380
(215) 692-6248

Ehlers-Danlos National
  Foundation
P.O. Box 1212
Southgate, Michigan 48195
(313) 282-0180

*National Marfan Foundation
54 Irma Avenue
Port Washington, New York
  11050
(516) 883-8712

*Osteogenesis Imperfecta
  Foundation, Inc.
P.O. Box 14807
Clearwater, Florida 34629
(813) 855-7077

COOLEY'S ANEMIA/
THALASSEMIA
AHEPA Cooley's Anemia
  Foundation
136-56 39th Avenue
Flushing, New York 11354
(718) 961-3666

*Cooley's Anemia Foundation,
  Inc.
105 East 22d Street, Suite 911
New York, New York 10010
(212) 598-0911
(800) 221-3571
(800) 522-7222 (within NY)

CORNELIA DE LANGE
SYNDROME
*Cornelia de Lange Syndrome
  Foundation, Inc.
60 Dyer Avenue
Collinsville, Connecticut 06022
(203) 693-0159
(800) 223-8355 (outside CT)

CRANIOFACIAL DISORDERS
(*See also* Cleft Lip/Palate)

*National Association for the
  Craniofacially Handicapped
P.O. Box 11082
Chattanooga, Tennessee 37401
(615) 266-1632

*National Foundation for Facial
  Reconstruction
33 East 34th Street
New York, New York 10016
(212) 340-5400

CRI DU CHAT SYNDROME
*5p-Society
11609 Oakmont
Overland Park, Kansas 66210
(913) 469-8900

CYSTIC FIBROSIS
International Cystic Fibrosis
  (Mucoviscidosis) Association
3567 East 49th Street
Cleveland, Ohio 44105
(216) 271-1100

CYSTINOSIS
*Cystinosis Foundation, Inc.
477 15th Street, Suite 200
Oakland, California 94612
(415) 834-7897

DEAFNESS/HEARING
IMPAIRMENT
(*See also* Neurofibromatosis)

*Acoustic Neuroma Association
P.O. Box 398
Carlisle, Pennsylvania 17013
(717) 249-4783

*Alexander Graham Bell
  Association for the Deaf
3417 Volta Place, N.W.
Washington, D.C. 20007
(202) 337-5220

*American Society for Deaf
  Children
814 Thayer Avenue
Silver Spring, Maryland 20910
(301) 585-5400 (Voice/TDD)

*American Tinnitus Association
P.O. Box 5
Portland, Oregon 97207
(503) 248-9985

*Helen Keller National Center
111 Middle Neck Road
Sands Point, New York 11050
(516) 944-8900

National Association for Hearing
  and Speech Action
10801 Rockville Pike
Rockville, Maryland 20852
(301) 897-8682
(800) 638-8255

National Information Center on
  Deafness
Gallaudet College
800 Florida Avenue, N.E.
Washington, D.C. 20002
(202) 651-5000
(202) 651-5976 (TDD)

SHHH (Self Help for Hard of
  Hearing People, Inc.)
7800 Wisconsin Avenue
Bethesda, Maryland 20814
(301) 657-2248

DENTAL CARE
National Foundation of Dentistry
  for the Handicapped
1600 Stout Street, Suite 1420
Denver, Colorado 80202
(303) 573-0264

DES
*DES Action National
Long Island Jewish Hospital
   Medical Center
New Hyde Park, New York 11040
(516) 775-3450

DIABETES
*American Diabetes Association
505 Eighth Avenue
New York, New York 10018
(212) 947-9707

*Juvenile Diabetes Foundation
   International
432 Park Avenue South, 16th
   Floor
New York, New York 10016
(212) 889-7575

DOWN SYNDROME
*Association for Children with
   Down Syndrome, Inc.
2616 Martin Avenue
Bellmore, New York 11710
(516) 221-4700

*Caring, Inc.
P.O. Box 400
Milton, Washington 98354
(206) 922-5680

*National Down Syndrome Society
141 Fifth Avenue
New York, New York 10010
(212) 764-3070
(800) 221-4602

DWARFISM
(*See* Short Stature/Dwarfism)

DYSAUTONOMIA
*Dysautonomia Foundation, Inc.
370 Lexington Avenue, Room
   1504
New York, New York 10017
(212) 889-5222

DYSLEXIA
(*See* Learning Disabilities)

DYSTONIA
*Dystonia Medical Research
   Foundation
9615 Brighton Way, Suite 310
Beverly Hills, California 90210
(213) 852-1630

DYSTROPHIC EPIDERMOLYSIS
BULLOSA
(*See* Skin Disorders)

ECTODERMAL DYSPLASIAS
(*See* Skin Disorders)

EHLERS-DANLOS SYNDROME
(*See* Connective-Tissue Disorders)

EPIDERMOLYSIS BULLOSA
(*See* Skin Disorders)

EPILEPSY
*Epilepsy Foundation of America
4351 Garden City Drive
Landover, Maryland 20785
(301) 459-3700

EPSTEIN-BARR VIRUS
SYNDROME
National CEBV (Chronic
   Epstein-Barr Virus) Syndrome
   Association Inc.
P.O. Box 230108
Portland, Oregon 97223
(503) 684-5261

EXSTROPHY
*National Support Group for
   Exstrophy
5075 Medhurst Street
Solon, Ohio 44139
(216) 248-6851

FAMILY PLANNING
*National Family Planning and
   Reproductive Health
   Association, Inc.
122 C Street, N.W., Suite 380
Washington, D.C. 20001
(202) 628-3353

*Planned Parenthood Federation
   of America, Inc.
810 Seventh Avenue
New York, New York 10019
(212) 541-7800

FAMILY SUPPORT
*Family Resource Coalition
230 North Michigan Avenue,
   Suite 1625
Chicago, Illinois 60601
(312) 726-4750

Family Service Association of
   America
254 West 31st, 12th Floor
New York, New York 10001
(212) 967-2740

*Military Family Support Center
4015 Wilson Boulevard,
   Suite 903
Arlington, Virginia 22203
(703) 922-7671
(800) 336-4592

*National Parent CHAIN, Inc.
933 High Street, Suite 106
Worthington, Ohio 43085
(614) 431-1307

*Parents Helping Parents
535 Race Street, Suite 220
San Jose, California 95126
(408) 288-5010

*Sibling Information Network
991 Main Street, Suite 3A
East Hartford, Connecticut 06108
(203) 282-7050

FANCONI SYNDROME
*Fanconi Anemia Support Group
2875 Baker Boulevard
Eugene, Oregon 97405
(503) 686-7803
(503) 686-0434

FRAGILE-X SYNDROME
*Fragile X Foundation
P.O. Box 300233
Denver, Colorado 80203
(303) 322-8223

FREEMAN-SHELDON
SYNDROME
Freeman-Sheldon Parent Support
   Group
1459 East Maple Hills Drive
Bountiful, Utah 84010
(801) 298-3149

GAUCHER DISEASE
*National Gaucher Foundation
1424 K Street, N.W., 4th Floor
Washington, D.C. 20005
(202) 393-2777

GENETIC DEFECTS
(*See* Birth Defects)

GLYCOGEN-STORAGE DISEASE
*Association for Glycogen Storage
   Disease
Box 896
Durant, Iowa 52747
(319) 785-6038

HEARING IMPAIRMENT
(*See* Deafness/Hearing
Impairment)

HEMOCHROMATOSIS
(*See* Iron Overload)

HEMOPHILIA
*National Hemophilia Foundation
Soho Building
110 Greene Street, Room 406
New York, New York 10012
(212) 219-8180

HEREDITARY HEMORRHAGIC
TELANGIECTASIA
Hereditary Hemorrhagic
   Telangiectasia Registry
RFD 3-Pratt Corner
Amherst, Massachusetts 01003
(413) 259-1515
(413) 545-2048

HUNTINGTON DISEASE
Hereditary Disease Foundation
606 Wilshire Boulevard, Suite 504
Santa Monica, California 90401
(213) 458-4183

*Huntington's Disease Foundation
of America, Inc.
140 West 22d Street, 6th Floor
New York, New York 10011
(212) 242-1968

HYDROCEPHALUS
*Guardians of Hydrocephalus
Research Foundation
2618 Avenue Z
Brooklyn, New York 11235
(718) 743-4473

*Hydrocephalus Parent Support
Group
1222 Tower Drive
Vista, California 92083
(619) 726-0507

*National Hydrocephalus
Foundation
Route 1, River Road
P.O. Box 210A
Joliet, Illinois 60436
(815) 467-6548

HYPOPIGMENTATION
(See Skin Disorders)

ILEITIS AND COLITIS
*National Foundation for Ileitis
and Colitis, Inc.
444 Park Avenue South
New York, New York 10016
(212) 685-3400

IMMUNE DEFICIENCY
*Immune Deficiency Foundation
P.O. Box 586
Columbia, Maryland 21045
(301) 461-3127

*World Hemophilia AIDS Center
2400 South Flower Street
Los Angeles, California 90007-2697
(213) 742-1354

INFERTILITY
*Resolve, Inc.
5 Water Street
Arlington, Massachusetts 02174
(617) 643-2424

Surrogate Parent Program
11110 Ohio Avenue, Suite 202
Los Angeles, California 90025
(213) 473-8961

INTRAVENTRICULAR
HEMORRHAGE
*IVH Parents
P.O. Box 56-1111
Miami, Florida 33156
(305) 232-0381

IRON OVERLOAD
*Hemochromatosis Research
Foundation, Inc.
P.O. Box 8569
Albany, New York 12208
(518) 489-0972

*Iron Overload Diseases
Association, Inc.
Harvey Building
224 Datura Street, Suite 911
West Palm Beach, Florida 33401
(305) 659-5616

KIDNEY DISORDERS
*National Association of Patients
on Hemodialysis and
Transplantation
5 Woodhollow Road,
Room 1-I20
Parsippany, New Jersey 07054
(201) 581-7327

*National Kidney Foundation, Inc.
2 Park Avenue
New York, New York 10016
(212) 889-2210

*PKR (Polycystic Kidney Research)
Foundation
922 Walnut Street
Kansas City, Missouri 64106
(816) 421-1869

LAURENCE-MOON-BIEDL
SYNDROME
LMBS Support Network
122 Rolling Road
Lexington Park, Maryland 20653
(301) 863-5658

LEARNING DISABILITIES
*Association for Children and
    Adults with Learning
    Disabilities
4156 Library Road
Pittsburgh, Pennsylvania 15234
(412) 341-1515

*Council for Exceptional Children
1920 Association Drive
Reston, Virginia 22091
(703) 620-3660

Foundation for Children with
    Learning Disabilities
99 Park Avenue
New York, New York 10016
(212) 687-7211

*Orton Dyslexia Society
724 York Road
Baltimore, Maryland 21204
(301) 296-0232

LEUKODYSTROPHY
(See also Adrenoleukodystrophy)

*United Leukodystrophy
    Foundation, Inc.
2304 Highland Drive
Sycamore, Illinois 60178
(815) 895-3211

LIPID DISEASES
National Lipid Diseases
    Foundation
1201 Corbin Street
Elizabeth, New Jersey 07201
(201) 527-8000

LIVER DISORDERS
*American Liver Foundation
998 Pompton Avenue
Cedar Grove, New Jersey 07009
(201) 857-2626
(800) 223-0179

*Children's Liver Foundation, Inc.
76 South Orange Avenue,
    Suite 202
South Orange, New Jersey 07079
(201) 761-1111

LOWE SYNDROME
*Lowe's Syndrome Association
222 Lincoln Street
West Lafayette, Indiana 47906
(317) 743-3634

LUPUS (Systemic Lupus
Erythematosus)

*American Lupus Society
23751 Madison Street
Torrance, California 90505
(213) 373-1335

National Lupus Erythematosus
    Foundation, Inc.
5430 Van Nuys Boulevard,
    Suite 206
Van Nuys, California 91401
(818) 885-8787

*Systemic Lupus Erythematosus
    Foundation, Inc.
95 Madison Avenue, Room 1402
New York, New York 10016
(212) 685-4118

LYMPHATIC AND
VENOUS DISORDERS
*National Lymphatic and Venous
    Foundation, Inc.
P.O. Box 80
Cambridge, Massachusetts 02142
(617) 784-4104

MALIGNANT HYPERTHERMIA
*Malignant Hyperthermia
    Association of the United
    States
P.O. Box 3231
Darien, Connecticut 06820
(203) 655-3007

MAPLE SYRUP URINE DISEASE
*Families with Maple Syrup Urine
    Disease
24806 SR 119
Goshen, Indiana 46526
(219) 862-2992

MARFAN SYNDROME
(*See* Connective-Tissue Disorders)

MENTAL DISABILITIES
*Association for Retarded Citizens
P.O. Box 6109
Arlington, Texas 76006
(817) 640-0204
(800) 433-5255

*Joseph P. Kennedy, Jr.,
   Foundation
1350 New York Avenue, N.W.,
   Suite 500
Washington, D.C. 20005
(202) 393-1250

*National Mental Health
   Association
1021 Prince Street
Alexandria, Virginia 22314-2971
(703) 684-7722

*TASH: The Association for
   Persons with Severe Handicaps
7010 Roosevelt Way, N.E.
Seattle, Washington 98115
(206) 523-8446

MUCOLIPIDOSIS
Children's Association for
   Research on Mucolipidosis IV
6 Concord Drive
Monsey, New York 10952
(914) 425-0639

MUCOPOLYSACCHARIDOSES
*National MPS Society, Inc.
17 Kraemer Street
Hicksville, New York 11801
(516) 931-6338

MULTIPLE BIRTHS
*National Organization of Mothers
   of Twins Clubs, Inc.
12404 Princess Jeanne, N.E.
Albuquerque, New Mexico
   87112
(502) 275-0955

Twins Foundation
P.O. Box 9487
Providence, Rhode Island 02940
(401) 274-8946

MULTIPLE SCLEROSIS
(*See* Neurologic Disorders)

MUSCULAR DYSTROPHY
*Muscular Dystrophy Association
810 Seventh Avenue
New York, New York 10019
(212) 586-0808

MYOCLONUS
Myoclonus Families United
1564 East 34th Street
Brooklyn, New York 11234
(718) 252-2133

National Myoclonus Foundation
845 Third Avenue, 4th Floor
New York, New York 10022
(212) 758-5656

NARCOLEPSY
*American Narcolepsy Association
335 Quarry Road
Belmont, California 94002
(415) 591-7979

Narcolepsy and Cataplexy
   Foundation of America
1410 York Avenue, Suite 2D,
   MB 22
New York, New York 10021
(212) 628-6315

NEUROFIBROMATOSIS
(*See also* Deafness/Hearing
Impairment)

*Acoustic Neuroma Association
P.O. Box 398
Carlisle, Pennsylvania 17013
(717) 249-4783

*National Neurofibromatosis
   Foundation, Inc.
141 Fifth Avenue, 7th Floor
New York, New York 10010
(212) 460-8980

NEUROLOGIC DISORDERS
*Center for Neurologic Study
11211 Sorrento Valley Road,
  Suite H
San Diego, California 92121
(619) 455-5463

*National Multiple Sclerosis
  Society
205 East 42d Street
New York, New York 10017
(212) 986-3240

National Spasmodic Torticollis
  Association, Inc.
P.O. Box 873
Royal Oak, Michigan 48068
(313) 649-5391

NEUROMETABOLIC DISORDERS
*Association for Neuro-metabolic
  Disorders
5223 Brookfield Lane
Sylvania, Ohio 43560
(419) 885-1497

ORGANIC ACIDEMIAS
*Organic Acidemia Association
1532 South 87th Street
Kansas City, Kansas 66111
(913) 422-7080

OSLER-WEBER-RENDU
SYNDROME
(See Hereditary Hemorrhagic
Telangiectasia)

OSTEOGENESIS IMPERFECTA
(See Connective-Tissue Disorders)

PERONEAL MUSCULAR
ATROPHY
National Foundation for Peroneal
  Muscular Atrophy
University City Science Center
3624 Market Street
Philadelphia, Pennsylvania 19104
(215) 387-2255

PORPHYRIA
*American Porphyria Foundation
P.O. Box 11163
Montgomery, Alabama 36111
(904) 267-2372

PRADER-WILLI SYNDROME
*Prader-Willi Syndrome
  Association
5515 Malibu Drive
Edina, Minnesota 55436
(612) 933-0113

PREMATURITY
*Parent Care
University of Utah Medical
  Center, Room 2A210
50 North Medical Drive
Salt Lake City, Utah 84132
(801) 581-5323

PROGERIA
*Progeria International Registry
Department of Human Genetics
1050 Forest Hill Road
Staten Island, New York 10314
(718) 494-5230

RADIATION
*National Association of Radiation
  Survivors
78 El Camino Real
Berkeley, California 94705
(415) 658-6056
(415) 652-4400, ext. 441

REHABILITATION
*National Organization on
  Disability
910 16th Street
Washington, D.C. 20006
(202) 293-5960

National Rehabilitation
  Association
633 South Washington Street
Alexandria, Virginia 22314
(703) 836-0850

Rehabilitation International
25 East 21st Street
New York, New York 10010
(212) 420-1500

RETINITIS PIGMENTOSA
(*See* Blindness/Visual Impairment)

RETT SYNDROME
*International Rett Syndrome
  Association
8511 Rosemarie Drive
Fort Washington, Maryland
  20744
(301) 248-7031

RUBINSTEIN-TAYBI SYNDROME
*Rubinstein-Taybi Parent Contact
  Group
414 East Kansas
Smith Center, Kansas 66967
(913) 282-6237

SCLERODERMA
*United Scleroderma Foundation,
  Inc.
P.O. Box 350
Watsonville, California 95077-0350
(408) 728-2202

SCOLIOSIS
*National Scoliosis Foundation,
  Inc.
P.O. Box 547
93 Concord Avenue
Belmont, Massachusetts 02178
(617) 489-0888

*Scoliosis Association, Inc.
P.O. Box 51313
Raleigh, North Carolina 27609
(919) 846-2639

SEXUALITY
SIECUS (Sex Information and
  Education Council of the United
  States)
32 Washington Place, 5th Floor
New York, New York 10003
(212) 673-3850

SHORT STATURE/DWARFISM
*Human Growth Foundation
4720 Montgomery Lane, Suite 909
Bethesda, Maryland 20814
(301) 656-7540

*Little People of America, Inc.
P.O. Box 633
San Bruno, California 94066
(415) 589-0695

Parents of Dwarfed Children
11524 Colt Terrace
Silver Spring, Maryland 20902
(301) 649-3275

SICKLE-CELL DISEASE
*National Association for Sickle
  Cell Disease, Inc.
4221 Wilshire Boulevard, Suite 360
Los Angeles, California 90010-3503
(213) 936-7205
(800) 421-8453

SIDS
(*See* Sudden Infant Death
Syndrome)

SKIN DISORDERS
*DEBRA (Dystrophic Epider-
  molysis Bullosa Research
  Association of America, Inc.)
Kings County Medical Center
451 Clarkson Avenue, Building E
6th Floor, Room E6101
Brooklyn, New York 11203
(718) 774-8700

*National Foundation for
  Ectodermal Dysplasias
118 North First Street, Suite 311
Mascoutah, Illinois 62258
(618) 566-2020

*NOAH (National Organization
  for Albinism and Hypo-
  pigmentation)
4721 Pine Street
Philadelphia, Pennsylvania
  19143
(215) 471-2265
(215) 471-2278

Xeroderma Pigmentosum Registry
c/o Department of Pathology
Medical Science Building,
    Room C 520
UMDNJ-NJ Medical School
185 South Orange Avenue
Newark, New Jersey 07103
(201) 456-6255

SPASMODIC TORTICOLLIS
(*See* Neurologic Disorders)

SPINAL MUSCULAR ATROPHY
*Families of SMA
P.O. Box 1465
Highland Park, Illinois 60035
(312) 432-5551

SUDDEN INFANT DEATH
SYNDROME (SIDS)
*National Sudden Infant Death
    Syndrome Foundation, Inc.
8200 Professional Place, Suite 104
Landover, Maryland 20785
(301) 459-3388

SYSTEMIC LUPUS
ERYTHEMATOSUS
(*See* Lupus)

TAY-SACHS DISEASE
*National Tay-Sachs and Allied
    Diseases Association
92 Washington Avenue
Cedarhurst, New York 11516
(516) 569-4300

National Tay-Sachs Parent Peer
    Group
92 Washington Avenue
Cedarhurst, New York 11516
(516) 569-4300

THALASSEMIA
(*See* Cooley's Anemia/Thalassemia)

THROMBOCYTOPENIA ABSENT
RADIUS SYNDROME
*TARSA (Thrombocytopenia
    Absent Radius Syndrome
    Association)
312 Sherwood Drive, R.D. 1
Linwood, New Jersey 08221
(609) 927-0418

TOURETTE SYNDROME
*Tourette Syndrome
    Association, Inc.
42-40 Bell Boulevard
Bayside, New York 11361
(718) 224-2999
(800) 237-0717

TOXOPLASMOSIS
Toxoplasmosis Interest Group
52 Edgell Street
Gardner, Massachusetts 01440
(508) 632-7783

TUBEROUS SCLEROSIS
*National Tuberous Sclerosis
    Association, Inc.
4351 Garden City Drive, Suite 660
Landover, Maryland 20785
(301) 459-9888

TURNER SYNDROME
*Turner's Syndrome Society of
    Sacramento
2246 Green Blossom Court
Sacramento, California 95670
(916) 635-8164

USHER SYNDROME
(*See* Retinitis Pigmentosa)

VENOUS DISORDERS
(*See* Lymphatic and Venous
Disorders)

VISUAL IMPAIRMENT
(*See* Blindness/Visual Impairment)

WILSON DISEASE
National Center for the Study of
  Wilson's Disease, Inc.
5447 Palisade Avenue
Bronx, New York 10471
(212) 430-2091

XERODERMA PIGMENTOSUM
(*See* Skin Disorders)

SELF-HELP CLEARINGHOUSES

Self-help groups offer people who face a common problem the opportunity to meet with others and share experiences, knowledge, strengths, and hopes. Self-help clearinghouses collect and disseminate information about local self-help groups and assist people in forming new groups; they differ in their ability to provide information and referrals. If you are unable to locate a service to meet your needs, contact your state mental-health association or the resource centers listed below.

National Center for Education in
  Maternal and Child Health
38th and R Streets, N.W.
Washington, D.C. 20057
(202) 625-8400

National Self-Help Clearinghouse
Graduate School and
  University Center/CUNY
33 West 42d Street
New York, New York 10036
(212) 840-1259

New Jersey Self-Help
  Clearinghouse
Saint Clare's Hospital
Pocono Road
Denver, New Jersey 07834
(201) 625-9565

Additional Common
Genetic Disorders and
Interesting Inherited Traits

Condition	Description/Explanation	Comments
Autosomal Dominant Conditions*		
Achoo syndrome	Sneezing stimulated by sudden exposure of a person in the dark or in semi-darkness to sunlight or other intensely bright light	Surprisingly common
Asparagus urine	Urinary excretion of a smelly chemical (methanethiol) after eating asparagus, and the ability to detect the odor	Very common; the ability to smell asparagus urine and to form it are probably transmitted by separate genes
Baldness, male-pattern	The common loss of scalp hair experienced by some men as they age	Females become bald through heredity only if they receive the gene for baldness from *both* parents
Beet urine (beeturia), absence of	Urinary excretion of beet pigment after eating beets does not occur	The dominant characteristic, which is much more common, is *not* to excrete this pigment

* See chapter 7.

Condition	Description/Explanation	Comments
Autosomal Dominant Conditions (continued)		
Calluses, hereditary (callosities)	Painful calluses over pressure points in the hands and feet	Can be successfully treated with the drug tretinoin (Retin-A)
Chilblain, susceptibility to	Injury to the skin — red or blue discoloration, swelling, and itching or burning sensation — caused by exposure to cold	Especially affects young women
Cold, hypersensitivity to (cold urticaria)	Exposure to cold results in skin wheals and possibly pain and swelling of joints, with associated chills and fever	
Color blindness, blue-yellow (tritanopia)	Affected individuals have problems seeing blue and yellow but no difficulty seeing red and green	Although this may be an autosomal dominant condition, a sex-linked form is also known
Dimples		
Acromial	Dimples on the back of the shoulder	
Cheek		
Chin (cleft chin)	A midline dimple or cleft in the chin	
Earlobes, structure of	Described either as present or absent, or, alternatively, as free or attached	Probably a dominant characteristic either way
Ears, ability to wiggle without touching		Probably dominantly inherited

Condition	Description/Explanation	Comments

Autosomal Dominant Conditions (continued)

Condition	Description/Explanation	Comments
Earwax (cerumen), moistness or dryness of		Earwax may be dry (as in about 85% of Japanese) or wet (as in most Caucasians and blacks); both are dominantly inherited traits
Eye dominance (ocular dominance)	One eye is favored over the other	About 97% of people have this condition, which is probably a dominantly inherited trait; about 65% show right-eye preference
Eyelashes, long		Unusually long lashes are probably inherited as an autosomal dominant condition
Fingers, hereditary cold	Intermittent attacks of numb and white fingers due to inadequate blood supply to the digits	May be caused by a disorder called Raynaud disease
Hair		
Curly		Curly hair (as opposed to straight hair) is probably dominant
Facial, excessive		Excessive facial hair in Caucasian females is probably inherited as an autosomal dominant condition
Finger (midphalangeal)	Hair on top of the middle segments of the fingers	

Condition	Description/Explanation	Comments

Autosomal Dominant Conditions (continued)

Hair (cont.)

Condition	Description/Explanation	Comments
Prematurely gray	Very early graying, beginning in the late teens, with almost all hair white by the midtwenties	
Red, very (*see under* Autosomal Recessive)		
Hayfever	Allergic sensitivity to ragweed pollen	
Hernia, double groin		Probably dominantly inherited
Insect stings, hypersensitivity to		Extreme sensitivity to insect stings may be dominantly inherited
Hip, dislocation of (*see under* Polygenic Conditions)		
Moles (nevi), pigmented		Multiple pigmented moles frequently occur in families, probably through dominant transmission
Nearsightedness (myopia)		One form of severe myopia is transmitted as an autosomal dominant disorder; autosomal recessive and sex-linked forms of nearsightedness are also known
Nipples		
Inverted		Affects more females than males

Condition	Description/Explanation	Comments

Autosomal Dominant Conditions (continued)

Nipples (cont.)

Supernumerary	Extra nipples, sometimes accompanied by extra breasts	May be associated with twinning
Sleep, obstructed (apnea)	Periods of no breathing during sleep, often associated with restlessness, loud snorts and snoring, and even sudden death	
Smell, absent or deficient general sense of (congenital anosmia)		Usually an autosomal dominant condition; sex-linked in some cases
Stammering (stuttering)		Dominantly inherited in some families; unusually frequent in Japanese, very infrequent in Polynesians, and almost completely absent in American Indians
Stretch marks (striae distensae)	Stretch marks in the skin, especially around the lower back	More common in males
Sweating, excessive (hyperhydrosis), while eating certain foods	Abundant sweating, limited mainly to the face — especially the forehead, the tip of the nose, and the upper lip — while eating spicy or sour foods	

Condition	Description/Explanation	Comments
Autosomal Dominant Conditions (continued)		
Teeth		
Oddly shaped	Peg-shaped, conical, variably-sized, and with small surface pits	
Present at birth		Occurs about once in 3,000 live births; also a feature of some very rare birth-defect syndromes
Protuberant	Overbite due to upper front teeth that jut out too far	
Supernumerary	Extra teeth at variable sites	
Thumb, trigger	A flexion defect of the thumb in which the digit is held in a "trigger position"	
Thumbs and big toes, stubby	Short, broad-ended thumbs and big toes	About 75% of the population have these features on both sides of the body
Toe, long second	A second toe that is longer than the first (the big toe)	Less common than a longer first toe; thought to be dominantly inherited
Toes, stubby (see Thumbs and big toes, stubby)		
Tongue, curlable or rollable	The ability to roll up or curl the edges of the tongue and thus form a trough down the center	Probably inherited as a dominant condition

Condition	Description/Explanation	Comments

Autosomal Dominant Conditions (continued)

Tune deafness (dysmelodia)	Inability to distinguish specific tones	Probably inherited as a dominant condition
Urine (*see* Asparagus urine; Beet urine, absence of)		
Varicose veins (*see under* Polygenic Conditions)		

Autosomal Recessive Conditions[*]

Aspirin intolerance[†]

Asthma[†]

Color blindness, blue-yellow (tritanopia)	Affected individuals have problems seeing blue and yellow but no difficulty seeing red and green	This disorder may be transmitted as a sex-linked condition, but also as an autosomal dominant one
Hair		
Blond, very		May be inherited as an autosomal recessive trait
Red, very		About 1.9% of Caucasians have strikingly red hair, which may be transmitted as an autosomal recessive trait; there is also evidence that red pigment in the hair (as opposed to its absence) is dominant, so the matter is still unresolved

[*] See chapter 7.

[†] Aspirin intolerance, asthma, and nasal polyps may variably be present in the same person. Individuals with two or three of these conditions may have parents who were blood-related or of similar ancestry.

Condition	Description/Explanation	Comments

Autosomal Recessive Conditions (continued)

Condition	Description/Explanation	Comments
Musk, inability to smell		About 7% of Caucasians are unable to smell musk; the characteristic is not yet reported in blacks
Nasal polyps*	Small lumps of tissue that grow inside the nostrils	
Nearsightedness (myopia)		Severe infantile myopia is probably transmitted as an autosomal recessive condition; autosomal dominant and sex-linked forms of nearsightedness are also known
Skunk, inability to smell		Probably inherited as an autosomal recessive trait
Teeth, absent (anodontia)	Complete absence of permanent teeth (primary teeth are not affected)	
Toe, long first	A first toe (big toe) that is longer than the second	More common than a longer second toe; probably recessively inherited
Tongue, foldable	The ability to fold up the tip of the tongue	May be inherited as an autosomal recessive trait

* Aspirin intolerance, asthma, and nasal polyps may variably be present in the same person. Individuals with two or three of these conditions may have parents who were blood-related or of similar ancestry.

Condition	Description/Explanation	Comments
Sex-linked Conditions[*]		
Color blindness, blue-yellow (tritanopia)	Affected individuals have problems seeing blue and yellow but no difficulty seeing red and green	Although this may be an autosomal dominant condition, a sex-linked form is also known
Cyanide, inability to smell		May be sex-linked, but mode of inheritance is still uncertain in most cases
Nearsightedness (myopia)		At least one form of myopia is transmitted as a sex-linked recessive trait; other, severe forms may be passed as autosomal dominant or autosomal recessive conditions
Smell, absent or deficient general sense of (congenital anosmia)		Although mostly transmitted as an autosomal dominant condition, sex-linked forms occur in some cases
Teeth, multiple impacted	Teeth that fail to emerge from the gum	Probably sex-linked in some cases
Polygenic (Multifactorial) Conditions[†]		
Celiac disease	A failure to absorb foods normally, especially fats; the malabsorption is secondary to intolerance to wheat gluten	The risk to the child of an affected person is about 3%

[*] See chapter 7.

[†] This group of disorders includes many different birth defects, most of the common diseases of midlife, the common mental illnesses (schizophrenia and affective disorders), and certain types of mental retardation. Many of these are discussed discussed in chapter 7 and elsewhere (see the index). A few additional examples of interest are included here.

Condition	Description/Explanation	Comments

Polygenic Conditions (continued)

Condition	Description/Explanation	Comments
Crohn disease	A noninfectious inflammation of the bowel that mainly affects the last portion of the small intestine (ileum)	The risk to the child of an affected person is probably between 0.4% and 1%
Eye color		Earlier ideas that blue eyes are inherited as a simple autosomal recessive trait are no longer accepted
Hips, dislocation of		Dominant inheritance may also be responsible; loose joints, which are more common in females, probably account for the higher frequency of congenital hip dislocation in females
Peptic ulcer	A break in the lining of the stomach or duodenum (a part of the small intestine) caused by exposure to certain digestive juices	May also be inherited as an autosomal dominant condition; the general risk to the child of an affected person is likely to be between 5% and 10%
Psoriasis	A chronic itchy and scaly skin disorder	The risk to a child of an affected person is about 1%
Rheumatoid arthritis	A chronic, multisystem disease that causes pain, swelling, and deformity of the joints; mainly affects hands, feet, wrists, ankles, elbows, and knees	More frequent in females; the risk to a child of an affected person is 3% to 5%

Condition	Description/Explanation	Comments
Polygenic Conditions (continued)		
Ulcerative colitis	A noninfectious inflammation of the bowel, that mainly affects the colon	The risk to a child of an affected person is probably between 0.4% and 1%
Varicose veins	Dilated, twisted, veins on the surface of the legs	Autosomal dominant and sex-linked dominant inheritance may both be responsible in some families

Questions You Thought of Asking

Innumerable questions must have crossed your mind while reading this book. The purpose of this text was not to provide an encyclopedic coverage of genetic disease, nor to masquerade as a genetic textbook or an introduction to biology. Its fundamental purpose was to alert you to the importance of genetic disorders as they affect you and your family, to provide a vital guide to securing the best health possible for those already afflicted, and to advise what steps you might take to preclude personal tragedies. This appendix is devoted to answering questions that may have occurred to you and that have, in fact, been asked of me repeatedly over the years.

My father's brother twitches constantly, especially with his face, head, neck, and shoulders. He also swears uncontrollably while twitching. I'm worried because I noticed my father also has some twitching, although much less marked. While he does not swear, he has very compulsive behavior. Is this a hereditary disorder?

This is indeed a hereditary disorder, first described in France in 1825. Over 100 years ago, in 1885, Dr. Georges Gilles de la Tourette recognized the familial nature of this syndrome, which now bears his name. Originally, the Tourette syndrome was characterized by recurrent uncontrollable twitching resulting in violent head jerking, shoulder movements, and tongue thrusting, all often accompanied by obscene shouts. This disorder affects more males than females and mostly becomes apparent between two and fifteen years of age. Doctors David and Brenda Comings of Duarte, California, have studied over 800 patients and report that the Tourette syndrome is one of the most common genetic disorders affecting man. Remarkably, about 1 in 100 individuals show one or more signs that they have the gene for this disorder.

The Comingses' extensive studies now show that a range of features may occur as part of the Tourette syndrome. These encompass *attention-deficit disorder,* including learning disorders, dyslexia, emotional disturbance, severe test anxiety, and stuttering; *disorders of conduct,* including vandalism, lying and stealing, fighting, starting fires, short temper, hurting animals, attacking others, and shouting at parents or peers; *phobias (fears) and panic attacks,* including fear of public transportation, fear of being alone, fear of crowds, fear of being in water, fear of animals, and fear of public speaking; *abnormal behavior,* including obsessive, compulsive, or repetitive behavior characterized by persistent unpleasant or silly thoughts, meaninglessly repeating words of others, touching things or others excessively, sexual touching, self-mutilation, exhibitionism, paranoid ideas, and hearing voices; *depression and manic-depression; sleep problems,* including sleepwalking, night terrors, difficulty in falling asleep, inability to take afternoon naps in early childhood, and early awakening as adults; and *problems in early development,* including delayed toilet training and bed-wetting. An affected individual may only have twitches (called tics) or may have one or more of the major features mentioned. The latest evidence is that the Tourette syndrome is caused by a gene carried on the long arm of chromosome number 18 and is transmitted as an autosomal dominant condition (see chapter 7). Hence, the child of an affected parent has a 50-percent risk of inheriting the gene, but the effects may be very variable.

A few people in our family have allergy problems, including asthma, hayfever, and eczema (skin allergy). Are these allergies inherited?

The predisposition to become allergic is inherited, not the specific allergy itself. Various modes of inheritance probably apply in different families, all involving the body's immune system (which makes antibodies). Studies from the Swedish Twin Registry of 7,000 like-sexed pairs of twins over forty years of age and 2,400 identical twins showed that both members of an identical twin pair had allergy 20 to 25 percent of the time. While there is clearly overlap between various types of allergic disorders, some idea of basic risks for the most common can be gleaned from the figures in table D-1. From the table, you can see that the risk of asthma for a child is 26 percent when only one parent has asthma,

34 percent when both parents are affected, and 10 percent if neither parent is affected but one sibling has asthma. Remember, however, that because of the inherited predisposition — most often passed through dominant or multifactorial inheritance (see chapter 7) — a person with one allergic disorder may often have another as well, or different types may occur within the same family.

Table D-1. Chances of Common Allergies Affecting a Child

Disorder	Only One Parent Affected	Both Parents Affected	Parents Unaffected and One Sibling Affected
Asthma	26%	34%	10%
Hayfever	12	uncertain	6
Skin allergy	34	57	14

Is migraine inherited?

Migraine headaches are probably inherited as an autosomal dominant disorder (see chapter 7). If both parents are affected, the offspring's risk of having migraine headaches is approximately 70 percent, while when only one parent is affected the risk is about 45%. Migraine is very common: it bothers some 5 to 10 percent of the population. Curiously, the frequency of migraine is higher than average among families affected by common allergic disorders.

Our first child has Rett syndrome. Is this condition inherited, and if so, what are our risks of having another affected child?

This devastating disorder of the brain was first described by Dr. Andreas Rett in 1966 in Vienna. Today we know that it is as common as phenylketonuria — 1 in 10,000 to 1 in 15,000 births — and that it affects over 8,000 children in the United States alone. Only females are affected, and development appears to be normal for at least the first six months of life. Sometime between six months and four years of age, normal development ceases. Typically there is deterioration in all areas of development, including behavior, personality, communication skills, and use of the hands, coupled with loss of ability to walk. Clumsiness, lack of balance, and hand wringing or flapping or clapping are typical. Autistic behavior and lack of emotional contact combine with developing mental retardation and seizures between four and seven years of age. Usually, between five and twenty-five years of age the affected female is confined to a wheelchair — unable to use arms or legs, possessed of severe multihandicaps, and growth-retarded.

Rett syndrome is thought to be a genetic disorder that is transmitted as a sex-linked dominant condition (see chapter 7) and that is lethal in all male embryos. Thus far, the existence of two affected sisters has been reported only twice in the more than 600 Rett families studied. Currently, every case is regarded as a new mutation; hence, the risk of having a second affected child is extremely small. No specific prenatal diagnostic test exists yet for this disorder.

Is childhood deafness inherited?

Deafness in childhood is common. It affects between 1 in 600 and 1 in 2,000 of all children. About 30 percent of patients directly inherit deafness, about 40 percent become deaf from various other causes (such as infections), while in about 30 percent the cause is unrecognized. When a child is born deaf and has no other abnormalities (for example, of the hair, bones, face, and so on), the two commonest forms of inheritance are autosomal recessive and autosomal dominant (see chapter 7). Parents who are related (such as first cousins) or who belong to the same ethnic group, and who are not deaf, probably each carry a corresponding harmful gene if they have a deaf child who has no other signs. In this situation, the risk of having another deaf child in any pregnancy is likely to be 25 percent.

We have one child with epilepsy of unknown cause. What is the likelihood that our next child would also have epilepsy?

Epilepsy, which is remarkably common, affects between 1 in 200 and 1 in 300 individuals. There are many known causes of epilepsy acquired after birth (brain injury and infection, for instance) or inherited as part of a genetic disorder (such as tuberous sclerosis). Frequently, however, no cause is discovered. In such cases, the observed risk of recurrence, given healthy parents without epilepsy, is about 5 percent. In cases in which one parent has epilepsy of unknown cause and one child is already similarly affected, the risk of recurrence is thought to be about 10 percent.

Is nearsightedness inherited?

Roughly 25 percent of individuals in the United States between twelve and fifty-four years of age are myopic, or nearsighted. The best evidence pointing to a significant hereditary component in myopia comes from a study of twins. A high degree of concordance (both twins nearsighted) in 78 pairs of identical twins was a striking finding, in contrast to no difference being noted between nonidentical twins and other children.

Nearsightedness can be inherited as a dominant, recessive, or sex-linked trait (see chapter 7). But disorders that affect various structures of the eye may cause or contribute to nearsightedness. The hereditary pattern is therefore not always easily understood or explained.

We lost our son to a condition called the sudden infant death syndrome (SIDS). Is this a genetic disorder? Do we have an increased risk that this would happen again if we have another child?

SIDS is a disorder in which apparently healthy babies are found dead in their cribs without any warning or prior illness. Sometimes called crib death or cot death, SIDS is probably the most common single cause of death in babies between one week and one year of age in the United States. Over 10,000 well babies die each year from this cause. The incidence varies in other countries but is thought to average about 1 in 500 live births.

The current view is that the electrical conduction of nerve impulses through the heart may be defective in some of these babies, predisposing them to sudden, fatal heart-rhythm disturbances. The risk of recurrence in subsequent children of the same parents appears to be about 2 percent in this disorder, which consistently occurs more often among boys. The mode of inheritance involved remains uncertain, but available data tend to point to multifactorial transmission (see chapter 7).

My mother has multiple sclerosis. Is this a hereditary disorder? What are my risks?

Multiple sclerosis (MS), a chronic disease of the central nervous system, is regarded as a multifactorial disorder that results from the interaction of an environmental agent, such as a virus, and a genetic susceptibility. The near relatives of an affected person are at increased risk, and age is a factor. Studies in Vancouver, Canada, show the following risks:

If a *parent* has MS:
- his or her *daughters'* risk is about 5% until about 28 years of age, when it approximates 2.5%, before becoming 1.5% at 33 years, and 0.5% at 43 years
- their *sons'* risk is much lower (exact figures are not certain)

If a *female* has MS:
- her *sister* has about a 5.7% risk until about 28 years of age, when it approximates 2.5%, before becoming 1.5% at 33 years, and 0.5% at 42 years
- her *brother* has about a 2.3% risk until about 25 years of age, when it approximates 1.5%, before becoming 0.5% at 38 years

If a *male* has MS:
- his *sisters'* risk is about 3.5% until 23 years of age, when it approximates 2.5%, before becoming 1.5% at 29 years, and 0.5% at 39 years
- his *brothers'* risk is about 4.2% until 26 years of age, 1.5% by 33 years, and 0.5% by 42 years

Nieces, nephews, and first cousins of an affected individual on either side of the family have between a 1% and 3% risk until 17 to 23 years of age, when the odds diminish.

Index

About the Author

Aubrey Milunsky, M.D., was born in Johannesburg, South Africa, and was graduated from the University of Witwatersrand Medical School in 1960. He is a triple board-certified specialist — in internal medicine by the Royal College of Physicians in London, in pediatrics by the American Board of Pediatrics, and in clinical genetics by the American Board of Medical Genetics. After thirteen years as a medical geneticist at Harvard Medical School and the Massachusetts General Hospital, he was appointed Professor of Pediatrics, Obstetrics and Gynecology, and Pathology at the Boston University School of Medicine, where he has also been Director of the Center for Human Genetics since 1981. He is a member of the editorial boards of several prestigious medical journals.

Dr. Milunsky has written more than two hundred scientific papers, and is the author or editor of eleven books, including two for the lay public: *How to Have the Healthiest Baby You Can* and *Choices, Not Chances* (originally titled *Know Your Genes* and published in nine languages).